VIDEO AND MEDIA SERVERS: TECHNOLOGY AND APPLICATIONS

VIDEO AND MEDIA SERVERS:
TECHNOLOGY AND APPLICATIONS

KARL PAULSEN

Focal Press
Boston Oxford Johannesburg Melbourne New Delhi Singapore

Focal Press is an imprint of Butterworth–Heinemann.

Copyright © 1998 by Butterworth–Heinemann

 A member of the Reed Elsevier group

All rights reserved.

No part of this publication may be reproduced, stored in a retrieval system, or transmitted in any form or by any means, electronic, mechanical, photocopying, recording, or otherwise, without the prior written permission of the publisher.

∞ Recognizing the importance of preserving what has been written, Butterworth–Heinemann prints its books on acid-free paper whenever possible.

Butterworth–Heinemann supports the efforts of American Forests and the Global ReLeaf program in its campaign for the betterment of trees, forests, and our environment.

Library of Congress Cataloging-in-Publication Data
Paulsen, Karl, 1952-
   Video and media servers : technology and applications / Karl Paulsen.
      p.   cm.
   Includes index.
   ISBN 0-240-80296-9 (pbk. : alk. paper)
   1. Digital television. 2. Client/server computing. 3. Video recording. I. Title.
TK6678.P38   1998
621.388—DC21   98-10222
                                                                                 CIP

British Library Cataloguing-in-Publication Data
A catalogue record for this book is available from the British Library.

The publisher offers special discounts on bulk orders of this book.
For information, please contact:
Manager of Special Sales
Butterworth-Heinemann
225 Wildwood Avenue
Woburn, MA 01801-2041
Tel: 781-904-2500
Fax: 781-904-2620

For information on all Focal Press publications available, contact our World Wide Web home page at: http://www.bh.com/focalpress

10 9 8 7 6 5 4 3 2 1

Printed in the United States of America

To my wife and partner, Jenny,
and to my family, without whose support,
patience, and caring,
this would not have been possible.

CONTENTS

	List of Figures	ix
	List of Tables	xiii
	Acknowledgments	xv
	Introduction	xvii
Ch 1	Video and Media Servers: The Journey Begins	1
Ch 2	Disk Recording: A Brief History	11
Ch 3	One Step Before the Server	25
Ch 4	Distinguishing the DDR from the Video Server	35
Ch 5	Selecting the Video Server	47
Ch 6	Video Server Systems	59
Ch 7	Server Configuration Basics	67
Ch 8	Media Server Architectures: The Front End Processor	77
Ch 9	Media Server Architectures: Video Server I/O	89
Ch 10	The Disk Dilemma	101
Ch 11	Where Servers Fit	117
Ch 12	Issues with Servers and Digital Systems	133
Ch 13	Living in the Digital Domain	141
Ch 14	Digital Storage Solutions for the Long Run	151
Ch 15	Storage and Information Retrieval Management	159
Ch 16	Media Management and Automation	173
Ch 17	Facilities for the Information Age	183
Ch 18	The Model Changes	193
Ch 19	Integrating the Server in the Facility	201
Ch 20	Shared and Distributed Server Environments	215
Ch 21	Network Basics	229
Ch 22	Principles in Network Protocol	245
Ch 23	Connections at the Network Level	255
Ch 24	Introducing Packets	263
Ch 25	Inside ATM	273
Ch 26	RAID Revealed	285
Ch 27	RAID by the Numbers	293

Ch 28	Parity in the Performance of RAID	303
Ch 29	The Highs and Lows of SCSI	313
Ch 30	SCSI's Second Step	327
Ch 31	Fibre Channel Disclosed	341
Ch 32	Serial Storage Architecture	357
Ch 33	Solid State Disks	373
Ch 34	Inside the Drive	379
Ch 35	Preventing Disaster	391
Ch 36	VOD: New Outlets for Content Producers	407
Ch 37	Approaches to Video On Demand	417
Ch 38	Interactive Television: Once an Emerging Industry	425
	Acronyms and Abbreviations	435
	Index	443

LIST OF FIGURES

Figure 5-1:	Two different types and uses for servers.	49
Figure 5-2:	Availability of data, on-line versus near-line.	52
Figure 5-3:	Common storage array with channel expansion.	53
Figure 5-4:	Conceptual bit stream server front end.	57
Figure 7-1:	Rear interface, two-channel video server.	70
Figure 7-2:	Conceptual RAID expansion chassis.	73
Figure 8-1:	Basic video server structure.	78
Figure 8-2:	Analog input to a video server.	78
Figure 8-3:	Advanced video server inputs.	79
Figure 8-4:	Server processing engine.	81
Figure 8-5:	Multiple port, single codec processing.	83
Figure 8-6:	Dual codec processing.	84
Figure 9-1:	Reference clock generator.	93
Figure 9-2:	Simplified audio and RS-422 flow.	94
Figure 9-3:	Timecode flow within a video server.	97
Figure 10-1:	Early file transfers from video to computers.	107
Figure 10-2:	Mapping of video, audio, and data for streaming.	111
Figure 10-3:	Streaming technology application spaces.	113
Figure 11-1:	Peer-to-peer network.	119
Figure 11-2:	Point-to-point video connections.	121
Figure 11-3:	NLE "islands" lack common shared library solutions.	123
Figure 11-4:	Store and forward for time-zone delay.	127
Figure 11-5:	In-flight entertainment video server.	129
Figure 12-1:	Coding grids for various $Y/C_B/C_R$ sampling.	137
Figure 15-1:	Simplified block for MPEG-2 coding system.	164
Figure 15-2:	Basic browse server network.	166
Figure 15-3:	Browse server search engine.	168
Figure 16-1:	Simple video server interface control interface.	174
Figure 16-2:	Elementary automation flow.	181
Figure 17-1:	Functional multichannel model for DTV.	187
Figure 18-1:	Proposed SMPTE 305M - SDTI structure.	197

x Video and Media Servers: Technology and Applications

Figure 19-1:	Two-channel server operations.	202
Figure 19-2:	Four-channel server for two-channel broadcast operations.	203
Figure 19-3:	Four-channel server feeds two transmitter systems.	205
Figure 19-4:	Mirrored server operations provide 100 percent protection.	206
Figure 19-5:	Interface for digital streaming tape backup.	209
Figure 19-6:	Multichannel automated servers with archive.	212
Figure 20-1:	Island connectivity.	218
Figure 20-2:	Collaborative server-based editing environment.	223
Figure 21-1:	Types of communications connections.	231
Figure 21-2:	Peer-to-peer and server-based networks.	238
Figure 21-3:	Distributed server model.	239
Figure 21-4:	Bus and star network topologies.	240
Figure 21-5:	Token ring network topology.	241
Figure 22-1:	Basic data frame construction.	250
Figure 22-2:	CCITT X.25 in comparison to the OSI model.	253
Figure 24-1:	Transmission time slot.	268
Figure 24-2:	Letter example depicts packet assembly process.	270
Figure 25-1:	ATM cell and transport packet.	274
Figure 25-2:	ATM cell header.	278
Figure 25-3:	Mapping of MPEG into ATM.	282
Figure 26-1:	Growth in drive capacity.	287
Figure 26-2:	Disk drive prices per megabyte.	288
Figure 27-1:	RAID 1—mirroring.	296
Figure 27-2:	RAID 3—striping with separate parity drive.	298
Figure 27-3:	RAID 5—block record striping with rotating parity.	300
Figure 28-1:	RAID 3 drive failure and data reconstruction from parity.	307
Figure 29-1:	Internal computer bus and devices.	315
Figure 30-1:	Time line of SCSI development.	329
Figure 30-2:	Disk drive structure.	340
Figure 30-3:	Generic sector format structure.	340
Figure 31-1:	Fibre Channel frame format.	345
Figure 31-2:	Arbitrated Loop with four nodes and five ports.	349
Figure 31-3:	Point-to-point topology.	350
Figure 31-4:	Fibre Channel layers.	351
Figure 32-1:	Connecting SSA nodes and ports.	363

List of Figures *xi*

Figure 32-2:	SSA string and loop topologies.	364
Figure 32-3:	SSA switched topology.	365
Figure 32-4:	SSA frame characteristics.	366
Figure 32-5:	Basic SSA-based networked system.	370
Figure 35-1:	Disaster tolerant video server and disk architectures.	395
Figure 35-2:	Data archive and four transport tape array backup.	403

LIST OF TABLES

Table 8-1:	Dual codec processes.	85
Table 12-1:	MPEG-2 profiles and levels.	138
Table 13-1:	HDTV standards references.	147
Table 13-2:	The FCC's famous "Table 3."	147
Table 20-1:	Common broadcast video bandwidths.	222
Table 21-1:	Common data rates for communications.	235
Table 21-2:	Network topology advantages and disadvantages.	242
Table 22-1	International standards organizations (ISO).	246
Table 22-2	IEEE Project 802 committees.	247
Table 22-3:	OSI model.	248
Table 24-1:	Information classifications.	271
Table 25-1:	ATM traffic classifications.	274
Table 25-2:	ATM base layers.	275
Table 25-3:	TC sublayer of the ATM physical layer (PHY).	276
Table 25-4:	ATM cell header functions.	277
Table 25-5:	Information services classifications.	278
Table 25-6:	ATM adaptation levels (AAL).	279
Table 27-1:	Common RAID levels and descriptions.	294
Table 28-1:	Exclusive OR binary functions, related to parity.	306
Table 28-2:	Parity interpretation.	306
Table 29-1:	SCSI transfer rates [1995].	321
Table 29-2:	SCSI structures and implementations.	322
Table 29-3:	Asynchronous SCSI transfer rates.	324
Table 29-4:	SCSI cables, connectors and uses.	325
Table 30-1:	SCSI Trade Association (STA) SCSI performance.	332
Table 30-2:	SCSI performance by configuration.	333
Table 30-3:	Open Systems Interconnection (OSI) model.	334
Table 30-4:	SCSI-3 Architectural Model (SAM) mapping schemes.	335
Table 30-5:	SCSI device classes.	338
Table 30-6:	SCSI commands, protocols, and models.	339
Table 31-1:	Fibre Channel port nomenclature.	343

Table 31-2:	Protocols, mapping, and interfaces.	352
Table 31-3	Fibre Channel classes of service.	354
Table 32-1:	SSA frame components.	366
Table 36-1:	Characteristics of video-on-demand.	409

ACKNOWLEDGMENTS

This book is the culmination of many years of working with, exploring, investigating, experimenting with, and living the vision of digital video and disk recording. Of seeing their value and of making them work with the tools available at the time.

There is a lot of practical application in this book. Some of the material you will find remains in its originally published form, which provides a perspective into how and possibly why things have changed.

The pages between these covers are a labor of love. They explain changes I see coming to the industry I've spent the past 25 years of my life working in. I have continually found personal satisfaction in sharing my thoughts, explaining this ever-changing technology, and conveying the guidance that has been contributed by many—not by just one.

Our industry's standard-setting organizations have provided an invaluable amount of information not only for this book, but to all our associates, friends, colleagues, and professionals. Without the tens of thousands of hours of effort given on behalf of the memberships of SMPTE, EBU, ANSI, and other worldwide standards organizations, we would be a boat without a rudder.

The computer industry has also contributed an overwhelming amount of technology and support for expanding the capabilities of disk drive storage systems, in turn making it possible for the broadcast industry to apply this technology on a cost-effective basis.

There are several individuals with whom I have spent the past ten years who are visionaries in this industry. Each of these engineers, managers, and friends has contributed in many ways. It would be impossible to name them all, so apologies go out in advance to those that should be part of this listing but may have been left out. My personal gratefulness goes out to Ken Ainsworth, Ray Baldock, Hilton Crete, Peter Dare, Geordie Douglas, David Fibush, Hugo Gaggioni, John Hennessy, Al Kovalick, Phil Livingston, John Luff, Tom McMahon, Bill

Miller, Bill Moran, Peter Owen, Bob Pank, Krishna Pendyala, Bruce Penney, Johann Safar, Tim Slate, Peter Symes, Larry Thorpe, Marc Walker, and especially to S. Merrill Weiss—all of whom provided materials, assistance, and support. I can't say enough. All I can say is thanks.

Other credits must go to those who instilled the concept, provided the opportunity, and continued the encouragement to keep the "Media Server Technology" column an ongoing part of *TV Technology*. I'd especially like to give recognition to those who have stood by me over the years I've been contributing. You've made this all worthwhile!

First, to Richard Farrell, who accepted my very first *TV Technology* writings and encouraged me to write the column beginning nearly four years ago. Then, to Marlene Lane, Mary Anne Dorsie, and Traci Sibalik, who've put up with my "just in time" contributions (especially during the assembly of this book) and who had the task of editing my work on a monthly basis. And to *TV Technology's* publisher, Steve Dana, and associate publisher, Carmel King—a heartfelt thank you for believing in me as well.

Next, I'd like to express my gratitude to the editors at Focal Press, in particular Marie Lee and Terri Jadick, who first persuaded me, then convinced me, and finally trusted me to do the work of assembling this writing into what I hope is a valuable and useful product. Moreover, to Mark Schubin—thanks for suggesting that this book could happen and for seeing my column exactly as it is intended to be read.

Many, many people have helped make this book this possible. I am grateful to those who, in their own—and possibly unknown—ways contributed not only to the contents of this book, but to the visions for future of our industry. If I've failed to mention you, it is not intentional—you know who you are!

Lastly, I must acknowledge all the readers of *TV Technology*. Without you, the "Media Server Technology" column would not have happened. Your letters, e-mails, and calls have helped immensely. I hope this work is of even greater value than the previous years of the series.

And with that said, let the journey begin. . . .

INTRODUCTION

The television industry has undergone some profound changes in the 25-plus years I've been making it my profession. All of us who work in this dynamic industry are witnesses to changes not only in the technology of television, but also in the business of television.

As a business, television has metamorphosed from a means of conveying entertainment and information to the most widely accepted method for communication on the planet. Television technology has migrated beyond discrete components for capturing, recording and displaying the image to software-based architectures capable of bridging many new mediums. Video, and the facility to produce it from, is changing from a isochronous to asynchronous environment. Information bandwidths now range anywhere from a few megabits per second to almost a gigabit and half. And video connectivity has grown beyond the bounds of the physical plant to that of the virtual facility.

Yet even with the advances we've seen so far, the broadcasting industry is just now preparing to undergo a profound revolutionary transformation the likes of which it has not seen in its scant half-century of commercialism. In the not-too-distant future, certainly within the lifetimes of most of the readers of this book, we will leave behind the comfort of analog and evolve to the complexities of digital. Ironically, most of the current analog video technology we've learned to live with will become ancient history in less than a generation.

The domain of the broadcaster has already expanded from a narrowly focused point-to-multipoint entity (as in over the air broadcasting) to a multichannel system aimed at a new audience and as of yet, an uncharted changing market. We already know that the next ten years of this technological transformation will culminate in the elimination of analog television broadcasting as we know it today. Television stations have already begun the rebuilding process so as to address the requirements of the future—a future that includes the forced obsolescence of present day technology on a time table that is fraught with decisions about systems that do not exist.

Television equipment manufacturing is also restructuring. With restructuring comes a new focus into areas that traditionally had been left to another industry altogether.

We've witnessed the convergence of television and computer. We've seen that this convergence is far more than just digits and disk drives. The convergence in reality is a collision—of ideas, of structure, and of turf.

The switch in the direction of digital television broadcasting, the acceptance of compressed digital video and storage, video-on-demand, pay-per-view delivery, and multichannel digital television are all elements in this universal shift toward servers for the storage and delivery of video media. This book discusses both operational and technical issues related to disk-based, digital recording and the implementation of video server– and media server–based technologies for broadcast and nonbroadcast applications. We will include developmental and technical discussions of optical, magnetic, and silicon-based solid state storage systems. We will look at the architectures of automation system interfaces, multichannel video server systems, archive and backup, and the all-important issues of fault tolerance and data protection.

This book contains visions of the future and forecasts of circumstances to come. The reader will find that the pages between these covers are laced with predictions. In some cases, you may think the statements are too progressive. In other instances, you'll see that the hype of a few short years ago is now covered in dust. Predictions are opportunities for magnification—and because this book was generated from over 40-some articles written for *TV Technology* over the past three and a half years, it was interesting to retrace the technology historically and perceptually.

The development of this book began in earnest well over 15 years ago. The concept for the *TV Technology* journal article series "Media Server Technology" began with my involvement in the Society of Motion Picture and Television Engineers (SMPTE) and our company's need to build full-bandwidth, non-linear digital editing into a post house that specialized in graphics, animation, and effects. In 1989, having placed Digital Post and Graphics, in Seattle, Washington, straight into the domain of component digital video—we needed devices that were effectively the simplest of video servers we know today. We endeavored to build and integrate multiple disk, full-bandwidth recording long before there were any significant server products available. We found ways to take existing products and force them to do the things we needed, sometimes working directly with manufacturers, often with only our own creative resources. We tried to get those who could make a difference wake up to what our industry really wanted—but hadn't quite realized it needed. At DPG we were not alone, as what we attempted to develop with stones and knives eventually became reality.

The "Media Server Technology" series, which began running in August 1994, was rooted in a desire to explore and to explain new ideas and changing technology. Over the years, I've tried to convey to the readers what is happening today and where we might all be headed tomorrow. In many cases, the concepts presented came from real and practical working experience.

Portions of this book remain as they originally appeared in the continuing column for the *TV Technology* series. There are statements made that were intended as predictions for tomorrow, and there are cases where tomorrow has already come and gone. You will find that some of the content remains as it was in the original context, in some cases now very much in the past. This was done purposefully as it shows some important points that have shaped the development of this technology.

It is my hope that anyone with a technical understanding of video, whether analog or digital, will find answers to many of the questions that surround the whys and hows of (video)tapeless digital disk and server-based video recording. This book provides insight into technologies in place today, and planning for the changes of tomorrow. Some of these concepts are already poised for tomorrow, and others are not yet on the drawing board.

The reader may use this book as a reference, as a tutorial, and as a guide. The pages within contain both history and technology. The intent of this book is to provide a different approach to that of traditional text or reference books. We intentionally avoid providing a lot of formulas and complicated details about MPEG or how to compute disk drive costs. Much of that type of material is already available from any number of other resources, including manufacturers and standards organizations–at any level of technology–and from books by prominent technologists in the industry.

What you'll find here is about potential—the prospects and considerations for implementing media file servers, streaming video servers and data file–based systems, into the television facility of the future.

CHAPTER 1

VIDEO AND MEDIA SERVERS: THE JOURNEY BEGINS

The emerging facets of digital video and recording have been constantly reshaping both the TV and the computer industries. It didn't take a rocket scientist to realize that broadcast companies like Abekas, Quantel, and even the former industry giant Ampex, had cleverly visualized the future of recording for the television medium. New companies, such as Avid Technologies, conceived the challenges that lay ahead. All the players now realize that since the invention of videotape recording there hasn't been a greater shift in the video industry, until now.

We are witnessing a paradigm shift in the way the video medium is stored and retrieved. With this shift comes random access, instant recall, and the potential elimination of videotape usage as we know it today. Video storage will forever be recast in both purpose and principle.

What follows is a reprint, with minor edits, of the original column this author contributed to kick off what has been a journey of more than three years through a constantly refining technology. You will see firsthand how initial perspectives have changed and how our predictions, for the most part, either have been fully realized or have radically departed from the state of the art.

2   Video and Media Servers: Technology and Applications

It was August 1994 when *TV Technology* began a series of columns focusing on new media and video server technology. The new media being discussed were not actually new, and the server *technology* employed had been around for years. What was new about media server technology was its impact and its integration. This period marks the dawn of the next major change for moving images and delivery systems bound for the 21st century.

As this new technology evolves, the video technologist, engineer and producer are each facing new issues and asking new questions. Never in the history of both the video and the computer industries have there been more choices and broader decisions to be made. While the information superhighway is under construction, many of us are probably seeking the answers and opinions to hundreds of new questions.

While sifting among the hype, the newness, the flash and the confusion, it will be our goal to provide insight and vision into the future of our distribution and production mediums, and how the new media server technology of tomorrow will impact all of our lives, both professionally and personally.

This first monthly installment [written for August 1994] will focus on some new terminology and a general look at purpose. We will start with a generic, nontechnical overview of what compression technology is about, then move into a summary of broadbased definitions that will serve as the foundation for future installments.

None of our definitions are ever cast in stone. In truth, the very manufacturers creating and implementing this emerging technology are all searching for appropriate definitions and applications.

IS ALL VIDEO COMPRESSED?

In terms of today's understanding, video signals are sets of samples made up of luminance and chrominance elements, and structured into interlaced fields that compose a series of frames. These frames are generally displayed linearly to produce the moving image.

It should be acknowledged that much of any digitized video signal contains redundant information that adds insignificant data to the bit stream. Deciding, in real time, what is redundant and what is actually required to reproduce an image or series of images, and at what level of integrity and visual accuracy, is the process of encoding/decoding.

This process is independent of the format, whether it be an analog NTSC signal, a full-bandwidth CCIR-601[1] signal or a 45-Mb/s compressed digital video signal.

The storing and retrieving of video and audio signals, to and from disk-based recording media, in real time, is possible because of at least two technologies. First is compression—the optimization of the digitized data stream applied to the intended application. This is valid for all modern compression technology, whether lossless or lossy, and is essential to understanding the applications in media server technology.

Second is the understanding of the principles being applied to massive data management, in terms of both software and hardware systems, including the various techniques and methods of media capture, storage, retrieval, delivery, error correction, integrity, availability, protection, and fault tolerance.

Over the next several installments, we will spend a proportionate amount of time reviewing the technologies associated with digital media and server technology. We will certainly be referring to the applications of this technology as "new media," even though in many instances these principles have been used in broadcast graphics and image processing for better than a decade.

DIVIDING VIDEO AND THE COMPUTER: NO MORE

Many of the heretofore "exclusive to the computer"—based technologies are now being applied to a new form of media communications—the moving image.

Until recently there was a definite dividing line between video and computer generated imaging. The motivation for the convergence of these two powerful technologies is deserving of more than just an editorial comment. Perhaps the emergence of the computer workstation for motion picture entertainment has been the best monetary reason to step up to the starting line and get in the race. While it might be argued

---

[1] CCIR-601 has become the generic reference for the D-1 format. The industry has connoted this to mean "D-1", or in general terms "digital video."

The proper notation is now Recommendation ITU-R BT.601 with subsequent "dashes" following the 601 to indicate certain revisions. Rec-601, as it is sometimes shortened, is the international standard for digital video, with a sampling frequency of 13.5 MHz with $Y'C_BC_R$ coding.

that "special effects" was a reason for this inevitable migration in the entertainment industry, most recently the Clinton Administration's astounding promotion of the "information superhighway" changed the direction and the flow of two major industries in a profound way.

While Advanced Television (ATV) was moving along at nearly the political speed limit, it was being passed in a blind lane by a much larger and more powerful computer industry. [At that time] ATV had been motivated with three prime objectives.

First, the desire to spawn a new manufacturing segment in television receivers meant that some 250 million television sets would need to be replaced over a 20-year period. That's big bucks!

Second, the intense desire to convince a mass audience that the only way TV can look better is if it has a wider picture, more lines and more closely approximates the movie theater. Arguably, this is of little interest to the average TV viewer who thinks VHS recordings are "just fine" and is just beginning to accept the concept of 15 frames per second with 256 colors on a computer screen at 352×240 ($H \times V$) pixels (give or take a couple hundred).

And finally, and probably the most important overall, next to the astounding political ramifications, is that advanced digital imaging technology might well find its way into every crack and crevice of both leisure and business—through the understanding that video does not necessary have to live only in the living room or the sports bar.

With a foundation for the reasoning behind new media technology, let's leave behind the (editorial) politics and hitch a ride on what's making this really happen. This relatively new technology has been evolving from a few distinct product lines into an all-encompassing storage, processing, and delivery medium. The inevitably expanding future is expected to be much more than a spinning disk and certainly more involved than merely converting to a "tapeless" industry.

OPEN AND SCALABLE GOALS

Some of the generic goals and tasks of this new media technology will necessitate open and scalable multiserver architecture. The media server systems of the future will need to have such inherent technical features as automatic load balancing, fault tolerant transparent changeover, automated real time error correction, and self-healing diagnostics and data repair capabilities.

Most likely, these systems will initially employ MPEG-1 compression. A transition to MPEG-2, once volume chip sets and subsystems become economical to manufacture, is inevitable.

A caution is warranted—yet the suggestion to "wait and see" is ill-advised. There are many sides to the issues; some are technological, others are purely business related. Remember, what you see today is most likely to cost half as much in 18 months, and it will be superseded by twice the computer horsepower to boot.

ARCHITECTURE IN SERVERS

In the technology-based sense, the term "architecture" is a descriptive word that basically means "how it's made up or put together." The meaning of architectures in the computer, server, and applications models will vary greatly as technology expands, guided by demand.

For example, the concept of a symmetric architecture, where systems will be capable of simultaneous read (playback) and write (record) processes, is something we will all expect from media servers. Other architectural specifications will include superior sound and video quality, which will be expected, and in most instances achievable, at far less cost than existing tape technology.

Networked architectures will be created as the technology and the applications improve. Entire systems will be built around the local and the wide area network concepts—although the development of those systems on other than private networks is expected to be a long time away.

MEDIA SERVER TECHNOLOGIES DIVIDED:

The industries involved in developing this new media storage technology are focused into a few broadly defined groups. One of these is the broadcast video server (BVS), which might be thought of as a family of "bullet-proof" servers that are intended for the broadcast environment. BVS applications may include broadcast commercial insertion, traffic management, overlays for promo insertions, time delay, and spot/program backup.

6 Video and Media Servers: Technology and Applications

In the case of live news, irrespective of future concepts for interactive television, the ability to instantly shuffle the entire news show for any reason exemplifies the integration of the BVS concept into daily broadcast.

While the BVS may be somewhat categorized on its own, it truly is another form of a locally managed, internally controlled video-on-demand (VOD) interactive media server. This logical extension takes the BVS concept and adds multiple streams of outbound video, selected entirely by the enduser according to some finite number of parameters defined in the "traffic" or control management software. Depending upon the size of the server, these number of streams may vary from a few (four or five) to literally tens of thousands.

One of the primary distinctions in the VOD system is in the BVS's level of fault tolerance, a factor that broadcasters have been expecting for years as an ongoing commercial revenue generating capability.

VOD systems, as mentioned, take the multiple VCR play concepts of the existing hotel pay-per-view (PPV) systems and create a more economical and flexible system that virtually takes the human interface and mechanical dependency out of the loop.

Preliminary markets for VOD systems now appear to be hospitals, cruise ships, hotels, and oddly enough, even the penitentiaries. The potential next-generation acid test may well be the home cable delivery system. Here, more extended and varied choices prevail and the physical number of customers (and choices of services) will be orders of magnitude greater than those required in private carrier distribution.

ELECTRONIC SLIDE PROJECTORS

Another area for this new media server technology is also not very new. Still store servers (SSS) are perhaps the oldest and most familiar of the disk based technologies to the broadcast industry. Within this same family, the electronic still store (ESS) systems have been used on the air and in cable messaging systems for over a decade. The ESS is a direct spin-off of the high-end character generation revolution of the early 1980s.

Today's technology has already extended ESS functions to include scalability and multiple user access in higher bandwidths than ever before. SSS/ESS systems may also be applied as a simple subset to the VOD or BVS, since there is nothing magical about displaying a single frame of a clip instead of a linear series of frames that compose a movie or commercial spot.

PRODUCTION QUALITY VIDEO DISK RECORDERS

In the higher-end media servers, the production and post production community will continue to demand the quality and performance of higher resolution full-bandwidth systems. This will extend to High Definition and Advanced Television for obvious reasons.

The digital video disk recorder (VDR), or until more recently, the digital disk recorder (DDR), has been evolving in both an extended feature set and extended storage capabilities. Storage capacities for VDRs/DDRs are now approaching the size, cost and capacity of modern digital videotape recorders. In the post production community, the VDR/DDR systems are already becoming as common as the second or third VTR in the edit suite. It is expected that VDR/DDR systems will eventually merge directly into the media server domain, expanding flexibility and multiplying user capabilities to that of the application-specific BVS or VOD.

INFRASTRUCTURE OF DESIGN

For the present, there seem to be at least three broadly defined schools of thought in new media server technologies. First, there are those companies (and their alliances) that promote the supercomputer system approach. Another school is looking at extending the capabilities of the "good ol' PC" or workstation concept. A third class is promoting a combination of dedicated drives and video engines, special-purposed for their particular market application.

Each camp has its benefits and/or trade-offs. Benefits in some cases include reduced price point at the expense of bandwidth and the benefits of increased storage. The opposite end of the spectrum is a higher cost with advanced features and moderate storage, but a superior image and performance.

Some approaches employ self-standing and closed architectures, where the manufacturer intends to produce the hardware and dictate the feature set and the control or operating software. This, in turn, permits a specialized application for a focused direct purpose.

Another approach appears to be a software-only concept, one where hardware is independent and scalable to suit the application (and the budget) of the end user.

In-between are companies that recognize their products as devices that can stand alone but are actually designed to be part of a subsystem (i.e., a media server). They expect to have their products integrate via control software so that any number of other manufacturers can implement full system designs. Such a system might find itself becoming a plug-in solution with a high degree of flexibility and application.

Hybrid systems might include a combination of such concepts as a virtual recorder—a device that effectively combines memory storage, magnetic disk, and magnetic tape. Data recorders and optical discs that are capable of recording data at rates faster than their playback speeds in essence will allow recording and archiving capabilities at greater than real time.

The principals of this concept will allow downloading to local servers, including future set-top home devices, at a rapid rate. It is even conceivable that both movies and other video streams might come to the home simultaneously without either of them influencing the other.

THE POST PRODUCTION INFLUENCE

There is a very strong interest in the post production communities in disk-based server and array technology. While it is expected that "tape" will be around for years to come, the fluid integration of optical and magnetic disks with that of a data streaming tape will change the meaning of "tape storage" as we know it today.

At least two of the major players in the videotape transport industry multipurpose their mechanisms for both video and data recording and retrieval. Automated tape systems (i.e., "library management" systems) are in heavy use as a replacement for the older nine-track reel-based tape drive systems of the last several decades.

The obvious extension to the post production community will be the networking of these larger database-structured imaging libraries for the repurposing and extension of existing stock footage and graphical elements.

The arts and photographic mediums will be able to store and retrieve incredible amounts of digital imagery and in turn will eventually solve the licensing and royalty issues through automated means rather than through the uncontrolled methods in place today.

And the ability to permanently store imagery to CD-ROM-based media and correlate it through libraries will change the face of the reference-related industries.

IT'S ONLY AS GOOD AS THE SOFTWARE

Software solutions may have the most significant long-term influence on media server development. At first, it seemed that hardware was the driving force behind expanding system capabilities. Now that off-the-shelf, generic components, such as SCSI drives, have proven to be reliable and pliable, hardware based systems have stabilized. So now it's software's turn to tax and grow the capabilities of the modern media server system.

Systems seem to evolve over time. New ideas and uses for the combined hardware and software technologies seem to wait in the wings for new engineering concepts to emerge.

For the longest time it was the hardware engineer who faced the product design problem. Engineering would create a new product based upon what the engineer thought could be accomplished given a certain new set of guidelines and hardware-based tools. Digital design, compression, and software applications have subsequently reshaped that concept—and now it appears to be the code writers' turn to make the technology really have a purpose.

Entire divisions within software companies are now devoting full-scale efforts to harnessing the power of these established and proven hardware structures. In order for a system to truly make money, which is the real goal of all of this technology, a system must be engineered to control, process, track, integrate and respond to all the demands of the customer, who ultimately pays the freight.

Since software, ultimately, is the most flexible and moldable item, it makes sense to look at the viability of a software architecture in the same venue as a hardware structure.

10   Video and Media Servers: Technology and Applications

Software and hardware must continue to live in harmony. As market testing, trials, demonstrations and continual feedback shape the growth of this "new media" server technology, we hope to clarify what's going on and how to use this technology in a productive and profitable venue.

That was the state of the industry in mid-1994. As this recitation moves forward, we will explore the functionality and the technology that began to emerge only a few years ago. Various new models for server environments continue to develop.

We—as users and consultants, designers and integrators, manufacturers and inventors—can all expect monumental changes in the applications of media and video servers as networking and digital television continue to expand.

CHAPTER 2

DISK RECORDING: A BRIEF HISTORY

In this chapter, we will recap, in condensed form, the history of the digital video disk recorder and its application to core industry technologies—as we understood it up through the early 1990s. The original article, published in December 1994, has been expanded to provide more detail surrounding the uses and the technologies that formed the mold for disk recording between the early 1970s and start of the 1990s.

Recording images onto a rotating surface has a long history. Looking back in time, the first records of digital magnetic recording on a disk surface were made around 1951. Random-access video-on-demand was predicted as early as 1921—and was said to be available as early as 1950. Magnetic recording was proposed in 1888, only about 43 years after magnetic polarization rotation was discovered.

The concept of recording video onto a spinning platter was demonstrated in the later part of the 1950s, nearly in sync with NTSC[2] (1954) and videotape recording, which became commercialized in 1956. The first video recording standards (1959) and the early solid state computer (1958) were all-important markers that would contribute to the development of digital recording on magnetic disk drives.

---

[2] NTSC: National Television Systems Committee, established in 1941, first developed 525-line, 60.00 Hz, 2:1 interlaced, monochrome television. Today, NTSC is often misstated as the National Television *Standards* Committee.

12 Video and Media Servers: Technology and Applications

For the professional broadcast industry, the introduction of the video disk recorder has functionally changed from a device used for sports replay to a common tool with extensions well beyond single purpose applications. The use of hard disk drives for still stores, animation and editing was just the beginning. Following the introduction of video compression in the early 1990s and the extension of the personal computer for graphics and editing, we've witnessed a fundamental change in the way video is recorded, stored and manipulated.

As we continue to reflect on the various stages that the development of video disk recording has transcended, we will see why the steady development since the 1990s has changed the way we will all do business. We will investigate only a few of the high-water marks in disk recording as they represent the major strides that took the concepts from a product to an industry. There have been literally hundreds of take-offs from each of the key major products in the history of disk recording, and it would be next to impossible to comment on them all.

FIRST-, SECOND- AND THIRD-GENERATION DISKS

The most remembered first-generation disk drive employed in the broadcast industry was most likely the Ampex HS-100. Introduced in the late 1960s, this device marked the entry point for slow motion and instant replay of recorded video images. By 1993–1994, technology for recording moving images to disk media had developed considerably and was considered to be in its fourth generation of development.

What had begun as a limited application with limited feature sets had grown in capabilities and respect. It was right around 1993–1994 that the disk recorder had matured and the division between it and the video server was about to begin.

Manufacturers continued expanding their quest for what would amplify the dimension of high-quality component digital disk recording in a new storage medium. Where once the broadcast industry saw the disk recording as a medium well suited for sports replay, it was now visualizing it as the backbone to its very operations.

The early video disk recorder (popular from the mid-1970s to the mid-1980s) remained in its own limited domain for quite some time. Around 1985, magnetic disk recording started its gradual move toward a second-generation stint primarily in video still stores and paint systems for graphic applications. Several specialized recording systems on proprietary hardware platforms were developed and sold. This period demarcated the move away from large-scale supercomputer disk drives to small form-factor conventional Winchester-type drives.

It would take several more years before the large-scale drives would vanish from the market place. The abundance of fixed and removable hard disk recording assemblies made it easy for those manufacturers who were serious about developing a product to carve a niche in the market. However, once the 5-¼" and 3-½" form-factor hard drives became producible in mass quantities, the door was shut for large form removable storage devices for the video industry.

The third-generation disk recorder saw a return to short-term, limited-length storage for the post production and editing environments. The early 1990s brought an even more important and broadened use for disks beyond exclusive applications seen only at the highest end of production. The change was demonstrated once the transition to non-linear and random access editing and storage was in full swing.

Disk recorders matured rapidly once the workstation, as a PC, Mac or RISC system[3], became a more prominent element in the industry. Video could now be stored on conventional form-factor hard disk drives in a variety of formats, using both compressed and uncompressed video data. In the years that followed, both the applications and the acceptance of disk recording matured well beyond adolescence—just in time for the video server to begin breaking new ground.

Industry envisioned the third-generation applications for the disk recorder as a supplement to accelerating the editing and post production process. As still stores, character generators, B-roll disks, and stunt recorder devices became increasingly widespread, so did the demands for the post production edit room. Even film-based motion pictures and animation special effects saw a significant use for the disk recorder as storage capacities grew economically.

Ever since the advent of the computer-assisted editor, the industry's craving for faster, more sophisticated editing has grown. In the mid-1970s, experiments and eventually real products began coupling

---

[3] PC refers to the IBM or Intel based Personal Computer; Mac refers to the Apple Macintosh; and RISC, for Reduced Instruction Set Computer, refers to workstation-type computers—such as Silicon Graphics Incorporated—that use processors of a streamlined nature that execute instructions more quickly and more efficiently than CISC, or Complex Instruction Set Computer, processors.

14 Video and Media Servers: Technology and Applications

computer disk drives with conventional linear tape machines. Under the control of a light pen or keyboard-based, computer assisted editor, disk recorders temporarily removed the requirement of a third, at that time quadruplex VTR. This was the original conception of video disk-based editing.

Acceptance was limited, although some products actually matured to exclusive disk-based non-linear systems; for the most part the expense and limited availability of these systems kept videotape as a primary source for editing.

As time and technology advanced, special effects and graphics began to move further away from film-based opticals and real time layering on videotape. Experimentation with disk-based, proprietary systems integrated with linear tape machines for source and output was beyond the drawing boards. As soon as video could be compressed acceptably, editing shifted from linear tape to non-linear random access editing—mostly for use in off-line production. The image quality of early compression systems was certainly not sufficient for release, but it proved to be an extremely successful adjunct and preparatory stage before the final on-line session was begun.

Due credit must probably be shared with the special effects graphics industry and the manufacturers who took risks to develop and promote the new technology. It was this group that stood up and let the world know that disk-based storage was needed and achievable—even though the early costs associated with the physical hardware were excruciating.

With a basis for growth in an emerging and maturing technology, the vision of video on a spinning magnetic platter was becoming acute. As mentioned, from a historical perspective the video disk recorder (VDR) or digital disk recorder (DDR) was a technology aimed originally at the high-end video post production industry. Nevertheless, the industrious minds behind it saw a future. They were persistent, and with the help of the exploding personal computer industry, they concocted the formula that would become the standard editorial process.

THE SLO-MO DISK RECORDER

In the professional broadcast industry, the disk era began with the birth of the replay in live sports. The most notable device of its time was Ampex's slow-motion replay system.

Looking some 25 years back, the household word among broadcasters for professional disk recording applications was the Ampex HS-100 series disk recorder. By end of 1976, television sports broadcasting without slow-motion replay was nearly unthinkable. The only accepted device at that time featured not just "slo-mo," but freeze frames and reverse motion. These commonly used features for program and spot commercials all relied on the SHS-100.

In 1976, the Ampex HS-100C, although only a few years old, touted some exceptional technical features. It included a derivative of the AVR series digital time-base corrector. Ampex advertisements described a subsystem that automatically lifted the heads for disk-surface protection purposes. Operational features included fast search and a twice-normal-speed playback. Search speed reached a viewable 4.5 times normal speed. At that time, only one device offered these kinds of high-end features. It would be a few more years before we'd see the birth of Type-A 1", the first production level, broadcast designed, stunt capable open-reel VTR. Ironically, as with most of technology, these "exceptional" features are things we all take for granted some 20 years later.

The HS-100 was considered a portable unit. It consisted of three separate units: a disk servo, an electronics unit, and an output-processing unit. Each unit weighed between 145 and 177 pounds and had a power requirement of 120 volts at 20 amps. The signal electronics, similar to the Ampex AVR-2, were already proven worldwide.

The digital frame buffer, capable of holding an entire frame of video, was not yet ready for prime time. This required the heads of the disk recorder to constantly step, or retrace over the same physical space on the drive surfaces, in order to play back a continuous still frame. One of the drawbacks to the disk recorder when constantly used in the mobile environment was that sometimes intense routine maintenance procedures were required to keep the stepper motors and end-stop adjustments aligned. Physical damage from head/surface contact was always a risk and there was always a possibility of damage due to dirt contamination.

16 Video and Media Servers: Technology and Applications

All told, the Ampex HS-100, with its ability to store up to 1800 fields NTSC (1496 fields in PAL), using analog recording and employing four steppers to cover the four surfaces on two rotating metal disks, was an amazing product.

There were other applications for the Ampex "disc"[4]. When not used for the live sports business, the disc could be found in the edit bay. In some of the early development for editing systems, the CBS-MemoreX (later shortened to CMX) editing system used the disc for temporary B-rolls. The disk-based recorder permitted A/B transitions such as dissolves and allowed still image reproduction using just a pair of quadruplex 2" videotape transports and the disc. One quad would be used as the master, or record VTR, and a second, which held the original material, was used as the A-roll. The disc recorder acted as the B-roll when transitions were needed. Other video sources would include a C-VTR or ancillary video sources as required.

Greater-than-single-frame transitions, usually dissolves or wipes, were possible by dubbing only the lap dissolve segment from the record VTR to the disk. This created just enough of the transition, a short B-roll segment, so that the disk could be used during the lap dissolve or other transition. At the start of the preroll, both 2" quads (record and A-VTR) would roll. If the transition was a 30-frame dissolve, at one frame before the edit in-point, the disk would roll and the record VTR would go into insert edit. The transition would trigger, playing back what was previously on the record VTR (and was now on the disk), and the dissolve to the A-roll source would take place. Once the transition was over, the disk would pause and the A-roll would continue until the edit out-point.

An outboard production switcher accomplished the electronic elements of the dissolve, wipe, or key transition. If the edit failed or was in the wrong place, only two VTRs needed to recue before the edit could be performed again. Since the original material on the record VTR still remained on the disk, it could be appended back onto the record VTR and nothing was lost.

---

[4] In Ampex's literature their product is called the "disc." To clarify the distinction, we will adopt the description found in *Wired Style* for this book. "Disc" shall refer to both read-only and read-write optical-storage devices and shall include optical-discs and laserdiscs. The term "disk" shall refer to the magnetic coated medium we relate to in hard disks, floppies, and the like.

Management of the B-roll track and edit decision list (EDL) became the task of the CMX editing system, whose founding principles would later become the mainstay benchmark for the edit suite. This principle is still used in modern edit systems for rendered dissolves (in non-linear edit systems) and in systems like Accom's RAVE, which uses disk recorders and linear transports for the same type of functions.

DISK RECORDING ALTERNATIVES

This infant video disk recording technology grew slowly over the next several years. The 1" Type-C videotape machine (circa 1977) temporarily stalled further development of the disk recorder beyond the 30-second analog-based disk recorder. A narrow range of products was adapted from the concepts of the disk recorder. These included a different form of analog, flexible media disk system intended as a lower-cost solution for sports and still image recording and playback.

In the early to mid-1970s, this flexible disk system employed analog principles similar to those used in the U-format video recorder, invented by JVC in 1969, which electronically used the "color-under,"[5] or heterodyne, system of recording. In this application, the recording surface was a replaceable flexible platter, approximately 12" in diameter, and made of material similar to today's floppy disk. A stepped motor and floating head arrangement, similar to that on a floppy disk, tracked the analog recording across the surface of the revolving media. The disk media itself had a limited life, and image quality was somewhere between VHS and 3/4" resolution.

Color-under is a form of composite video where the chroma is heterodyned onto a subcarrier at a frequency under (below) the subcarrier frequency of NTSC or PAL. The subcarrier is stabilized with a crystal lock but is not coherent with the television line rate. Heterodyne video recording separates the luma (Y) and the chroma (C), recording the luma in frequency modulation (FM), as in direct recording, and the chroma mixed down to a subcarrier frequency somewhere between 600 kHz and 1 MHz. A time-base corrector is not required to see a picture on a conventional receiver but is generally suggested for playback in a professional and broadcast system.

---

[5] Color-under is the recording technique used in Betamax, VHS, Hi-8/8mm, and best known in the ¾" U-Matic videotape recorders.

Even though the low-cost flexible slo-mo disk was not widely recognized, this device provided a more economical alternative to the more expensive quadruplex-recording principles of the HS-100 disk recorder.

Optical formats using laserdisc recording technology matured, and the principles of magnetic recording would be sidestepped for a while as the optical CLV and CAV community began to grow. From the professional applications of laserdisc recording came the home laserdiscs we know of today. Giving credit where credit is due since its introduction, there have been literally hundreds of various applications for laserdisc recording. Laserdisc technology has proved to be quite satisfactory up through and even beyond the advent of video on the CD-ROM for the personal computer.

DIGITAL RECORDING FOR STILL IMAGES

One of the next steps in advancing toward digital video storage on computer hard disks came around 1984–1985. This second generation of ancillary recording and image storing equipment for video editing and on-air applications began to change dramatically in the early 1980s. Still stores came from companies such as Ampex, ADDA, and Harris, among others, and continued to grow as a fledgling personal computer industry developed. The development of the personal computer thrust storage technology development into warp speed and imaging on magnetic disks into a whirlwind of proprietary formats.

As serious professional digital video production began, the UK-based company Quantel produced the first true non-linear random access digital video editor and effects compositing system. Quantel called the product "Harry," and it was a revolutionary piece of technology. Harry, which grew out of the industry standard Classic Paintbox and its drive storage principles, was truly digital and brought so many new functions, features, and approaches to video production that one could spent volumes describing them.

Quantel recognized that still images, which had been stored on computer disks since the early 1980s, could be recorded sequentially and played out in the same fashion. Their entire concept was based upon total and completely latency-free access to any individual or multiple series of fields stored over their four high-speed, OEM-provided and unmodified parallel transfer disk drives. Until Harry, no one had seen the true potentials of digital disk recording.

Harry used these highly reliable parallel transfer drives, something of a convention in mainframe and supercomputer drives at that time, to store over 80 seconds of run-time digital video. The video frames could be sequenced in clips that were edited or manipulated by the operator and could playout in real time. Complex processing (layering, keying, or manipulative effects) was typically done in non–real time and stored back to the disk drives. Video play out was always in real time. Harry could also make appropriate adjustments in the 3/2 pull-down properties of film to tape transfers, making it the first video product of its time to disentangle 30 frame video to 24 frame film and then reassemble (or stretch) the 24 frame properties back to 30 frames per second NTSC video. Only a disk recorder could accomplish this feat as well as reverse the interlace properties of video–both are tasks essential to the animation or rotoscoping processes for video graphics or special effects work.

The Harry was a premium priced system that employed four large Fujitsu disks running in complete synchronization to provide a continuous series of discrete video fields (frames). Each video field was always stored in exactly the same location on the appropriate disk drive surfaces. The data would be extracted off the disks to a multiplexer/combiner system inside the Harry chassis, where it was assembled into a bus-based, parallel digital bit sequence on a proprietary internal high speed bus (the "Q-bus"). Inputs and outputs could be in RGB/YUV[6] or composite NTSC. Later a CCIR-656 interface upgrade was available, but at the introduction of Harry, there was no standard in place.

It took all four disks running in sync to produce each full-bandwidth digital video field. Because each field was independently addressable, reordering any field in any sequence became as simple as "tapping down" on the pressure-sensitive pen and tablet combination, whose menu was displayed on a "heads up" GUI in front of the artist.

Harry was so revolutionary and contained so many features that it was rarely understood for its underlying technological principle–high

---

[6] The YUV term is used generally to define a component system of luma, and two color difference signals B−Y and R−Y, respectively. Y'UV is scaled for encoding into composite video, such as NTSC or PAL.

quality, random access, digital disk recording. Harry is still recognized far more for its effects abilities than for its disk recording concepts. The core feature of Harry, embedded in its disk recording capabilities, became the fundamental backbone for the sophisticated video and media server systems of today.

Even though Harry was limited to about 80 seconds of continuous running D-1 (component digital) quality 8-bit video, the idea of true random access recording and editing, coupled with infinite internal copying or layering, became the quintessential axiom behind the effects post production process we know today. Quantel found ways to make 8-bit video every bit as good as many 10-bit systems by perfecting a look-ahead, look-up, and recall principle that dithered only the appropriate bits of the signal. Quantel called this "Dynamic Rounding" and has since licensed its concepts to many makers of digital video products.

Quantel's Harry was targeted at the top of the scale of high-end production facilities. This is evident from the purchase price and more importantly, from the results achieved. There essentially was no other way to produce the multitude of effects possible in Harry without resorting to film.

Shops that purchased a Harry had or were going to specialize in graphics or special effects. Yet, over time, some of these same shops viewed Harry as more than just an effects/graphics device. Artists recognized that products other than commercials, such as short-form programs and laserdisc interactive, could be produced and edited on Harry in rapid fashion. Even though time line editing had not been perfected in the conventional edit suite, the Quantel concept of visual editing gave way to a new and completely different method of assembling program content. Many couldn't wait for the technology to develop—and for prices to go down.

By the end of the 1980s, Harry had become the benchmark for high-end video graphics and compositing. In 1992, Harry's successor, Henry, was revealed. Henry added features such as a full five minutes of expanded storage and multiprocessing capabilities with an expanded user interface. Henry would use Quantel's advanced disk system, called Dylan, coupled with RAM in the disk interface to act as a buffer (this process was referred to by Quantel as Chatter Disk Management). Multiple Dylan drive arrays could be added to extend storage, and these same drives became the fundamental storage device for all Quantel disk storage systems.

Although most Harrys have been either replaced with Henry or seriously upgraded, the concepts in both products remain in widespread use worldwide. The concepts have been applied to other Quantel products in editing, compositing, and effects manipulation in formats from 525/625 up through high-end film scanned digital optics.

GETTING TO DIGITAL

At the same time Harry was introduced, a new evolution in disk recording was born. A spin-off/start-up company, Abekas Video Systems, was developing a technology that would radically change the landscape in digital video disk recording.

In 1985, Abekas unveiled its A42 Video Slide Projector to address a series of issues related to image capture and storage. This affordable still store system incorporated conventional 5-¼" form-factor Winchester drives in the WD100 (40 MB) and WD350 (140 MB) capacities, to store 100 frames (200 fields) and 350 frames (700 fields), respectively. Each Winchester drive had three platters and used both the upper and lower sides of each platter for recording. Additional storage drives could be added to bring capacity to 2010 fields of video online. A streamer tape drive could store pictures off-line, holding 50 frames (100 fields) on a conventional 20 MB computer data cassette.

The Abekas Video Still Projector incorporated somewhat conventional computer drive storage technologies and adapted it to video signal systems, using binary coded data, or digital video coding, as its means to capture and store still frame images.

The next major change in recording technology was two more years away. It was 1987 when digital video recording on linear tape would be introduced. Sony's entry into digital video came with the DVR-1000—the first component digital (D-1) video transport. The CCIR-601 and CCIR-656 digital video standards were still on the horizon. Once those standards were in place, a framework from which manufacturers could conform for developing digital disk recorders was solidified. From there the disk recorder I/O and data structure would allow for universal conformity.

Although the success of the D-1 recording format was to be shadowed by both the cost of the D-1 and the introduction of the D-2 composite digital recorder, the structure of the D-1 format would eventually win out. The D-1 format would eventually be proven superior in image quality to other forms of recording in the 525I/625I-line[7] structure. Over time, the component digital video format would become the basis for all serious coding of video in component formats.

However, before the 601 or D-1 format firmly took its place as the next generation in video, a somewhat unfortunate and probably unnecessary side step occurred. Composite digital or D-2, the 143 Mb/s, 4f_{SC}-sampled (four times the frequency of 3.58 MHz subcarrier) NTSC digital video format, the *substitute*, was introduced.

By substitute, we mean that in 1988, manufacturers recognized a set of rationales for the phasing out the 1" Type-C format open-reel videotape transport. The cassette was becoming more accepted as a storage medium. The D-1 format digital recorder was far too expensive for the average production house. Betacam and BetacamSP, at the time an acceptable analog recording technology, still suffered from the same types of effects as 1"—most notably dropouts and generational copying losses. In addition, manufacturers needed an alternative solution that would put the post production industry into digital without having to immediately rebuild their entire infrastructure.

Both Ampex and Sony developed their own solutions to digital. They brought out composite digital video and then fought feverishly to promote these new D-2 products. All the while, the only two other digital VCRs—one D-1 model from Sony and one from BTS—followed a different path. The concept of digital was here and despite the hardships and trials, there would be no turning back to analog recording.

For the early adopter being digital connoted perfection. However, the magic and uniqueness of the digital concept would have a relatively short life span. Only a brief period would pass before the marketing efforts of D-2 took hold for the production and post production arena. The technical problems associated with composite digital, which in effect only digitized an analog signal, would be overshadowed because many believed that as long as you had D-2, you were digital, and that was all that mattered.

---

[7] The representation of "I" (either upper- or lower-case) will be indicative of "interlaced" scanning. "P" will symbolize "progressive."

Disk Recording: A Brief History   23

Digital was new, it was exciting, it offered new possibilities, and it was expensive. Once digital video recording was established there also needed to be a disk recorder to round out the complement. While a few recorders actually had composite digital inputs and outputs, most of the disk recorder manufacturers shifted to component digital recording, even before the demise of D-2 and long before the acceptance of compressed digital video.

Not long after the transition to D-2 was underway, the production industry and its delivery medium, the broadcast community, began to understand and accept what digital was about. Seen mostly by example, the confusion and the cost of the transition to digital were apparent. It would be a while before digital would be accepted in all areas.

Toward the end of the 1980s, the broadcast community as whole would find it difficult to invest in the new digital videotape recording technology. In the late 1980s, broadcasters were still having a hard time getting agencies to deliver spots on anything other than 1". Betacam cassettes were holding strong but were not accepted universally by agencies. Quad 2" VTRs, last exhibited at the 1983 NAB show, were still in use, although in just a very few places. The ¾" tape remained "just fine" for hundreds of stations in the smaller markets. These facilities had made that shift years before and looked upon S-VHS and 8mm as alternatives before they'd accept a new digital tape format.

PRODUCTION COMMUNITY GETS CREDIT

It was indeed the production community, generally credited with developing visions into reality, that saw very quickly what digital could do for them. As soon as D-1, followed shortly by D-2, was introduced, multi-generation, high-quality recording and compositing (layering) began to be requested at post houses. Production effects, using 1" tape for layering, would all but disappear once the multiple-generation capabilities of digital were demonstrated.

Digital tape mastering and digital disk recording would eventually become the quintessential format for the future. But by the end of 1989, even with two new videotape recorder formats in place, many operators still did not fully understand digital. There just wasn't enough ancillary support equipment around to make effective value out of digital.

We had all become familiar with digital from the time-base corrector and the font in the mid-'70s. The many flavors of still stores had also been around for at least half a decade. Still, it would be the start of 1990 before the next generation of digital products, following Harry and the Abekas A62/A64 disk recorder line, would really be put into the perspective it needed to be in.

Beginning in 1990 or so, industry predominantly selected JPEG, which stands for Joint Photographic Expert Group, as the solution for integrated editing and storage of digitally compressed images. JPEG, which is a lossy compression format, was developed as a method for compressing photographic images for electronic imaging uses. Non-linear editor manufacturers capitalized on motion JPEG, and it has since become the most widely accepted method for compressing and storing video on smaller form-factor hard disk drives.

MPEG, which stands for Moving Pictures Expert Group, is the acronym that describes both spatial and temporal video and audio compression. In the mid-1990s, MPEG-1 was perceived to be a staple for CD-ROM storage, especially for the playback of video and audio streams in multimedia applications. Other multimedia formats, such as AVI (audio-video interleave) and Indeo (from Intel), have been developed for compressing digital audio and video images. MPEG has hence become accepted by the PC and multimedia industry as a valid method of delivering moving images on computer platforms. MPEG is also the basis for DVD (digital video or versatile disk) the forthcoming promise for delivery of higher quality motion pictures for home uses.

Motion JPEG, MPEG-1 and MPEG-2 applications for the storage and delivery of video are now well entrenched in many parts of industry. For the studio and the future of digital television, the MPEG-2 Main Profile at Main Level (MP@ML), and its studio type counterpart, MPEG-2 4:2:2 Profile at Main Level (4:2:2P@ML), are here for the long run.

There are further developments in expanded flavors of MPEG, such as MPEG-4. There are methods of moving compressed video at faster than real time and soon in the form of FTP file transfers instead of strictly video streaming. No matter where these concepts head, we can with reasonable certainly postulate that most of the advanced principles associated with professional digital video storage, including compressed and full-bandwidth video, have come from the persistence in development of the early digital video disk recorder.

CHAPTER 3

ONE STEP BEFORE THE SERVER

> There were several early product developments for disk recording technology that preceded video servers and cache systems for library based VTR cassette systems. Between 1986 and 1994, significant engineering developments were made in the areas of digital recording transports, standards for digital recording, and the embryonic stages of digital disk recorders.
>
> In this chapter, portions that originally appeared in the "1994-NAB Review" for *TV Technology* are as they appeared in that issue.
>
> Servers were not quite a reality in 1994, and ATV, HDTV, and a whole lot more was still in flux. Yet, looking back over history, an important fact to place into perspective is that many of the features we take for granted less than four years later were at that time revolutionary concepts for industry.

There was great speculation in the video industry in mid-1994. Concerns that the disk drive was replacing the tape transport and that tape was dead were echoed by many a writer. What is important to recognize is that none of the developments in digital transport and digital recording technology (with the possible exception of D-2) has diminished any of the impact on the future of media server technologies.

Even though formats, such as quad, have fallen into disuse or been discontinued, videotape has and will remain an important part of the video industry for some time. The tape transport and linear tape industry will evolve and formats will go by the wayside. The announced discontinuation of Sony's U-Matic transport manufacturing in late 1997 isn't going to end the use of ¾" tape—it's just going to diminish it. These things are inevitable and necessary as old products make room for new ones.

A NEW DIMENSION

Digital video has brought a new dimension to industry. When you talk to a computer hack about "digital video," they think of it mostly from the perspective of streaming video on a computer or network. They don't seem to understand a thing about D-1, D-2, D-3 or D-5. The very existence of ITU-R BT.601 is probably so far from their minds that if you told them you could put 270 Mb/s video onto a disk drive they'd probably just laugh.

On the other hand, if you talk to a broadcast professional about the same questions, you'll get a wide variety of answers. This industry is beginning to feel the impact of digital video in many more forms than the computer industry. Moreover, their impacts will be felt just as strongly.

Even though computer video has made great strides in the past decade, broadcast digital video has made even greater ones. As history unfolds, we will eventually forget that the early recording formats of digital tape were understood by only a handful of brave pioneers. Strangely enough, even today, with multimedia, video servers, and a plethora of disk recorders, confusion in D-1 versus D-2 formats remains.

Early digital recording brought little to no equipment that could take advantage of the features and principles of digital technology. Most switchers, effects and picture manipulation units still had analog I/O. There was no serial digital interface[8] (SDI), and the only parallel digital routing switchers were expensive and complicated to install.

For the broadcaster, accepting digital made little economic sense. Those who made any transition did it primarily in the commercial playback machines. These operators either needed to replace their dying quad cart machines or they wanted to satisfy their buyers that the image

---

[8] Serial Digital Interface, or SDI, although in standards committees during the late 1980s, was not finalized so that equipment could be manufactured for interoperability until approximately 1991.

quality would be improved—so they might consciously be able to charge more for their spots.

COMPOSITE DIGITAL REVISITED

Encoders and decoders were designed around low pass filters that considered more of the television transmitter's requirements than the capabilities of the developing physical plant. The market for distribution of the final product was becoming constrained to a quality level that matched analog 1" or Betacam tape specifications.

For production, effects work continued to increase. Layering or compositing became a buzzword. With digital, the ever-familiar dropout was on its way out, at least in production. Yet, over time, D-2 proved to be less than satisfactory for high-end effects and layering. In composite digital it was shown that after only a few, usually between eight and ten generations, or copies, the artifacts from sampling and resampling the same essentially NTSC ($4f_{SC}$) signal became quite evident.

The quantizing structure of the D-2 format resulted in an inconsistent scaling of black levels, due in part to the fact that there was no defined bit boundary level for the 7.5 IRE setup level. Chroma levels were still difficult to manage and most facilities had only analog interfaces to other equipment, making the benefits the benefits to digital transports of little value. These principles added to the multigenerational degradation handicap at almost as serious a level as those found in poorly maintained oxide-Betacam transports.

Even having a D-2 composite digital disk recorder did not help the matter significantly. Until there was a means to keep operations totally within the box and drives itself, outside influences on digital were unforgiving. The compositing DDR concept was better than external VCRs and analog switcher/effects equipment because it kept control within a closed set of disk arrays, coupled with an integral keyer/compositor (e.g., the Abekas A62). The addition of an A62 (composite DDR) or an A64 (component DDR) to an edit suite made multi-layering and effects keying simpler and cleaner than it was with external equipment.

All the confusion of early D-2 digital was indeed unfortunate. Besides the low level sampling error associated with composite digital, the D-2 transport and head structures were also plagued with numerous problems, essentially branding the format with nearly as many negatives as its analog counterparts. In addition to the potential for serious video degradation, audio editing problems in some early machines would result in the destruction of all video data, rendering the recording useless.

On the other hand, for the highest-end production, component digital could be layered ostensibly forever. The very nature of component digital recording, sampled at 4:2:2[9], removed the NTSC footprints and kept the constantly degrading properties of encoding and decoding the chroma subcarrier away from the video image. You could not visually recognize the degradation of the video after upwards of 10, 20 or more digitally rendered, D-1-to-D-1 copies. This was truly revolutionary and was soon recognized as the benchmark for digital quality that remains today.

DRIVE LAYERING SOLIDIFIED

The driving desire for effects-based work in video drove the first successful stand-alone, digital video disk recorders—such as in the Abekas A60, A62 and A64 product line—steadily upward.

Abekas Video Systems, a relative newcomer, produced the industry's first self-contained, stand-alone 25- and 50-second 525 line (30- and 60-second products in 625 line) disks. The drive structures could also be interfaced with a keying and compositing module to perform the basic layering functions of Quantel's Harry at a price point well below that of the Quantel technology. The Abekas products were also the first step in nondedicated semi-open architectures in digital video storage. These disk recorders provided the first links between video and computer platforms via Ethernet connection.

Abekas paved the way for advances between second- and third-generation disk recorders. These models of disks offered in the late 1980s and early 1990s what is still an industry standard profile for today's modern disk systems. The Abekas feature set was comprised of an SCSI-based interface for Exabyte tape streaming archive, digital video input and output, analog monitoring and recording I/O, timecode addressing, Ethernet (for computer graphics interface) and random

---

[9] 4:2:2 sampling is composed of two chroma components (C_B and C_R) that are subsampled horizontally by a factor of two with respect to the luma (Y) sample, making up the component digital system of ITU-R BT.601.

One Step Before the Server 29

access, continuous play video including stunt-motion, still imaging, looping and segment play.

The drive system allowed for connection to existing edit controllers in Ampex (SMPTE), Sony and CMX protocols. This made it possible to deploy these devices into on-line editing suites for a new application of an early technique in reducing the requirements for linear videotape machines. The extension of these control interfaces also made it possible to adapt what is now nearly eight-year-old technology into modern Mac, Amiga, PC and UNIX workstation environments as the mainstay render workhorse of computer generated imaging in 525 and 625 formats.

Abekas also provided a relatively unsuccessful connection interface between the Quantel Paintbox, the A60 disk and external transports for production in the graphics, rotoscoping and animation domains. The TouchUp software/tablet interface began to marry both "closed" and "semi-open" devices to give productive flexibility and far more on-line storage than had been achievable before. The TouchUp product was termed "unsuccessful" probably because it was a device too far ahead of its time. It was also plagued by the difficult task of adapting a closed-architecture Quantel Paintbox into an open-minded digital production world. Even after the 601 digital upgrades were available directly from Quantel, the attempt at a digital video disk interface between the two competing companies ended in failure.[10]

Just as products and technology mature, companies and employees do also. Some time after the overwhelming success of the A6x recorder line, the time came for a new product line that offered the next transformation in digital video disk recording. The recording length of DDRs (digital disk recorders) was still relatively short. Most held less than 30 seconds of full-bandwidth, 601 video. Although compressed motion-JPEG digital was already on the early Avid Media Composers, the output was still not acceptable for air-quality release material.

The migration to longer recording lengths at lower costs was started as the A66 and A65 lines of Abekas recorders came into being.

---

[10] Ironically, the two companies (Quantel and Abekas) were eventually acquired by the same parent company, Carlton. This still made little difference in resolving compatibility issues between the two products.

The product lines intended to extend the record length by many times. However, the promises of chaining drives together for longer record times were beaten down by several factors. One of these was the change in the disk's form-factor technology. Another was the difficulty in getting sequential drive chains to synchronize for glitch-free overlap from drive to drive[11]. The concept of linking multiple drives together for extended play would have to wait for an entirely new architecture of DDR to arrive.

One of the features desired in editing dated back to the root concept of the early CMX editing system. The concept of reducing expensive transports and generation losses that resulted from copying a B-roll for dissolves or edits would be answered at least partially with the D-2 transport. Composite digital videotape recorders had a feature call preread which was used as a marketing factor over D-1 in the early years. Preread allowed a single D-2 transport to play out a segment and then make a permanent recording immediately after, and on top of, the original recording. Of course, this rendered a re-record impossible, as the source material had just been recorded over, leaving a new recording with no original material remaining.

The preread concept also permitted, within certain technical confines, processing of external material (a key or insert), typically from an effects switcher, without having to copy B-roll video to another transport. Since the digital disk recorder was a rare commodity in most post houses, preread seemed to be the answer to fewer transports, and it sold like hotcakes. In reality, it was not an answer at all. Facilities with only one D-2 edit suite found themselves still needing the third D-2 or an external DDR, just to satisfy the risk of loosing or ruining a client's master.

This also contributed to a reduction of sales of D-2s and development of the component digital tape recorder, forcing other products to take a temporary backseat. Market confusion over D-1/D-2 was further complicating the decision of which format to manufacture for—D-1 or D-2, or both? And Sony had another idea on the horizon for 1994–Digital Betacam!

---

[11] Attempts to use external editing controllers to match frames continuously from one disk to another were unsuccessful as well. This author worked closely for nearly a year with two major equipment manufacturers to develop this integration, but the attempts ended as newer technologies prescribed alternative methods to achieve the same results.

The market brought out other problems associated with making multiple drives work together. The third generation of DDRs brought the birth of many new players in the 1991–1992 digital disk market. The HDTV revolution was on the forefront, and already HD-DDR technology was sprouting. Yet the market for HDTV and associated recording systems was so small, that mass-market acceptance of products would be a long time coming.

GROUND BREAKING WITH A NEW DDR

Accom, a spin-off founded by management and engineers from Abekas and other prominent video equipment manufacturers, took the steps to expand the disk recorder to a new limit. Intrigued by what could be accomplished using the new SCSI-2-based drive technology and by what was rumored to be a new five-minute-plus recording system by Quantel (their Henry), Accom set out to produce their version of a realtime disk recorder, which became the RTD-4224.

The functional feature set of the DDR had to be extended. Editors, with a ferocious desire to edit faster and with less linearity, drove the development of the RTD. Audio had not yet been successfully integrated, although that was desired, on disk recorders. The ability to do a 30-second spot on one disk necessitated a longer-than-25-second (NTSC) disk recorder. The requirement to expand the drive beyond a base record length of 32 seconds was also a necessity.

All the desired feature sets were expected in the next generation of Abekas DDR products. The desired features of Ethernet, SCSI, timeline edit control, timecode and eventually audio, were set in stone, but had yet to materialize in totality.

The real problems that needed addressing in edit environments included the deployment of features including instant start or zero frame pre-roll, extensibility, and roll-through-midnight[12]. Few of these problems were solved or even realized until disks had become more integrated into edit suites.

When a manufacturer added high-quality slow-motion, motion compensated imaging and feedback related digital effects—plus two 32-second drives in one chassis that could be separated or linked together for up to 64 seconds of record or playback, you now had the ideal

---

[12] "Midnight" is the point where the disk or time code number crosses 00:00:00.

32 Video and Media Servers: Technology and Applications

product of the day. Accom was about to introduce just such a product in their RTD-4224 Real Time digital Disk recorder.

In the early 1990s, other new players were stepping into the playing field as well. Grass Valley tried a feeble attempt to produce a digital disk recorder that integrated with their Kadenza/Kaleidoscope switcher/digital effects product lines. GVG's idea of recording video and depth information for three dimensional image manipulation of their DVE seemed like an interesting concept, yet the fact of the matter is it was never delivered, and a Grass disk recorder never became a product. Still others were attempting storage on small platforms using the early stages of video compression. While useful for security applications, the quality of the images was unsatisfactory for broadcast or postproduction applications.

Nevertheless, for high-end postproduction, Accom had done it! They'd taken the quantum leap into drive technology that has produced an expandable, economical system structured around multiple drives and dual channel operations. When coupled with other Accom products, including their compositing module and later the Axial 2020 editor for sophisticated control, the edit suite of tomorrow became a reality.

Abekas continued to develop outstanding lines of new disk-based products employing uncompressed component digital recording technology. Accom also expanded their disk product line into server-based architectures and workstation edit decision systems. Quantel created a product line employing their proprietary Chatter disk technology—and developed an entire new diversified line of servers, editors, compositors and long playing digital video systems. Quantel also moved into their own grid-compression technology with Micro Henry, an economical means for off-line subsets with the high-end features of the Henry product.

Relatively new advances in multiple resolution recording technology became evident in many of the newest companies building disk recorders. Companies including Recognition Concepts Incorporated, Hewlett-Packard, Sierra Design, Ciprico and a growing many others had begun integrating their own technology applications for both real time component digital recording and non–realtime graphics workstations.

Accom successfully entered the stand alone workstation disk product line with their WSD (Work Station Disk), and Abekas developed

One Step Before the Server 33

new products aimed at workstations, effects work, and editorial composition.

The early developers of servers during 1994 included BTS, who adopted a unique approach by developing a server-based network system that allowed both full-bandwidth and user definable variable compression-based recording technology. In 1994, nearly all disk-based products employed 10-bit component digital recording, a demonstrated requirement if one did not dither or apply Dynamic Rounding; Quantel developed digital a scaling feature for 8-bit component digital.

In 1994, the concepts were new and the needs were still being defined. This was the start of the race for the future of digital video recording, storage, forwarding, and networking.

Since the first sale of a videotape recorder in 1956, it had taken some 38 years, but in 1994, the light came on. Random access, non-linear, high-quality video recording and playback would be cast as a critical part of editorial composition, production and broadcast delivery for untold years to come.

CHAPTER 4

DISTINGUISHING THE DDR FROM THE VIDEO SERVER

Today, with the acceptance of disk-based storage for a variety of applications, the boundaries between the modern on-line editing system and any of the vast number of off-line editing devices are in actuality just a fuzzy line.

Most of the definitions that are now household terminology came straight out of the manufacturers' marketing efforts. Manufacturers coined phrases like "non-linear," "digital," "random access," and the like, turning them into broad definitions that really pointed to their own specialized products.

This was not necessarily bad, as the production industry ate it up. Operators could provide new services, have "gee-whiz" tools, and in turn expand business in directions previously not possible. All they had to do was convince the paying producers that this saved time, was *cool*, and was the "wave of the future."

"*Avid*" eventually became *the* household name for non-linear editing systems, following in the footsteps of "ADO", which was patented by Ampex and came to be the name used for nearly every digital picture manipulating device. How about the Chyron, a universal word for all electronic character generators—yet does anyone remember the impact Vidifont had? Looking back, it seems obvious that the word "non-linear" referred to anything other than a linear editing process. As capabilities grew and systems began to come to market, some manufacturers put a steep price on defining just how far the term non-linear could be stretched. Some looked at the application perspective first, rather than the physical system that might be one self-contained unit or a series of discrete components. Some systems were all disk-based and others were a combination of tape and disk.

OFF-LINE OR ON-LINE

In the early years of the non-linear editor (NLE), the process was always assumed to be off-line only. Perhaps because of the quality issues presented from the final output standpoint, this was all it could be. It was not that long ago when the primary purpose for the NLE was *only* to produce an edit decision list (EDL). Today, with the possible exception of high-end post and film, technology has completely reversed that position.

Some of the earliest commercial entries into non-linear editing were strictly tape-based, using only still frames to produce in- and outpoint references. One privately developed system used a screen to temporarily store single frame images as stills—without the use of videotape. The small still frame images (later named thumbnails) were either cached to a disk drive or held in memory. The locations of the thumbnails pointed to timecode positions on the worktapes. An editor could shuffle and reorder the still frames, on the monitor, creating a series of cuts (later renamed clips) of the desired visual content, prior to final assembly on videotape. Moreover, the system existed years before today's dual screen, GUI-based, time line editor was a real product.

These early concepts became the foundation of today's non-linear editing systems, but made little headway in the market because of cost, the complexities of manufacturing, and reliability. There was also little desire by conventional production facilities to take away from their big-ticket video edit bays. The other side of the same equation was that edit bays fed the growth of other products such as edit controllers, switchers, and multiple 1" and even 2" tape machines. Not until a proliferation of ¾" and VHS off-line edit systems took hold would there become a real market for what would be the future disk-based editing system.

Today, there is little doubt that the term "NLE" refers primarily to a disk-based editing system. Credit must be given to Avid Technology for its profound success in changing the industry. Yet after nearly a decade of non-linear editors based around Macintosh and PC

Distinguishing the DDR from the Video Server 37

platforms, there have been only a few radical changes in the methods of time line, disk-based off-line editorial production.

ADVANCES IN NLE

Only in the past couple of years have some manufacturers begun to reintegrate videotape transport as part of the software functionality side of editing. The Axial RAVE editor was really one of the first to take an on-line room and transpose it into a non-linear on-line environment. However, Accom's Axial product line had to incorporate its own digital disk recorder products to achieve this.

Similar but more self-contained and proprietary functionality came as Quantel expanded its product lines from the Harry to Henry, then on to Edit Box. Although systems employing tape and disk editing have always existed, we're now seeing more sophistication added to the applications—especially where the integration of traditional on-line production hardware is a part of the mix.

The Lightworks VIP system combines contemporary external equipment and Profile video server equipment from Tektronix in a fashion that bridges the gap between digitizing linear tape and finishing on disk-based media. An extension of the Lightworks NLE, VIP attempts to decrease the dependence on predigitizing all the raw media before beginning the edit session. Transferring the video material only once, and only if it is the shot the editor really wants, practically eliminates the processing time that precludes an actual edit session.

These hybrid systems have gone beyond the process of predigitizing. The true hybrid NLE systems have put smarts into the software and made it so that the operator needs to see a field recorded videotape only once. Every piece of the tape can be cached to the disk in any order that the editor wants. The tape could be shuttled well into the pack, viewed (and cached), then wound back, viewed (and cached), etc. With sophisticated digital disk recorders and servers integrated into the edit control system, there is no requirement for predigitizing before editing.

Now the editor needn't wait for all the material to be cached to disk before the process starts. Unwanted material can be discarded from the record database as soon as it is viewed, thus providing significant increases in virtual disk space. This speeds up the editing process and adds more functionality to the edit session.

Going one step further requires bringing in the concept of the video server. The server concept adds multiple channels. It allows multiple feature sets to be added without the need for prerendering. Beyond that is the opportunity for collaborative editing, where others gain access to the same materials simultaneously and for other purposes—without necessarily tying up another storage system or another non-linear editor.

Complete NLE systems today consist of many components that may be discrete devices or may be collected into a single unit device offered as an integrated turnkey solution. In any case, most NLE systems consist of at least an edit controller made up of software or hardware plus some degree of local storage. The storage is usually, but not entirely, disk-based, and may include off-line and/or near-line storage. The species of storage media might consist of data tape, removable magnetic disk drives, optical drives, or videotape.

There are varieties of input and output subsystems, some are integral to the editing system, and some are not. Compression is usually used, although the purist may insist on full-bandwidth, uncompressed disk storage with off-line storage on full component digital videotape (D-1 or D-5).

RANDOM ACCESS DISK RECORDERS

Random access devices for the storage and reproduction of video and audio media can incorporate a number of types of devices and systems. Disk-based recording devices may operate as stand-alone or as integrated editing systems. They may be used in surveillance systems, commercial playback systems, and even full-fledged program distribution systems for video-on-demand or kiosk presentation displays. Store and replay devices can be broadly classified in two categories: the video disk recorder (VDR) or digital disk recorder (DDR) and the server. We'll discuss the disk recorder first, as it lays the foundation for a return to the server definition.

For the purpose of explanation, we will classify both the VDR and the DDR as "disk recorders." We will assume that these disk recorders could record strictly video, video and audio, or a combination of video, audio, timecode, and other user data. We will further assume that the format in which they record the media onto the physical disk drive is left to the manufacturer. The format could be full-bandwidth,

Distinguishing the DDR from the Video Server 39

compressed, streaming media, files, or the like—structured in a manufacturer-specific method that essentially inhibits direct "disk-format-A" to "disk-format-B" replicating without some level of translation or I/O between them.

Disk recorders are considered general-purpose devices that emulate the basic functionality of a videotape transport with a number of added stunt features, depending upon what the manufacturer has designed into the recorder. The disk recorder is further characterized by its ability to handle higher data rates that permit full-bandwidth recording or compressed recording.

Control software is integrated into the disk recorder (VDR or DDR) such that it makes the recorder act like a tape transport. The devices typically incorporate standardized protocols familiar to other external controllers, such as editors or graphics systems. The control software might be extended to include file transfer protocols for topologies such as Ethernet and, of late, Fibre Channel, and soon, IEEE 1394 (FireWire).

Control functions connect over conventional RS422 or RS232 interfaces, integrated SCSI control, direct data bus interface, external 10BaseT Ethernet, or a combination of any or all.

Encoding and decoding for input and output may or may not be incorporated into the disk recorder. An example is a disk recorder, optimized for media purposes, which has no video-digitizing codec integrated into its product. Such disk recorders might be added to 3-D graphics systems, render engines, scanning input recorders, etc., with the full intention of outputting the final data product as a *video* medium—but employing an external encoding process.

One other form of disk recorder package includes packs of special purpose drives that by themselves have no outside functionality, but are the foundation for the storage portion of integrated professional graphics and editing systems. Examples include storage arrays such as Quantel's Dylan drives or Scitex's storage cubes that are specifically attached to only their product line of editor or graphics systems.

Indeed these devices act and appear as *disk* recorders, but the user must incorporate these storage systems into a specific manufacturer's system, or they will have nothing. These modular units also provide sole-source storage expansion for specific applications or groups of applications unique to one and only one manufacturer's product.

Some disk recorders offer a variety of input and output alternatives. They generally will include their own (or an OEM-provided) encoder and decoder. The coding method is for the most part motion-JPEG (sometimes written MJPEG), MPEG-1 and, most recently, some flavor of MPEG-2.

Some systems use Wavelet compression, which is a form of sub-band coding where successive filters are applied in vertical and horizontal directions, and only the data within those sub-bands is separated and then extracted for storage. Other systems use manufacturer specific software codecs, such as Quicktime, Cinepak, Indeo or AVI (which, incidentally, have all been adopted as industry accepted de facto "standards").

Today, the most widely used form of coding remains motion-JPEG. The original JPEG was devised as a method for the compression of still pictures and is based on the use of the discrete cosine transform (DCT) to provide image compression of between 5 and 100 times. Motion JPEG, or intra-frame (I-frame) coding, grew out of JPEG for editing systems because it offered ease in coding and it resolves the complexities of editing and producing still frames on a random access basis. JPEG techniques eventually formed the foundation for I-frame compression in the MPEG standards.

DRIVING THE INDUSTRY

The professional video disk recorder industry, for the most part, remains entrenched in using its own proprietary media structures and selection processes for determining which drives can be used with which products. There are several good reasons for this and some not-so-wise ones.

The broadcast manufacturing industry has a strong record of accomplishment for bringing to market products that will perform continuously and reliably. Engineer and producing products such as videotape machines for a narrow market has required that manufacturers develop designs for only one use. With the possible exception of the D-1 and D-2 transport, which found acceptance as a data recorder and a video recorder, the videotape machine formats are single purpose devices. Hard disk drives, on the other hand, have a multitude of uses for a variety of industry applications.

When developing any product for a mission-critical application, especially where it is part of the final on-air chain, the choice of materials is very important. By using narrowly defined constraints for media, including hand-picked drives and application-specific microcode, critical fine-tuning is possible. This assures that the drive substructure and all other peripheral components will perform exactly as intended.

This situation comes with its own set of problems. Disk drive manufacturers must respond to the marketplace for their principal products. Technology then forces a high turnover resulting in a relatively limited product life cycle.

To develop a reliable and tested product, the disk recorder manufacturer exerts its effort to fine-tune a particular 4.3 GB drive. Over a one-year or more product development cycle their particular hard drive may actually be irrelevant by the time their product is introduced, delivered, and accepted by the market. If the drive is not discontinued, its availability might be limited simply because a newer, higher storage capacity drive has already been introduced at a similar price point.

To the end user, the street price for the same 4.3 GB drive the recorder manufacturer uses is now half of what it was during product development. The prospective buyer knows this and finds that for the same disk drive price, the next-size hard disk drive would offer twice the storage capacity. However, it probably will not be incorporated into that particular disk recorder, and the user can't understand why.

DRIVE STRUCTURES WITH AN UPGRADE PATH

Most disk recorders that handle uncompressed, full-bandwidth component digital signals will generally utilize their own form of data distribution for the internal disk drives. These data formats again narrow the alternatives for selecting off-the-shelf replacement drives, as in many cases the disk recorder manufacturer has optimized some element of the drive microcode or other control software for its own specific application and performance.

With the forward thrust in disk recorder usage and new product development, changes are on the horizon. Those manufacturers that cannot provide a clear upgrade path, or do not follow similar parallels to the personal computer industry, may find their market position diminishing. This is a double-edged sword. The manufacturer that positions its products to serve multiple purposes, such as a disk drive element for disk recorders and server operations, has a better likelihood of providing upgrade paths and storage capacity increases.

On the other hand, this concept will also require enhanced support and the evaluation of a multitude of drives. In addition, the manufacturer must start the design with an expansion plan utilizing nonproprietary subsystems. All of these and many other factors must be equated to cost and to the anticipated volume for the product line.

A good example of the growth potential in a full-bandwidth, digital disk recorder is the Accom Real Time Disk (RTD 4224) recorder. In the early 1990s Accom recognized the shortcomings of its competitor Abekas Video Systems and set out to develop an expandable disk recorder system. Accom included feature sets such as multiple channels, Ethernet connectivity, file transfers, and add-on options such as motion processing, effects, and an integral audio recording scheme.

The biggest plus, however, was that Accom provided a field-installable storage upgrade path consisting of an add-on chassis and any number of drives that the user could add when and as desired. This concept set the tone for both the PC disk recorder and the modern server architecture of today. The principle can be applied to compressed or full-bandwidth systems, depending upon the particular architecture.

FROM DDR TO SERVER

The question then becomes "When does a disk recorder become a video server?"

When the functionality of the disk recorder goes beyond that of strictly record and playback functions, then the device should be placed into the server category. This broad distinction is the first defining line between a single-ended, single-threaded disk recorder and the entry-level video server.

A server should be capable of supporting non-linear or hybrid mass storage. A server might store data or media-based material or both. Besides just disk-based mass storage, a hybrid server system might consist of videotape, data tape, and optical or removable media—generally under control of another host device such as an automation system or software control features.

The video server should address multiple channel control, multiple access, and certainly multiple, simultaneous input and output operations. At least in broadcast operations, the server generally provides support for backup systems, data protection, redundancy, and a level of fault tolerance in the form of both software controls and hardware protection.

SERVICES TO OTHERS

A server, by definition, *services* one to many users or hosts. Servicing might be in the form of multiple playback channels linked via external media, including baseband video and component digital interfaces.

Inputs and outputs might include a high speed, independent transfer mechanism to move pure data from server to server or disk array to disk array. Transfers from host to host might be possible in faster-than-real time and managed under the rules of network topology architectures.

Another identifying characteristic of the server is its ability to record and immediately play back one to many channels independent of each other. This functionality permits recording on one channel with playback on many others, recording on many channels with playback on one channel or a combination of both—depending upon the architecture of the server.

Most video servers are built to operate with some level of compression. Often the bit rates can be set by the users, and others are set automatically by the type of input material being recorded. Some servers allow for bit rate selections on a clip-by-clip basis—allowing for quality variations and increased storage capacity management.

Until recently, inputs to the video server have essentially been raster-based, interlaced signal formats. The signals are typically NTSC, PAL, component analog or digital. Very few currently produced servers are utilizing composite digital input or output formats.

The next-generation digital video servers currently making their way into production are taking the computer server model and extending it to address the convergence of professional video and computer generated animation and graphics file formats. Devices in this domain now seem to be a cross between the server and the disk recorder. They are being built around the conventional Intel processor on a PCI bus structure, with the predominant operating system being Windows NT.

Many of these server models are simply board sets that the user configures or an integrator assembles. The storage units for many of these products employ user selected, off-the-shelf RAID storage units for flexibility and user definable expansion.

Servers of this flavor can address files as discrete rendered frames connected over a network or direct from host to disk. The connection medium can be a mixture of SCSI, Fibre Channel, SSA, or even some form of baseband video interface.

Rendered file sequences are stored in directories, on the server's disk array, for use in several forms. The typical use is streaming from the disk to an integral encoder. Sequenced files are encoded into a meaningful series of raster-based video frames, usually at a 60 field per second, 525 line NTSC or some flavor of component digital video—generally at 720 samples × 485 active lines. Synchronizing and timing reference data are added as a final step in the encoding process.

WORKSTATION SERVERS

Another use of this type of server is as a storage base for workstation-based functions. Again, this server may not be much more than an array of disks, usually in some level of RAID, that appears to the workstation as a single disk. Software for compositing or effects production takes individual files, converts them to the file format that can be most easily used by the software, performs the transformation function, and then returns the files to a native format on the server. The final set of frames can take the same path previously described and output real time video in full-bandwidth, digital format. In this example, the actual images are never true video until the final output, at which point the format might be presented in progressive or interlaced raster structures or any other file format suitable for the actual application.

It is common during the picture transformation process to review the sequence of test files as video. In this instance, only the needed files are called from the server, sent to an encoder, and displayed in a format suitable for proofing. This lets the original files always retain their integrity for other uses.

Eventually, the server may become the higher-resolution repository for production purposes, with the files ultimately being compressed to user assigned bit rates in either MPEG or JPEG formats.

This opportunity would allow for the absolute preservation of the original material on a mass storage array but permit several levels of distribution bit rates, depending upon the final transmission medium.

Film effects studios are already utilizing large-scale servers to store scanned 70mm film frames in order that they may be transferred to workstations for special effects, color correction, and even preparation of archival material. Since the resolution of these images remains extremely high, the storage arrays are typically very large and hold only a limited amount of actual run-time footage. These storage systems rely heavily on data storage tapes for preservation and conservation of physical drive space.

To summarize, we've described the digital disk recorder or video disk recorder as a single element with essentially a limited degree of functionality. The DDR, whether of high bandwidth or high compression, essentially provides a medium to store a relatively limited amount of content for short-term uses by individual users. The control of the DDR is such that it emulates the conventional videotape recorder and may incorporate stunt features such as slow motion, freeze/step frames, reverse motion, and near instant-access to any frame.

The server, on the other hand, must provide for multiple inputs and outputs, control schemes for both baseband and file manipulation and transfer, some level of redundancy and file protection, and expansion of storage space. The server can attach to multiple hosts, either over a network or under control of an automation system. It generally is built around either non-linear or hybrid mass storage systems. Finally, the server contains integral codecs that allow for a common I/O and a file-based I/O—structured around known industry recognized transports.

CHAPTER 5

SELECTING THE VIDEO SERVER

It is 1995 and the station's general manager has just returned from a management seminar at a major trade show where the headlines read "Video Servers to Replace Videotape." You've read a little about servers, seen one or two at the trade shows, and seen where non-linear editing systems have headed over the past two years. The facility has a dozen or so linear transports, all in need of replacement or major repairs, and your cart machine for commercials is an orphan unto itself.

Suddenly you hear the miracle bells being chimed. "I can rid myself of transports, of tape, and of maintenance headaches related to the last-second repairs on the library management system" is a dream waiting to happen. And so the server is destined to become your savior . . . or so you think!

This is the concept—unlikely, but a good one if it's true. Still, the handwriting is on the wall. You know that over the next five years major changes will take place in the facility that will require tape format upgrades, digital conversion, and the implementation of a video server system to support the infrastructure you are about to change.

So the planning begins. Here, in 1995, there aren't as many options available. Products are just being introduced and the pressure of DTV is not upon us. Jump to 1998, and some of the same questions and requirements are squarely upon us. The server is now a reality, with at least four or five major suppliers capable of providing equipment and systems.

The one question that still remains, however, is "Will the implementation of a server in your facility eliminate the need for tape?" The answer still seems to be "Probably not," but if that's your ultimate goal there's a lot more to consider than just the reduction of transports and a shrinking tape budget.

The general statement about the video server replacing tape seems is similar to the claims of a few years ago when we all heard "We'll [soon] see 500 channels in the home." We still don't have 500 channels and we still have videotape. So taking the first steps toward full video server implementation requires a lot of important decisions be made. Consider the alternatives carefully and ponder the answers to these questions as you make choices for the future.

- How does one decide the best-suited server for your application?
- How does the user optimize the server for its own applications in its own environment?
- Will the server be used for short- or long-term storage?
- How important are bandwidth and I/O performance?
- What levels of redundancy or data protection are of most concern?
- Is there a cost/performance ratio that is established and measurable?

It is difficult to determine if the implementation of one server might lead to the need for a different server for a similar application a few years down the road. This is especially true as we begin to hear about requirements to handle various levels of compression for different types of bit streams and program delivery.

SERVER OVERVIEW 101

Selection of a server architecture may best be defined by carefully understanding the application in which it will function. Later, in chapter 8, we will go into greater detail on the internal architecture of the video server, but first let's look at the overview.

Servers do not all provide similar operations, functionality or quality, so one video server will probably not satisfy all of the facility's applications. To define what the server should look like, you should determine what the primary, secondary and tertiary objectives are. From

Selecting the Video Server 49

these guidelines, you can determine what type of server may be required and how to configure it.

Video servers with only a few inputs and outputs are now being used for time zone delay, spot insertion, spot-playback assist, editing, and computer graphics rendering. Servers can also provide resources for the short-term storage of timely programming, without the need for archival or tape storage.

Other types of servers, those with multiple streams or channels, are used for near-video-on-demand (NVOD), in-flight entertainment systems, and intracampus distribution of training courses. Cable head ends are also using MPEG-1 servers, with lower bit rates and multiple streams, to deliver local commercial insertions. These servers typically aren't broadcast quality, but they certainly produce reasonable-quality images with much higher compression rates.

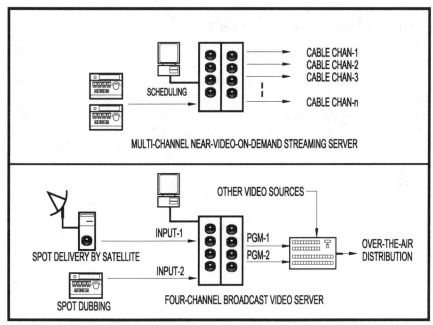

Figure 5-1: Two different types and uses for servers.

Preservation of your investment in a video server is critical. Technology changes too fast not to consider how your server will fit into the facility one, three or five years from now. You will need to understand several portions of the server architecture to make a qualified judgment. Some of those questions include:

- Is the server is scalable?
- How does adding additional nodes or workstations in a given manufacturer's architecture work?
- How do you determine or specify the amount and types of off-line, on-line, and near-on-line storage?

If you are considering a video server, you must do more than play the twenty questions game. You need to include today's and tomorrow's questions and expect that most of the answers will come in a variety of flavors and mixes.

WHAT'S AVAILABLE ON THE OUTSIDE

As broadcasters, we understand that professional video equipment usually is made of proprietary parts and that they just aren't available at the local parts house. With computer and video server technology, things are changing because many of the elements involved in a server could be generic to the industry. Windows NT is a readily available operating system, as should be the systems drive in the computer. Memory, display boards, modems, and CD-ROMs are also familiar and therefore might be prime candidates for computer discount warehouse purchasing.

If the vendor manufactures the server using a conventional PCI bus architecture, it may be for a good reason. On the other hand, some manufacturers have found very good reasons to use VME or other bus types to get better performance for their applications. Finally, there are some video server-based manufacturers that build 100 percent of their product from the ground up, and you couldn't get a second source for the parts if your life depended upon it.

The point here is to understand what elements make up the server you're considering. Some servers do combine off-the-shelf components with their own board sets. In some instances, this makes parts readily accessible and to some degree less expensive to replace should the need arise. Be sure, however, to understand what "off-the-shelf" really means. It probably doesn't mean you can call the local computer parts house and order a replacement 9.1-GB drive, but you might get that system drive there and use it as a spare.

Understand thoroughly if you will be locked into the original equipment vendor for their supply of OEM drives, which might be configured to work only in their particular hardware. With disk drive costs constantly decreasing and capacity continually growing, you might think there are benefits to using off-the-shelf components. Remember, though, with changes in technology, those replacements may not only be obsolete in three years; they may not be cost-effective because the price of that 4.3-GB drive today will get you a 9.1-GB drive tomorrow.

In some cases, you may be able to get replacements only through the original hardware vendor. This depends upon the carrier that the drive sits in, any special coding that they've done, and so forth.

FUTURE PLANS

Look at your entire enterprise's future desires for server implementation. Then consider if the manufacturer plans or will offer a program to keep the drive arrays used in today's system in step with the growing drive technology of the future. The computer disk drive market is much broader than the broadcast equipment market. This should be good news for the end user from a hardware standpoint, but if the broadcast equipment manufacturer locks you into just their set of options, you may find this difficult to deal with as this technology grows and grows.

With this in mind, consider what happens when the manufacturer's physical drive configuration jumps from 4.3-GB to 9.1-GB capacity for the same form factor. Can you take advantage of it, or will you be locked into 4.3-GB drives forever?

Will you be able to mix and match drive capacities in the future as the market drives the media costs down? What is the manufacturer's solution for future proofing, and will they back it in writing? Are they planning to add a drive expansion chassis that will handle different drive sizes? Are you going to be locked into the footprint of the server, from a storage capacity issue, forever?

What happens if you do run out of space? If you need more storage and you're currently running at 48-Mb/s motion-JPEG for compression, you could off-load your entire library to videotape and change the compression ratio to 24-Mb/s, then re-encode the entire library. This is not a suggested method, as it has two complications.

First, you've gone through two or more decompression and compression cycles, which will probably be noticeable, at least in some material. Second, the quality of the second compression may not be as satisfactory as that of the first, but that is a subjective property that hinges on how good the original material was in the first place, how it was originally coded, and if it was subjected to NTSC coding anytime along the path.

It is better to understand up front what growth capabilities are possible from the server you're selecting before the unfortunate surprise hits you squarely in the pocketbook.

EXPANSION OPPORTUNITIES

Expansion of a server system is crucial to the growth plan of any facility that is considering implementation on a total facility scale. And expansion can have different meanings as well. Figure 5-2 shows where adding near-line storage did not necessarily improve system performance.

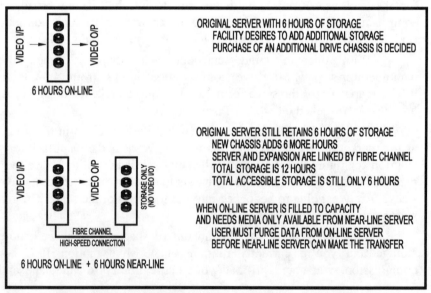

Figure 5-2: Availability of data, on-line versus near-line.

Selecting the Video Server 53

Some entry-level servers contend that you can "get into a server at a lower cost and then grow from there." The theory is good; the practice might not be. The users should be aware of the potential position they might be placed upon the architecture of the server itself.

Some servers are locked into only the amount of storage that can be placed in the principle drive chassis. The manufacturer may tell you there are expansion options, but if those options don't allow on-line access to the media—that is expansion requires that the entire data set be transferred from one array to another before on-line access is possible—that may not be in your best interests.

Manufacturers of video servers generally will offer scalability of their systems at varying levels. If the server utilizes a common storage array, accessible to all channels at all times, adding channels is simplified. The base server system (seen in Figure 5-3) consists of an I/O chassis including the compression engines. Adding channels is a matter of buying just the chassis and linking it, usually by Fibre Channel, to the common storage array.

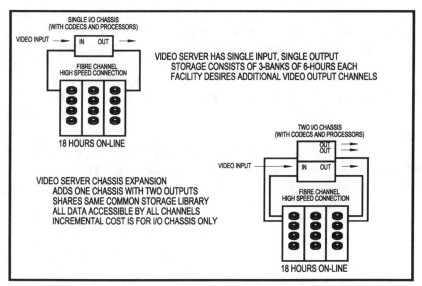

Figure 5-3: Common storage array with channel expansion.

The incremental expansion costs are typically less than the cost of a comparable linear videotape recorder, and the capabilities are greatly

54 Video and Media Servers: Technology and Applications

expanded. The storage arrays, in this example, are typically Fibre Channel-based providing for redundancy and, if the chassis are RAID-based, data protection as well.

Another form of expansion also deals with storage growth. Some manufacturers philosophically look at storage expansion by adding additional chassis with arrays of drives, which are usually RAID-based. RAID provides protection of data as well as certain elements of redundancy, depending upon how the RAID and the servers are configured.

BASEBAND OR BIT STREAMS: STORING THE DATA

Consider next if you are expecting to store signals delivered as bit streams rather than as baseband video. In time, we expect this option to be a requirement rather than an option, but there is little commonality between server architectures to make this a viable universal alternative at this time.

At least one company presently states that their server can accommodate different compression bit rates to be recorded on the same media without having to restripe and reconfigure their entire disk array. This is an important concept, as explained in the previous paragraph, but it has a different meaning for present-day delivery of video to the broadcast facility. Today, when deciding to encode video into data files for storage on a video/data server, the coding bit rate is selected during setup. This is usually set via the GUI (graphical user interface) and it generally will remain fixed.

When compressing with motion-JPEG, input from a SMPTE 259M, 4:2:2 sample structure, the industry has almost universally accepted 24-Mb/s as the comparable quality equivalent of BetacamSP. Subjective tests have shown that visually, when a like signal is input to a server (employing motion-JPEG) in component analog (Y, B–Y, R–Y) or in component serial digital (SMPTE 259M), the image, when output to either a component analog or a digital monitor, will match the quality of the original recorded and played back first-generation from BetacamSP.

Test signals are seldom used to make such subjective comparisons because they do little to bring into play compression artifacts that arise as a result of using DCT—which is the universal first step in all JPEG and MPEG compression. Since the DCT equation will always result in the same level of compression for the given image and bit rate, only altering the bit rate will cause relevant image quality degradation or improvements.

BIT STREAMS ARE THE FUTURE

In the applications for the future, we may be more concerned with the *type* of data that can be exchanged either server to server or server to interface—and less concerned with recording and playing back baseband (component digital) video.

To make this explanation palatable, let's first define how manufacturers store the data on their systems. One method is quite different from the other. We will not get into the details on how data is physically distributed over the drive surfaces, as the distribution of the data (interleaving, shuffling, etc.) is not the intent of this topic.

A server's disk array is typically formatted and stripped for the application in which it is intended to perform. If the intent is to establish one compression level, say 48-Mb/s, then this is the level that the compression engine will understand and it will format the array accordingly. That is how some servers function.

Other servers take on the appearance of a bit bucket, a device that stores data without regard for what that data is. This argument is taking on new dimensions in the march toward DTV. For it is this device that will be the one we will want to have available, universally, for the storage of bit streams ear marked for variable compression bit rates and different resolutions. The server that records just the bit stream, without concern over what is in the bit stream, will be well suited to recording packetized data that resides inside the envelop of 270 Mb/s or 360 Mb/s SMPTE 305M (proposed) SDTI[13] signal transports.

Another reason that some servers might not accept mixing of compression rates on the same server platforms has to do with whether and how fast their codecs can recognize and change decoding schemes. In order to address receiving one compression bit rate followed immediately by a different file compressed at a different bit rate, the

---

[13] SDTI, Serial Data Transport Interface, is currently under consideration by SMPTE. As of this writing, it is expected the standard will be approved paving the way for transport of packetized signals over convention SMPTE 259M signal systems.

codec reconfigures itself for the different bit rate. This is referred to as the "setup time" and for many codecs in production today, this setup time is anywhere from several frames to several seconds. Changing coding functions is a complicated process, requiring the codec to internally recognize and then reload the decoding software into the JPEG chip sets so that different streams can be seamlessly played out back to back. Most servers cannot or do not handle this operation without placing certain constraints on the total system.

One constraint might be that no clip could be less than three seconds in length to guarantee back-to-back sequencing. Another constraint might be that more than one decoder must be resident in the chassis so that preloading of the different bit-stream can be accomplished. Every one of these constraints affects the system as a whole and should be understood by the user before deciding to implement one platform over another.

Codecs usually function in a dual mode, utilizing common circuitry for similar tasks. The dual mode allows a codec to be used as a record channel and a playback channel. Depending upon the board and chip sets, this can take on different forms and is explained in greater detail in chapters 8 and 9.

Let us shift back to the concept of recording bit streams. The conceptual flow for a multipurpose, bit bucket input section is shown in Figure 5-4. Future packetized television will consist of MPEG bit streams that are sorted into other transports for recording on linear (data) tape or for exchanging between one device and another. A good is example is DVCPro where the format is such that the bit streams can be exchanged between like devices at four times real time for 25 Mb/s DVCPro and two times real time for 50 Mb/s DVCPro-50.

The data from the devices could be recorded as native DVCPro direct from a camera to a storage device, such as a tape transport or server. Should a server be capable of accepting native DVCPro, might also contain the codecs to convert the signal to baseband SMPTE 259M digital video. However, it is probable that servers will, in 1998, be capable of accepting an SDTI signal and either (a) record it as pure digits to the server (i.e., the bit bucket concept), or (b) strip the packets from the SDTI transport and record them as either native or another format.

Selecting the Video Server 57

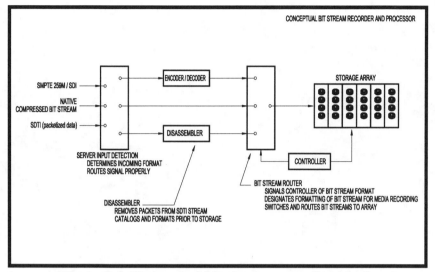

Figure 5-4: Conceptual bit stream server front end.

While as of this writing, in early 1998, no such product is currently on the market, a system could be constructed to address the variety of possible bit streams that will be in compressed digital systems before the middle of 1998. The possibilities exist of recording much higher data rates, including current proposals for mezzanine level compression such as MPEG-2 422P@HL (proposed for high definition television studio quality recording). Here, the concept to be understood involves making a system extensible and flexible to address the myriad of data bit streams we expect will be prevalent over the next five years.

The April 1998 NAB Conference will reveal the introduction of studio quality 50 Mb/s MPEG-2 422P@ML by at least two well-established names in broadcast video servers. This will bring the start of a gradual transition away from motion JPEG for server storage as industry heads into the first stages of DTV deployment. It is expected that MPEG will dominate the next generation of servers, just as soon as the internal architectures of the servers are completed such that we can allow seamless and non-seamless splicing of bit streams. This is an important advancement in digital servers that it just beginning to make its way into the equipment compliment of tomorrow's broadcast facility.

CHAPTER 6

VIDEO SERVER SYSTEMS

In the last chapter we dealt with general issues surrounding the selection of the video server. We will continue with a deeper investigation of the considerations you'll be facing when you begin the shift to server technologies. We'll start with video compression.

The vast majority of video servers use motion-JPEG codecs as the front-end compressor and back-end decoder. Motion-JPEG or MJPEG, as is sometimes abbreviated, has developed into a solution that meets the demands of both non-linear editing and video storage for commercial broadcast applications. This is mainly because of the benefits in all intra-frame DCT coding, which is the foundation of MJPEG.

We are not going to repeat the in-depth, nitty-gritty details of video compression, as there are several good tutorials on the subject, each spanning several dozen pages. One of the best sources for a complete look into digital compression, as it applies to Advanced Television, is S. Merrill Weiss's *Issues in Advanced Television Technology*. A thorough examination of digital video compression can be found in Weiss's book, Chapters 8 through 12.

In general, we understand image compression as "shortening or condensing," which for video images requires that we make two basic assumptions. First, successive video images contain redundant information—repeated, unnecessary, and extraneous. This information can be encapsulated in such a way that nearly every piece of redundant content need only be expressed once, and then the system will report to the decoder, "Repeat what I've already given you x times over the next y frames." Second, the human vision and auditory systems, by nature, have certain limitations in what is perceived. Taking proper advantage of these limitations, and applying them to the process of coding an image, you find you can discard or rearrange information to such a degree that the brain never misses it.

It is important to reduce the amount of information that must be recorded, stored, and transmitted if we're ever going to achieve the requirements necessary for digital video transmission and recording. We've already become accustomed to viewing suitable pictures at MJPEG rates between 24 and 48 Mb/s. If the same images are compressed using MPEG, the data rates can fall to between 12 and 18 Mb/s.

What we're really doing is taking a standard definition television image, at some 200 million bits per second, and compressing it by a factor of between 10 and 20, using MPEG IBP coding with a long Group of Pictures (GOP). For high definition, with over one billion bits per second, that rate when comparably compressed will be 50 to 100 Mb/s. The final goal, however, is to put the HDTV signal into the space of a 6 MHz television broadcast channel, and to do that means the signal will be reduced to around 19 Mb/s.

Videotape systems have been utilizing compression techniques for recording for over a decade. The D-1 format used 4:2:2 coding to reduce the chroma channel bandwidths. Digital Betacam also uses mild 2.5:1 compression. The latest Digital-S and DVC and DV-CAM systems utilize the next generations in compression as well. Even the analog tape formats of MII and Betacam used an analog compression scheme to reduce bandwidth and conserve videotape media.

Video servers for the most part all use some form of compression—be it MPEG, MJPEG, fractals or Wavelets. When you begin the process of selecting the video server for your facility, it is important to know some fundamentals about the internal systems you'll be supporting and living with for years.

SCALABILITY

We have talked about this in many forms. Scalability relates to the abilities of the system to adjust performance levels, including varying the amount of data recorded and varying the amount of storage that can be added over time.

In terms of imaging, adjusting chrominance and luminance resolution will affect performance. Other areas include changing the frame rate and varying the pixel depth of the image. In short, if you can predetermine what the image will be used for or displayed on, then you can design a system that fits those parameters. Unfortunately, that is not always the case and in many situations an image displayed in one form may look entirely different when conveyed in another manner.

EXTENSIBILITY

Obsolescence is a fact. Technology continues to evolve and with it comes obsolescence. For a product or a feature to be extensible means that it can continue to be of value, that it will not be rendered useless.

Designers and engineers strive for this when developing a product, but unfortunately, the product must eventually be produced and sold. Sometimes this is before all the features and desires are ready or completed. Nonetheless, the goal of making a system "future-proof" requires having a vision of what will be needed and what can be economically achieved in the future. Extensibility also involves building portions of systems that can address today's needs but also be prepared to address tomorrow's wants.

This is no easy task and it requires that the system be capable of discerning data that can be interpreted and used at that instant, while ignoring other information for which it has no use or value. It may seem difficult to apply the concepts of extensibility to a video server system without knowing a great deal about not only the technical inner workings but the marketing forces behind the product. The general concept to understand is that the devices used in the server, which were built before later definitions were established, will accept and process all the variations in materials they have been designed to handle.

INTEROPERABILITY

Video servers will be required to receive, store, and share images, signals, bit streams, and metadata across many applications and boundaries. For a server to be interoperable, it should be capable of conveying its information over interfaces it currently uses and those that have not yet been defined.

We have a fairly clear understanding of what lies ahead for packetized television, DTV and compressed digital video. Unless the servers of tomorrow are built to exchange information at a fundamental bit stream level, we will be without interoperability.

As evidenced by the lack of interchange on motion JPEG files, we have recognized that future video servers should establish a common subset of file formats and protocols with which to carry and translate data from manufacturer A to manufacturer B. We are continuing to watch as new standards unfold providing the ability to transport compressed digital video packets over existing router and cabling infrastructures. This is the first step to interoperability, and after years of privacy and protectionism, it is being met with open arms.

APPLICATIONS, COMMUNICATIONS, AND OTHER DETAILS

Application software is an important factor and is often taken for granted in considering the server of today and how it will be scaled upward tomorrow. The software application program interface (API) is the heart of performance once the drive arrays, compression codecs and input/output interfaces are fixed.

Examples of software that may influence your particular application include system and machine control, database tracking, archive management, monitoring, and other embedded applications. Most servers lack more than some basic fundamental controls, diagnostics, and monitoring routines. There has been a cry by users to provide more universal and sophisticated applications as part of fundamental video server systems.

Giving credit where it is due, those who are providing the next generation of storage systems are realizing that the user, after spending $150,000 for a video server, does not want to be told they must spend an additional $50,000 to $150,000 for an outboard, add-on software application before they can make substantial use out of their investment.

Another of the underlying elements in the video server is its operating system (OS). Today it is rare that the OS is completely proprietary, due to the enormous expense in developing code to control what are basically standard computer CPUs with sophisticated peripheral interfaces. At the very least, be aware or understand if the operating system is based upon UNIX, X-Windows, Windows NT, Mac, or OS-2, or if it is completely proprietary.

You may not need some of the functionality today, but knowing how the server networks and what its peripheral communications protocols include could become important pieces that will help identify the flexibility and longterm durability of the software control architecture.

How the server communicates with the host will determine the physical interfaces necessary to install during integration. Most servers use some flavor of Ethernet. Token Ring is all but vanished for new installations but may be part of an existing network environment you may need to interface. The server may communicate over TCP/IP protocol or it may be based only upon RS-422 for machine control.

Fault monitoring may be internal, external or a combination of both. You should understand what is monitored and how it connects with the monitoring host. Can the server be monitored or even controlled with traditional off-the-shelf monitoring applications such as HP-OpenView or SunNet Manager? Will there be ATM or ATM/AAL5 high-speed network IP implementation, and for what reasons? If the video server is intended for multipurpose operations, such as being shared between an on-line editing system and a news department's off-line or air-playback system, you should know how it communicates and what the expected throughput will be at peak times. In essence, will the server be adaptable to your enterprise's computing network so that sole sourcing is minimized?

UPGRADES

Chances are, if the server manufacturer is a computer-oriented company, many of the more traditional interface, monitoring and control functions will be implemented. This will have advantages as the future facility architectures become more integrated with network-based operations. Knowing what to expect from future upgrades of, say, Windows NT, may be of benefit should a custom control interface or the addition of a different manufacturer's server be contemplated within the facility.

If the server is to be controlled by an automation system, has the vendor supplying the automation system been staying in step with the latest implementations on the video server? When an OS upgrade is sent from the server manufacturer, do you know how to verify if their current automation application has been thoroughly tested using the actual new version(s) supplied by the server manufacturer?

Of course, this requires that someone with real computer savvy be employed to take advantage of these feature sets—but I think we're all in tune to the fact that the TV engineer of the future must be versed in networking as well as video.

INPUTS AND OUTPUTS, VIDEO OR DATA?

The generic term "digital video" carries with it a lot of confusion and ambiguity. To some, "digital" may mean any video created, captured, displayed, edited, stored or transmitted on a computer or over a computer network. To the broadcaster, the postproduction facility or the videographer, "digital" usually means video that conforms to Recommendation ITU-R BT.601.

The studio encoding parameters for digital television, in 4 x 3 and wide-screen 16 x 9 aspect ratios is referred to as ITU R BT.601-5. The standard that describes a serial digital interface for 525/60 and 625/50 digital television equipment, operating with either 4:2:2 component signals or $4f_{SC}$ composite digital signals, is ANSI/SMPTE 259M-1997. The bit rate for the resulting component serial data stream is defined for both 4 x 3 and wide-screen 16 x 9 aspect ratio pictures, nominally 270 Mb/s for 13.5-MHz luminance sampling and 360 Mb/s for 18-MHz luminance sampling. The nominal bit rate for serial $4f_{SC}$ composite NTSC is 143 Mb/s and for PAL serial $4f_{SC}$ composite signals it is 177.3 Mb/s.

Thus, the serial digital interface (SDI) is that methodology used to transport the above described digital video around the studio for input or output to such devices as switchers, monitors and video servers.

Knowing the details of the video server's I/O will be important from the very beginning. In chapter 9, we will deal in much greater detail with the I/O on today's video servers.

We fully expect to see servers with the capability to record more than just video. Today, most new products in the video server line accept SDI inputs and outputs. However, consider this: Will the video server allow for more than an SDI interface? Consider what is around the corner and how it will impact your facility's future—as Fibre Channel, ATM, FireWire (IEEE 1394), or some other transport technology under development that might come along.

In the future, it is likely that data, in many cases stored to the drives in less than full video bandwidth, would need to be moved around the facility at a data rate two or more times the real time playout rate. Will the video server, which may only have an SDI interconnect, permit alternative data dumps at greater than real time to another server or another location via a digital satellite link or a future intercity 1.2 Gb/s SONET fiber loop?

Will the server accept, process and/or store data delivered over SDTI? In other words, consider how, besides real time playback, you can get the data in and out of itself, or another server or possibly convert it to a DV bit stream.

STORAGE: TO RAID OR NOT TO RAID?

To the layperson, the ins and outs, whys and why-nots of the storage array remain a confusing element of the media storage subsystem. Today, the buzzword remains RAID, which stands for redundant arrays of inexpensive drives (or disks). RAID comes in many flavors and features, so consider what type of storage architecture options are available.

Know and understand which RAID-level is being implemented on the server you're looking at and why. Know the questions to ask, such as:

- Is the system's RAID level set for a specific reason, such as parity protection or improved access time?
- How many drives make up the RAID array for a given amount of storage?
- Does the system employ fast and wide, deep or Ultra SCSI drives?
- How deep and how wide are the drives striped?
- What are the form factors and where can I get off-the-shelf replacements?
- Will buying a replacement drive, drive pack or entire array become as painful as replacing a Digital Betacam head drum?

66 Video and Media Servers: Technology and Applications

We will hold further discussions on RAID as the subject is presented thoroughly in chapters 26 through 28. What is fundamentally important in this overview of selecting a video server is how the storage structure grows and adapts to the future.

The more prevalent systems in use today incorporate a SCSI protocol drive array. The interface is predominately parallel with multiple SCSI buses possible. The newest video servers to come out, as late as September 1997, are incorporating Fibre Channel drives and Fibre Channel–Arbitrated Loop interconnects between the server engines and the common storage arrays.

We are also finding RAID now implemented in software as a counter to the expensive and fault-prone hardware-based SCSI RAID controllers. Software implementation for RAID is another solution that arguably has no greater nor lesser values other than cost and changing the single points of failure or control. In some instances, implementing RAID in software makes good sense because as drives become faster and cheaper and can store more. A software change addressing a node instead of another piece of hardware has got to make better application to extensibility.

CHAPTER 7

SERVER CONFIGURATION BASICS

Placing a video server into service involves more than connecting video and audio inputs and outputs to the rear panel and turning it on. To effectively utilize the server, a thorough understanding of the underlying architecture is necessary.

This chapter is the predecessor to a series of in-depth chapters on the internal architectures of the video server. We will begin by discussing the ground level basics of the server as it applies in the broadcast environment.

Every manufacturer has its own particular way of making its server function. Some offer a very basic one or two channel device that is intended for the simpler functions such as record, library, cache, and playback. These servers have limited expansion and are marked for a price niche rather than from an expansion perspective. This is not to say that this type of server is any less capable of producing images from disks, but it was designed for streamlined operations without a heavy initial price point.

BASIC CONCEPTS

We are not going to delve deeply into the specific manufacturer configurations for video servers, for many reasons. To do so would mean we'd have to leave several servers off the list, which wouldn't be fair. Alternatively, we'd have an oversized book that would read like a catalog and would be obsolete in only a year or two.

Another reason for not covering all the flavors of servers is that the technology and the internal architectures of servers are changing regularly. If we covered the way one server is fabricated today, it might be meaningless only a few months later. So apologies are provided in advance.

We will, however, look into some of the basic concepts developed by specific manufactures because they have taken either very unique (at least for today) or more common approaches to building their storage subsystems.

Video servers runs the gamut from full-bandwidth disk recorders for storing uncompressed ITU-R BT.601 video to high compression, off-line editors that are used strictly for rough cutting with no intention of airing the resultant product.

To provide an understanding about how much drive storage is required, a full-bandwidth recorder is going to cut into storage capacity at the rate of 22.5 Mb/s for uncompressed, 8-bit video data only. If you record everything, including 10-bit video, embedded audio, and ancillary data you would consume 33.75 Mb/s.

A conventional 2-GB computer disk drive is capable of recording about 60 seconds of video at 10-bits ITU-R BT.601, 4:2:2 component digital sampling and eight channels of AES audio. For broadcast operations, this would be an impractical use of hard disk storage space for daily commercial operations. It would also be pointless to stack multiple disk drives into the hundreds and put only 60 seconds of video on each one. For this reason, we have seen compression technology flourish for video server applications, providing additional storage space with no sacrifices in visual quality.

We have established that the practical video server uses compression for the storage of video, audio and associated data. So for the remainder of this chapter, we'll dispense with the specifics and cover the general. Later, we will cover some broad-based applications in systems that will give examples of how larger-scale systems are utilizing video servers as a complete solution—in news, production, on-air/commercial playback, and program delay.

MAIN SERVER COMPONENTS

Most video servers are comprised of two main components. The first component is a record/play unit that is the interface to and from the second component, the primary disk array. Each of these components generally provides for some level of expansion.

Server Configuration Basics 69

The record/play unit, or video engine, as some have called it, consists of several elements:

Selectable video inputs—Either serial digital (SMPTE 259M) or analog flavors in composite and component. Professional video servers in broadcast applications are moving away from the analog composite inputs and outputs, or are offering them only as options. The inputs may accept either 8-bit or 10-bit, allowing the user to select the depth during server setup.

Selectable audio inputs—Similar to the video inputs, these are either analog or AES3 standard digital audio. Interface options include the 75-ohm BNC (unbalanced) input or the 3-pin XLR input for 110-ohm twisted pair, balanced. Most servers accept only 48 kHz AES sampling (20-bit resolution) in at least two sets of AES inputs, yielding four channels of audio, generally identified as two stereo pairs.

Embedded audio—This option allows the AES audio streams to run on the same media as the video input. This would be available only on devices that provide serial digital interfaces (SDI). We may see SDTI (SMPTE 305M) as an alternative method of interface for packetized compressed digital bit streams, which may also embed audio or other data into the transport structure.

Outputs—Those provided are generally in the same formats as the inputs.

Analog video reference—Generally video black burst from the facility's house reference generator. The user can select between external reference, internal reference (free run), or one of the video inputs (either digital or analog).

Control interfaces—RS-422 machine control, for emulation of common VTR modes, such as the BVW-75 protocol. Most servers will supply at least one RS-422 machine control port for each "channel" of the server. The terminology "channel" has no common definition and may vary depending upon the manufacturer's implementation. It is expected

70 Video and Media Servers: Technology and Applications

within the next few years an international control code will be presented for standardization. The goal of such a control code will be to normalize all the various control interface parameters (e.g., Sony, Ampex, CMX) between devices, to one common set of protocols. The protocol presumes to include specialized functions, some specific to video servers, in order to bring increased performance and capabilities to disk-based systems.

Figure 7-1 depicts a generic, two-channel rear interface panel for a video server with both analog composite/component inputs and serial digital interface (SDI). The inputs are shown as analog, but could very well be AES digital audio, internally selected with switches.

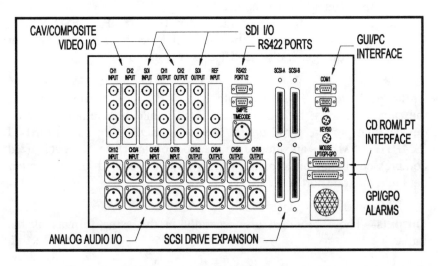

Figure 7-1: Rear interface, two-channel video server.

General purpose input or interface (GPI)—Servers usually have at least the same number of inputs and outputs as there are channels. GPI functions may be dedicated or set into software and assigned by the user.

Timecode—Digital time and control code, as referenced by ANSI/SMPTE 12M-1995, for use in systems at 30, 25 and 24 frames per second. Timecode is generally referenced to house clock. In most cases, timecode data, specific to the video clip recording, is carried and

Server Configuration Basics 71

conveyed to the video server from the source VTR on the RS-422 serial control interface.

Network connection—Generally this is Ethernet and is used for whatever the manufacture intends it for. Sometimes it is for connection to a LAN, sometimes only to another PC that acts as a GUI for the setup and operations of the server.

The balance of other connections to the server are manufacturer-specific. Some of these connections might include:

SCSI interface—Depending upon their drive structure, the physical drives may be mounted internally or may be in secondary enclosures, usually RAIDed. If the drives are internally mounted, then you probably won't find a usable external SCSI connection. Should the server use external drive chassis, the number of SCSI ports on the rear of the primary enclosure will depend upon the number of SCSI buses (sometimes several) that the system supports.

Fibre Channel interface—Most of the servers manufactured at the time of this writing were using FC over twisted copper pairs. The need for an optical fiber connection has not surfaced and manufacturers are reluctant to add them until demand warrants it. If more than two servers or hosts are networked, the twisted pair media is connected to an outboard managed Fibre Channel hub that controls the distribution of data between devices. Depending upon the specific server and storage architecture, Fibre Channel may be used to connect between a server and its Fibre Channel disk drives or arrays, either internally or externally. Fibre Channel may also interface between other servers and peripherals.

Alarm ports—These may be in the form of either GPO contact closures, or other independent ports, which allow the user to configure signaling as they desire. The signals that trigger alarms might include primary power supply failure, fan or airflow failure, over temperature, or a fault in the computer or compression engines themselves.

User interface—If the server is built around a PC-based board set it most likely has the normal PC-like I/Os. You will generally find a VGA output for display, a keyboard, and a mouse port. These three items are lumped into the term GUI (or graphical user interface).

Peripheral ports—Generally these consist of at least a 9-pin serial port and a 25-pin parallel port. Some servers make use of the 25-pin port to drive an external CD-ROM drive, used for loading software. Others have built-in floppy or CD-ROM drives for loading software or archiving data.

Modular expansion—At least some expansion should be expected both internally and externally to the server. Inside, additional slots destined for new or additional codecs should be available. Outside, modular access to breakout panels, control panels, or other interface devices should be available if the server does not currently populate all the input and output ports.

MODULAR ARRAYS YIELD VERSATILITY

Server storage subsystems are generally modular in makeup. We've already pointed out that some servers install the storage drives, called the array, internal to the video compression and I/O chassis. This provides for a finite amount of storage, regulated by the individual capacities of the installed drives and the compression rates selected during coding.

The more common approach for broadcast video servers is to add expansion chassis, usually in a RAID configuration, that can grow as the user decides they need added capacity. How these arrays are constructed and how the bandwidth of the system is kept high are the basic principals in disk drive storage technology for media applications.

Today, at least, we find three fundamentally different approaches being used for drive storage arrays. The first is the conventional RAID-3 system that is being used by the majority of manufacturers (each includes a few of its own modifications). RAID drive subsystems can reside internally to the server or externally. In our application, the convention SCSI interconnects the slave drive enclosures to a common video processing and I/O chassis.

The second generic form of storage array is the Fibre Channel disk array, which would use serial connections of shielded twisted pairs

Server Configuration Basics 73

in a loop arrangement (optical fiber is an alternative, but is not yet implemented on video servers in production). This form of array makes interconnection simpler, but generally requires an external Fibre Channel hub. The hub does become a single point of failure if it quits altogether, but the looping arrangement of this type of hub provides for internal bypass and changeovers to a certain degree. As of this writing, there were no hot-changeover Fibre Channel hubs being implemented, but it is expected that a product like this will be available soon.

The third type of array is the switched fabric and concentrator/commutator arrangement. This device physically resides between the I/O video server engine(s) and multiple drive array controllers—generally supporting outboard RAID chassis configurations.

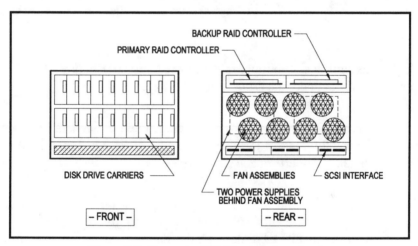

Figure 7-2: Conceptual RAID expansion chassis.

An example of a RAID expansion chassis, Figure 7-2, shows the components generally found in a RAID array. For storage expansion additional drives may be added to the chassis in balanced groups or banks. Arrays are made up of groups of drives, generally from four to five per group and consist of four data drives and one parity drive.

The generic RAID expansion chassis might consist of the following components:

74 Video and Media Servers: Technology and Applications

RAID controller—The hardware-based circuit card that manages the disk drives to make them appear as one drive to the system chassis. RAID can also be implemented in software, in which case that application may also be referred to as the "controller".

Backup RAID controller—A secondary, standby card, typically an option, that can be placed into operations either manually or, in some systems, automatically.

Power supply systems—You will find at least a primary and one backup supply specified in most RAID expansion chassis. These are hot-standby, automatic fault tolerant devices that also provide performance status back to the system controller and the RAID controller.

Cooling assemblies—Multiple fans designed so that a failure of one or more fans will not take the drive system off-line.

Drive carriers—Depending upon the chassis, these are generally single slots that typically house 3-½" form-factor 4.5 or 9.1 GB Ultra SCSI or Fibre Channel, high-performance disk drives.

Fault tolerance—Part of the RAID 3 specification is automatic detection and identification of failed or failing drives; plus, upon replacement, automatic rebuilding of the data on the new drive from information remaining on the other sets of drives. This should be a transparent operation, although there might be a slight but generally unnoticeable degradation in performance during the reconstruction.

Automatic formatting—When a new set of drives is added, thereby expanding storage in what was previously a less than full compliment of drives, the system will automatically format the new drives. A new parity channel will be created and any necessary striping will be accomplished, generally unattended.

Fibre Channel redundancy—The storage array is based upon the FC-AL, and the manufacturer will most likely provide dual ported nodes. Each node has a second port that may be connected to an entirely electrically isolated and separate loop. Both loops are concurrently active and are designed for automatic transfer of data traffic between them. Loosing one port on the node does not cause a system failure.

Fibre Channel Loop Resiliency Circuits – LRCs are high-speed bypass circuits used in arbitrated loop hubs. The LRC detects the presence or loss of the connected port. It acts as a switcher, placing the port on-line, if active, and switching it out if inactive or nonresponding. This LRC system is designed to eliminate a total system crash in the event a node, port, or a loop failure occurs.

In the next chapter we will move into the internal architecture of the video server, covering the details of general system I/O and codecs.

CHAPTER 8

MEDIA SERVER ARCHITECTURES: THE FRONT END PROCESSOR

All media server architectures can be looked at as being constructed from elements, or subsystems, of an overall system. In this chapter, we will be concerned with the front end of a specialized and unique system, the video server. While referring to this type of media server as simply a "video server," we understand and accept that video servers generally have inputs and outputs that include a media makeup of video, multiple audio channels, and usually at least some form of control interface.

A video server must essentially allow for the input, storage, and retrieval of video data in real time. Although the internal architectures of video servers will vary significantly from manufacturer to manufacturer, the basic principles of the video server remain essentially the same.

We can see in Figure 8-1 that the video server is comprised of an input side, a processing and storage section and an output side. Video and audio inputs and outputs can be of any flavor, including composite and component, analog or digital. The control and data I/O can be a mixture of timecode, machine control, or network interfaces.

The makeup of the video server subsystems, including the interface to and between each subsystem, is the major factor in a video server's performance. The more a video server system is flexible in each of the subsystems, the higher the degree of applications it has.

78   Video and Media Servers: Technology and Applications

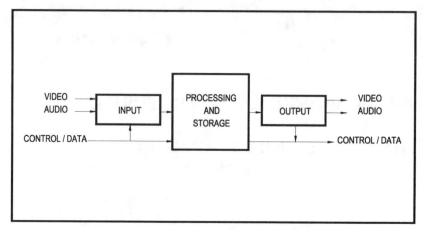

Figure 8-1: Basic video server structure.

INPUT AND OUTPUT

The formats typically found in video server inputs and outputs vary depending upon the intended application. We will first be concerned with the video aspects of input and output.

Over the years, as video has transgressed from analog composite to analog component and then to digital, manufacturers have tried to stay in step with the marketplace in deciding just what flavor of I/O to include in their products. This is true for videotape transports, display monitors and video servers.

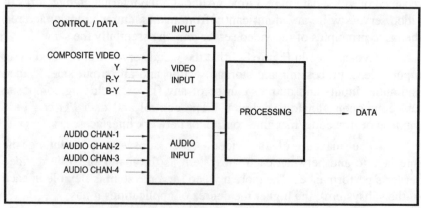

Figure 8-2: Analog input to a video server.

Media Server Architectures: The Front End Processor 79

By the end of 1997, the video industry was facing more potential and established formats than it had seen in its entire history. The inputs we would expect to see on video servers might well resemble those found on any analog videotape transport (Figure 8-2), such as BetacamSP or MII.

Following the input section, the video and audio information is processed in any number of ways, depending upon the choices the manufacturer has made in the video server's architecture itself.

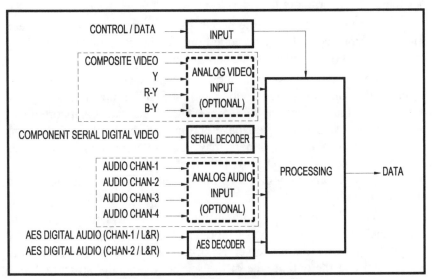

Figure 8-3: Advanced video server inputs.

As technology keeps moving further toward component serial digital video as the standard for physical plants, many of the input options seen on the video server (Figure 8-3) will be either eliminated or supplied only as options. We expect to see the format of inputs migrate entirely away from expensive analog, which requires conversion to at least some type of digital domain. The inputs to future video servers may need to include a variety of compressed digital bit stream inputs supported by some level of serial digital interface on coax, twisted pair or optical fiber media.

As digital television continues to evolve, we will recognize that the world of the digital I/O will be far more than what we now know as "601" or ANSI/SMPTE 259M, component digital video in a serial form. We fully expect that the video server of five years from now will incorporate I/O from a mixture of digital transports, including full-bandwidth, compressed bandwidth, and data.

Many video servers already provide ANSI/SMPTE 259M inputs and outputs in place of or in addition to component and composite analog video. Audio is also appearing as AES inputs on some models, with input options including 110 ohm or 75 ohm, but AES is not as readily accepted as the serial digital video I/O.

Embedded audio in the serial digital transport has so far seen limited acceptance and may be added only if the concept of distributing embedded audio and video catches on in the facility.

For simplicity, we will show the next several iterations of the video server using only digital video and AES audio inputs. It should be remembered that the options shown in Figure 8-3 could be supported in a number of ways. Some servers have option cards built into their systems that are merely plug-ins. Others may suggest converters be added as outboard devices (glue products). Others show the serial and AES digital inputs as options and only offers composite and component analog video and analog audio.

It is up to the individual integrator or system engineer to determine which combination of inputs and outputs is necessary for the environment in which the server is to be placed. It should be further understood that the output side of the server is generally structured as a mirror of the input side. We shall also assume the user would provide essentially the same output format as input format for the majority of applications.

BASE CODING CONFIGURATIONS

The next portion of the server's front-end to be considered is processing. This is the heart of the video server architecture ahead of storage. Processing may include the functions of preparing the video and audio signals for compression or for the assembly of the digital data into a structure appropriate for storage on hard disk drives.

Media Server Architectures: The Front End Processor 81

One distinguishing point in the server that makes it function uniquely from that of a digital disk recorder is that it can handle multiple inputs and outputs, generally simultaneously. In order to do this, some form of front-end is necessary to channel or route the signals to appropriate encoders or decoders either before or after the storage system.

We will first look at the simplest form of video server. A single channel in a video server can be thought of as an input converter, a coder, and an output converter. The input converter sets up the video ahead of the coder so that it is in the correct format for the coder to accept. For simplicity, we look at the coder as being a compressor-decompressor system, as opposed to a full-bandwidth transcoder or formatter that would essentially be used only in higher-end video disk recorders. Figure 8-4 shows the fundamental portions of input-output processing in a video server.

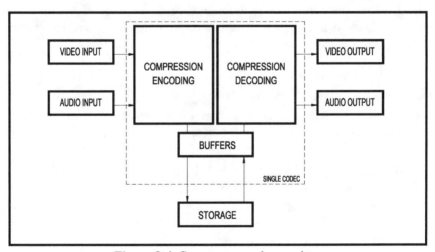

Figure 8-4: Server processing engine.

Video and audio enter the server, are converted from its native format and then are fed to a compression encoder. The output from the encoder is sent to a buffer that controls the flow and timing of data from the encoder and onto the storage system (hard disks). When instructed, data stored on the hard drives is sent back to a buffer, to the compression decoder and on to the audio and video output converters.

There is a variety of assemblies available for the encoder-decoder components of the video server. Some manufacturers use off-the-shelf PC type-hardware, similar to those used in non-linear editors and PC-capture systems, as their codec[14]. The circuit cards used in this application are generally ISA, EISA, or PCI bus-based. Some manufacturers may use VME cards and others may bridge between circuits cards with a higher-speed "over-the-top" bus-interface used strictly for digital video data transfers. Generic PC-type circuit cards have limited capabilities, optimized primarily for the available space and considerations, such as the bus-speed, of the PC-industry.

The broadcast environment's video server is generally designed for the rigors and performance expected in a mission-critical application. You will find some designs employ proprietary codecs built around more advanced compression chip sets even though the edge connector and bus architecture may strongly resemble that of a PC-type card. Since the codec in the server is the essential component in determining image quality, the selection of the compressor format, the chip set, the software tool set implementation, and other associated algorithms are paramount to the performance of the system.

CHANNELS AND CONFIGURATIONS

One of the architectural design factors in any server is its capability of simultaneously handling multiple streams of media (video and audio) going both into and out of the server. This issue is a common point of confusion and is the subject of the next portion of our discussion.

A simple server like that shown in Figure 8-4 would typically offer a single input and a single output. This would be the classic definition of a video disk recorder. This implementation allows only two functions to occur—a single record in and single play out.

An extension of this concept (see Figure 8-5) includes a slightly more advanced implementation that allows for more than one output operation to be delivered at the same time. This implementation is used in most video-on-demand systems, but the approach has only limited application in professional broadcast servers today.

---

[14] Codec, meaning "encoder plus decoder," is a universal device with complimentary, combined functions that will encode (compress) and decode (decompress) on the same physical hardware subsystem.

Media Server Architectures: The Front End Processor 83

In Figure 8-5, you can have two (or more) different playbacks simultaneously from the same storage system. You might use one output for confirmation when inputting data into the server, then switch so that two unique outputs are available simultaneously for playback of two different programs, still leaving an audio/video input for recording with an E-to-E[15] output confirmation.

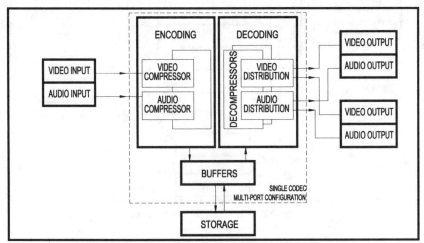

Figure 8-5: Multiple port, single codec processing.

Since broadcast operations tend to require more flexibility in daily operations, more inputs and more outputs may be desired in a user-assignable configuration. One way to do this is to use multiple codecs and have a steering device or router that would simply route the input signals to the proper codec. Another method would be to have a dedicated compressor and a dedicated decompressor for each expected stream in the server chassis. This concept is more expensive and may be viewed as a waste of resources, depending upon the architecture chosen by the server manufacturer.

Besides routing the signals to the correct encoder/decoder, more advanced coder arrangements permit the handling of more than one process at the same time. Using high-end codecs, larger buffers, data switching and bus management, the time-sharing of codec elements

---

[15] E-to-E is "electronics-to-electronics" where the signal is monitored without passing through the processing paths.

84 Video and Media Servers: Technology and Applications

within the server architecture is possible. The encoder-decoder software elements may then use the same chip sets for the complimentary or opposite processes and allow for "turning the signal around" inside the codec subsystem itself.

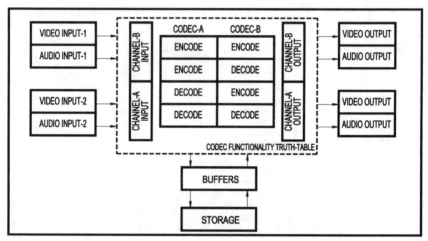

Figure 8-6: Dual codec processing.

A channel in a video server is considered a subsystem element made up of an input processor, a codec, and an output processor. The processors may include a discrete physical input/output section or may simply be an input point for preprocessed video media before or after the codec. A port is either an input or output access point (including a converter) that connects either video or data streams into or out of the channel. Figure 8-6 depicts the process whereby two channels can be set up to allow for multiple functions without the need to duplicate some of the codec circuitry.

The processing in a codec is complimentary; that is, it will be able to both encode and decode (but generally not simultaneously). By providing an outboard switching method, usually as a software steering device, the two codecs can process four different and discrete functions independently. The four functions are shown in Table 8-1.

Media Server Architectures: The Front End Processor 85

CODEC-A	CODEC-B	INPUT/OUTPUT FUNCTIONS
ENCODE	ENCODE	Two unique inputs — no playback outputs
ENCODE	DECODE	One record — one playback
DECODE	ENCODE	Same as above (one record — one playback)
DECODE	DECODE	No record inputs — two unique outputs

Table 8-1: Dual codec processes.

Larger video servers may have several channels that are fed to or from multiple input or output ports. The channels may feed several buffers, a network access port, or one to many disk arrays for storage of the video media data.

Other video servers have elected to use an architecture that keeps only one or two channels in each chassis. There may be many reasons for the decisions; most are based upon the target market the manufacturer sought to attract at the time their design philosophy was in early development.

CODECS

The codec, as mentioned, is an important element of the video server architecture. Some video servers have already been designed for extensibility, which means they provide an avenue to either upgrade or add on to the existing codec board set to address expected improvements in compression and other technologies. Others have determined that new circuit cards with new features, developed and provided by third party OEMs, will be their course of extensibility.

In proprietary systems, we find that some codecs have been built around mezzanine boards[16] that handle audio processing and compression separately. This concept might well be the answer for a server that presently does not directly handle AES digital audio or that uses only motion-JPEG but intends to upgrade to some flavor of MPEG-2 in the future.

There are many designs and implementations of codecs for video servers. Often the design is based upon a manufacturer's intent to grow the product well beyond what they saw as an entry-level concept into the

---

[16] A mezzanine board may be a daughter board or full second level add-on board designed as a subset to the original I/O processing board or compression engine.

server domain. Sometimes these designs are used to compare one server to another. Other times the design could only progress so far because standards were not in place that would have made it possible to market the server with an industry stamp of approval behind it.

Still, there always seems to be a degree of negative marketing between competing manufacturers that is used to substantiate one particular architecture over another. Some would like users to believe that because a codec plugs into an EISA bus, it actually carries the video data over the EISA bus—which is inherently a slow device. Others would like you to believe that using one type of codec over another is a reason not to buy the competing product. Others like to tell you that a particular proprietary architecture inhibits compatibility between coding processes and therefore means that that server is not compatible with other products.

Since video and data processing bandwidth (or speed) is another important element in server architectures, we have seen additional means implemented to enhance performance and data bandwidth. Some codecs have intentionally added an over-the-top bus connection as a means to carry higher data rates independently of the computer's built-in bus on the motherboard. The over-the-top concept consists of either a header or a ribbon cable connection from one codec card to another codec or I/O processor. It is called "over-the-top" because it rides between cards above or on the end opposite to the edge connector that plugs into the motherboard.

This technique, similar to the original Intelligent Resources VideoBahn bus, is quite effective in getting around the interrupt-based architecture of the PC bus and thereby using the PC motherboard as a control bus rather than a video data bus. Other high-speed data servers have used similar techniques, such as Raceway, for the connection of multiple processors on multiboard servers.

The codecs selected for most video servers are generally compression-based devices that use either motion-JPEG or one of the two more prominent MPEG-2 profiles. Both "PEGs" use DCT for spatial compression of each image, although MPEG uses both spatial and temporal compression consisting of I, B and P frames. Experimentation with other forms of compression, such as fractals, is ongoing with only a tiny fraction of non-PEG–type implementation being used at this time, and nearly zero in the broadcast environment.

Codecs often allow for multiple data rates or compression rates that give the user flexibility in the grade or degree of image quality they want. There are some considerations to understand about descriptions and applications for user definable, multiple data rate codec.

Some servers allow the data rate to be changed on a file-by-file or clip-by-clip basis. This is valuable when considering the quality of the image you wish to maintain during the encoding process only, as it is the bit rate (and GOP structure in MPEG) that determines how much storage will be required for a given unit of recording run time.

Some servers' codecs will allow for the dynamic processing of the bit rate based upon what the front end of processor determines as adequate to maintain the image quality based upon the type of content. The variables considered in dynamic bit rate adjusting include the amount of motion from frame to frame, the relative amount of detail in a given scene or sequence of scenes, how hard the processor has to "work" to get within a certain defined range of bit rates, and many others.

In MPEG-2, where bit rate and GOP structures can affect the number of total bits required to store a clip, an entirely different set of considerations must be used for dynamic bit rate adjustment. Since this process, referred to as variable bit rate, is relatively new and is in only minor use, we will leave that discussion to other experts on the subject.

CHAPTER 9

MEDIA SERVER ARCHITECTURES: VIDEO SERVER I/O

The media servers discussed up to this point remain within the "video server" definition because they deal essentially with video and audio, vertical blanking interval (VBI) data, and the timecode and frame location identification, in either the analog or the digital domain. For the video server, the major concern, for the immediate pre–high definition television generation of digital, is its ability to reduce the dependence upon conventional videotape as a sole and exclusive storage medium.

Justification that leads to the eventual divorce from videotape includes issues impacting the cost and maintenance of tape transports, linear videotape inflexibility, complexities in automation and tracking systems, physical media costs, and the reliability of the transport and the preservation of image quality over time. In most of the broadcast facilities in 1997, the professional media server seemed to be confined principally to video-based uses. This has been of benefit to designers because it has essentially constrained the timing and resolution of the video data to a defined and understood realm.

As the move accelerates toward DTV, we will be dealing with entities far beyond 525/59.94 video. How the facility and the server handle those issues will be an entirely different matter.

Many of the today's concerns will take on new dimensions once the recording and storage of higher bit rate signals (HDTV) become reality. Signals with data rates greater than 270 Mb/s and a host of other compressed signals at less than 100 Mb/s will begin to creep into our predicted implementation of DTV. These signals may be carried within

SMPTE 259M, but using a different *transport* such as the proposed SMPTE 305M (SDTI) for serial data transport interface.

In this chapter, we'll keep in tune with conventional 525/59.94 video as we know it, leaving the future for a later discussion.

BRIDGE TO THE OUTSIDE WORLD

We recall from the previous chapter that the input and output (I/O) portions of the video server form the bridges to the outside world. For several years it will remain necessary to link analog composite, component, and SMPTE 259M (digital video interface) signals to the internal world of the video server's codec and storage subsystems. The I/O cards that bridge the external world to the inside dimensions of the video server are selected to configure the video server processing for a specific manufacturer-selected application.

If the server's application is designed for a motion-JPEG compression codec, then the I/O card must prepare the video signal (either analog or SDI) so that it can be delivered to the input side of the codec. If another codec is used, then the output side of the input card may require another structure altogether. For the broadcast video server, the end user seldom has any say as to what these selections and configurations are.

Video server products from various manufacturers now offer several video format I/O options. Most new video server I/O options conform at least to SMPTE 259M for 10-bit 4:2:2 component and $4f_{SC}$ NTSC composite digital interface[17]. There are few implementations of either composite digital video or component digital bit-parallel (SMPTE 125M) interfaces for servers in current manufacture. At least two reasons exist for the demise of these interfaces. First, new support products in composite digital video, such as D-2 or D-3 capable devices, are just about nonexistent. Second, almost all bit-parallel (12-½ pair twisted cable) component digital interfaces have been replaced or converted to serial digital interfaces.

However, at the end of 1997, analog composite and component video is still in abundance throughout most broadcast facilities. For this

---

[17] $4f_{SC}$ is a composite digital system with versions in both 525/59.94 and 625/50 scanning standards. This system utilizes a sampling frequency of 4 times the respective sampling rates. For 525/59.94 this is 4×3.58 MHz = 14.3 MHz.

Media Server Architectures: Video Server I/O 91

reason, many video server products still have at least an option for these inputs, but that is changing rapidly. In most cases users are now being encouraged to use external converters (glue products) rather than purchase internal board sets for the server.

INPUT FILTERING AND PREPROCESSING

For broadcast applications, the server's analog composite input subsystem is expected to preprocess or filter the incoming signal for a variety of image-related purposes. Filtering is necessary prior to digitization so that aliasing does not occur in the digital domain. Noise reduction is another preparation element of compression that is performed in a dynamic or adaptive nature.

Composite analog decoders generally will provide four possible forms of filtering. A notch filter will tailor a signal to separate, remove or otherwise suppress portions of the signal that might otherwise overextend the amount of work necessary to compress the signal for little or no improvement to the image itself. Notch filters, when improperly applied, can introduce artifacts into pictures that have luma (Y) details at frequencies near the color subcarrier.

A temporal filtering process is good for still images but not for motion. The filtering templates detailed in ITU-R BT.601-4 have been standardized for studio digital video systems, and define both a passband and a stopband insertion gain that shape the signal so that it can be digitized properly and without overshoot or aliasing.

Spatial adaptive filtering is good for motion-related images but not for stills. The fourth filtering process is a combination of both spatial and temporal adaptive filtering that offers a compromise between the two.

The I/O subsystem should be capable of providing automatic gain control (AGC), including the ability to track noisy signals such as those from microwave or satellite feeds. Fine control of gain will prevent excessive excursions of the signal from falsely triggering the compression algorithms into coding erroneous or superfluous information into additional data of no value.

Much of the filtering described is less critical or unnecessary when the signal originates as a proper component digital signal. However, many signals in the current broadcast facility do not originate as wide-bandwidth component digital video. These analog signals must be converted to digital before feeding the input to a video server. If not, the video server will be expected to perform this function itself.

Care should be used in determining the type and quality of any analog-to-digital converter. An inexpensive converter will do little to improve performance and image quality when tacked onto the front end of even a modest-quality digital video server.

Most of the composite analog inputs to video servers should accept NTSC or PAL and should be able to read and process VITC or other VBI signals. If the server itself does not externally genlock (which is rare) it should permit genlocking in order to generate 27 MHz (ITU-R BT-601), $4f_{SC}$ (NTSC) or $8f_{SC}$ (PAL) clocks for internal server functions.

If the composite analog signal contains VITC or other data located in the vertical interval, it may be desirable to preserve that data in some form. Depending upon the codec structure of the video server, this data may be stripped from the analog signal and distributed as a form of ancillary data over the video server's internal bus architecture. The VBI data or VITC identifications should be reinserted on the output stream or available for use as ancillary or metadata of some form.

VIDEO REFERENCE

Another portion of the video input subsystem is the system timing reference for the video server. Typically, this is an analog composite reference signal that comes from the master video black burst signal. The reference should be synchronous to the various video processors, sources and signal systems in the facility.

The genlock signal is decoded by the front end of the video server to provide an internal reference clock for the various subsystems of the server.

The clock signal generated after the A/D converter is placed on the server's system bus and fed to appropriate devices that require video-based clock references for operations. All output video from the server is locked to the reference input. Should the facility wish to operate on a different standard, such as PAL, an appropriate video reference signal would need to be switched to the server's reference input.

Server outputs can be offset from the reference by fields, lines, or pixels. These offsets allow the server to compensate for timing errors that may occur in the analog video domain.

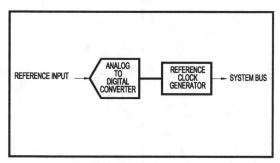

Figure 9-1: Reference clock generator.

When the server operates exclusively in the digital component video domain, the reference is used essentially for clock generation and offset purposes as necessary (see Figure 9-1). Typically, digital video signals are all reclocked (or retimed) just prior to the server's output card.

As time goes on and standards become developed for bit stream delivery of digital signals within the studio, it is expected that many debates will arise around reference signals. Working Groups within SMPTE Engineering Committees have been looking into needs and solutions for advanced reference signals aimed at the all-digital plant for several years.

AUDIO I/O

The process of handling different audio input and output formats brings similar options to those in video. Nearly all video servers will have at least two audio channels, predominantly in analog formats, at the input and output of the server.

An increasingly high number of servers are already using four-channel inputs, in either analog or AES digital. Some manufacturers are providing more complex solutions to the number and uses of audio inputs.

94 Video and Media Servers: Technology and Applications

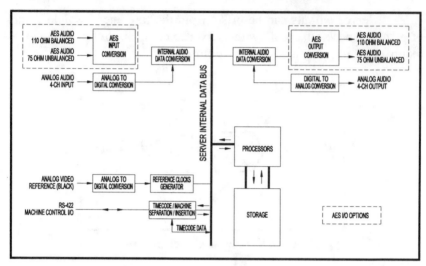

Figure 9-2: Simplified audio and RS-422 flow.

Figure 9-2 identifies audio and machine control data distribution within the video server.

Some video server manufacturers have uses targeted at editing-based functionality. Here there are reasons for having front-end routers integrally installed that can accept upwards of 8 to 16 audio channels, with flexibility to assign them to various record channels internally. Monitoring and confirmation paths are usually included in most video servers.

Servers that accept analog-only inputs will typically transcode the analog audio signal to some form of PCM[18] audio. Either the process will have at least 16-bit resolution with a selectable clock, or at the very least it should be 48 kHz sampling locked to the video reference input. Most server applications in the U.S. professional broadcast arena use either analog or AES/EBU audio inputs. Consequently, most servers have not offered the variety of sampling rates you might find in sound cards or prosumer and consumer equipment.

Once digitized, the audio bits are interleaved with the video data during the process of buffering and transferring to the disk storage

---

[18] PCM (pulse code modulation) is the sampling process that produces pulses whose amplitude is an analog quantity. The amplitude of the pulses is quantized into a number code which is called PCM.

Media Server Architectures: Video Server I/O 95

devices. How this is accomplished is strictly up to the manufacturer and each uses its own method to achieve its own results.

When the server is principally concerned with recording and playout clips, such as short-form commercials, promos and IDs, the added features, such as inserts or audio-only for voice-overs and tags, may not be important. Should the function of the server extend beyond this, then it becomes meaningful for the operator and the engineer to understand just how the video server implements its audio recordings—or if insert editing is even possible.

In multiple channel recording for broadcast applications, such as for SAP[19] or ProChannel, you will need to understand what the server's flexibility and usability factors are. If your facility regularly adds tags or overdubs spots and interstitials, or when overdubbing is required, you should determine a plan based upon the number of audio and video ports and channels your server may have.

For example, a server might require that all audio recordings be accompanied with an equal length of video. In this case, even though the physical space that the audio segment consumes in terms of bits is far less than the equivalent video run time, the same amount of storage may be required.

To carry this concept a few steps further, it is possible in servers that employ MPEG-2 coding, that of you are required to record video with a designated audio-only clip, you should use video black as the video input. The reasons are straightforward when you take into account that with MPEG you can express an indefinite length of black by coding only the first instance of black. On playback, MPEG simply instructs the decoder to "repeat this same frame for duration x." In this way, the bit count goes way down, although the audio may change radically throughout the length of the clip.

As we get closer to DTV, certainly the prospect for additional audio channels and coding will need to develop. If the facility plans to incorporate the elements in Dolby AC-3 coding, a core ingredient in DTV, the audio channels may want to be recovered, preserved, decoded,

---

[19] SAP or Second Audio Program, is used for adding second languages, continuous station promotions, or weather alerts to the stereo audio program material. Sometimes facilities only deal with SAP at the final stages of master control and others may actually place the third channel audio on the individual spots or program IDs themselves.

and/or carried onto server devices in some form. It is uncertain whether this would be at the system layer of the streams or in discrete channels. At present, no equipment is being marketed to address those issues. However, a proposal has been presented by Dolby Laboratories for an open model for audio elementary stream encoding in a DTV distribution chain. This concept would facilitate the multiplexing lightly compressed discrete AES3 audio bit streams and carry them within SDTI, extending the capabilities of four-channel AES3 digital video transports to as much as 24 discrete and recoverable audio channels on one video tape. The concept might be extended further, placing the transport stream at the input or output ports of a video server where it could be recorded as bits or disassembled and stored as discrete AES3 elementary streams.

TIMECODE

Those video servers that are designed for broadcast facilities, in particular as supplements to conventional linear tape transports, will generally need to employ some method to catalog or frame count the media being loaded or retrieved. Timecode[20] is the general and universally accepted method that assigns a number to each frame so that it can be uniquely identified in a clip or videotape program.

Variants for timecode systems include 24, 25, 29.97 and 30 frames[21]. The 32 bits used for timecode account for eight digits that represent hours:minutes:seconds:frames (HH:MM:SS:FF). There are at least three options for the carrying of timecode either to or from linear videotape.

Vertical Interval Timecode[22] (VITC) is carried in the VBI on one or two scan lines of the vertical interval, usually on line 14 (line 277) for 525/59.94 systems. For videotape, VITC should be carried on two nonconsecutive lines. Tradition has placed VITC on line 16 (279) and line 18 (281), with a field mark bit added for field 2 (even). VITC can be ingested into the server via the conventional analog composite signal input. The VITC data is usually automatically detected, extracted from

---

[20] Timecode for 525/59.94 television is standardized by ANSI/SMPTE 12M-1995. For 625/50 line standards, EBU Tech. 3097-E (1980) provides the standard.
[21] SMPTE RP 136-1990 gives guidelines for magnetic recording of timecode on film in these frame rates.
[22] SMPTE RP 164-1992 describes a recommended practice for locating VITC.

the baseband signal (often ahead of digitization) and saved so that it can be recovered once the respective clips are returned to their native condition at the output of the server.

Longitudinal timecode (LTC) is recorded on a longitudinal track on the original videotape. Each frame carries 80 bit cells and results in a bit rate of 1.2 kb/s. The LTC signal is brought into the server as an analog signal (0 dBm into 600 mW) in the same fashion as a conventional videotape transport. Depending upon the architecture of the server, LTC is saved in a similar manner to that of VITC. LTC is then regenerated and output at the time the specific clip is played back.

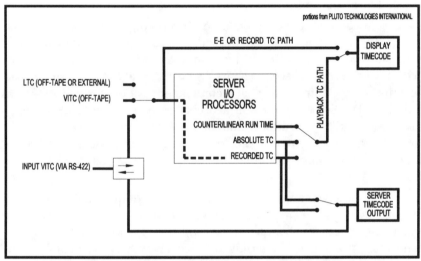

Figure 9-3: Timecode flow within a video server.

The third way that SMPTE timecode can be delivered to the server (see Figure 9-3) is in the same method that it might be delivered to an external device—such as the computer assisted edit controller—in a conventional editing environment. This timecode data is embedded in the control data stream and transported into or out of the server, using the RS-422 data protocol on 9-pin D-connectors. The control protocol (such as Sony BVW-75) will determine where that data is placed in the serial control bit stream.

Timecode is an important element in machine control and in automation systems. Many transport functions, such as recording into the server, play, cue, stop, and reverse rely heavily on existing established tape transport controls. Having a machine control interface that includes carrying timecode information like that of a tape transport control stream is almost a necessity.

Most facility automation systems rely on timecode information for addressing videotape segment locations. Once the program is transferred into a server for delay broadcasting, time-shifting or other purposes, the communications relationship between tape and clip must be preserved. Some servers may not preserve the original timecode but simply begin a new count sequence at the start of each clip. In most cases, however, the server will recreate the timecode sequence on the RS-422 machine control line when played back.

In addition to the 32 bits used for timecode, an additional 32 bits per frame are designated as user bits. User bits are often used for separate purposes specific to the controller or controlling device for which they were generated. There may be a second timecode stream as a real time offset from an original recording, or a set of characters (such as ASCII/ISO). Some automation and control systems build their cueing and tape identification structure around a dependence on user bits. The user and the vendor would need to be aware of what the server does or might do to this information if it is important to be preserved for other uses.

Timecode may be conveyed as either drop frame or nondrop frame. In 525/59.94 systems, the number of frames per second is no longer an integer number and an adjustment must be made for the extra numbers. A prescribed pattern, which occurs every $66\text{-}^2/_3$ seconds, is the frequency that frame numbers must be dropped, in pairs, in order for timecode to stay related to the video color frame. This is drop frame timecode.

The rule to remember for which frame numbers are dropped in drop frame timecode is:

Frame numbers 00:00 and 00:01 must be dropped at the start of every minute except at the tenth minute, when the frame numbers are all retained.

We will not go into the mathematics for, but the impact of carrying the proper information on a server should be understood. Each frame in a server using JPEG intra-frame (I-frame) encoding has a unique identification. This is not generally timecode-based. The clips stored in a server are not necessarily numbered with a "one frame to one timecode number" fashion. The timecode data is generally recomputed as the video stream plays, rather than being retained on a one-to-one relationship. With MPEG-2, however, time stamping may be used for similar purposes to what timecode is used for today (the details of the various time stamps of MPEG-2 are beyond the scope of this discussion).

Timecode data can be conveyed within the server in a number of ways. Sometimes the server will reconvert the data to a frame/field count and display it on the server's display GUI. If the video and audio coding will be MPEG, timecode like data may be embedded into the elementary levels of the data stream, signifying a time stamp that is preserved throughout the system regardless of where it is transported.

The methods used for timecode and timestamping are left to the respective server manufacturer. From a system level, the end user should understand what is carried and how it is preserved or transmitted to and from the server. As long as we retain the various videotape formats and the machine control methods we are most familiar with, the video server will most likely keep these interfaces just as they are today.

VBI

In analog television, the portion between frames or fields is referred to as the vertical blanking interval (VBI), an ambiguous term (per Charles Poyton) because not all this region is actually "blanked". The data contained in the blanking interval for analog video conveys synchronization information necessary to reconstruct the images once transmitted or when displayed on a conventional video monitor.

Digital video has little use for the blanking information and effectively does not bother with encoding that portion of the analog picture. The information needed to reconstruct the vertical blanking portion of the picture can be carried in the digital bit stream without consuming the nearly 8 percent of video frame in the analog domain.

In modern broadcasting, certain lines in the VBI typically carry important information; including such signals as VITS (test signals) and VIRS (a reference signal for transmission). Most of these signals are seen as being transmission-related, and they generally are not present in advertising spots delivered to broadcast facilities for air.

Certain digitization processes might actually obliterate some portions of the VBI. It may be valuable to capture some areas of the VI (vertical interval) and preserve them for return to the analog video domain. There are 12 lines (lines 10 through 21) of vertical interval video in the odd field and 11 lines (lines 273 through 283) in the even field. Within this region closed-captioning information, VITC, teletext, market identification codes, and in some cases even copy protection coding could all be placed in this nonpicture area. If you are using the server for program pass through and time delay, misunderstanding what the server might do to this signal region could cause operations difficulties that could surprise the unaware should they be altered or eliminated.

In the digital domain, the treatment of these various signals and references takes on a new dimension altogether. As the ancillary capabilities of the digital transport take on more sophistication, it is expected that legacy signals, such as those in the VI, may be carried as segments of the digital data stream. We further expect that the rather limited definitions and data space currently available in the analog VI will be expanded to entirely new services.

CHAPTER 10

THE DISK DILEMMA

This chapter will look into disk recording from various perspectives and somewhat differently from what we have been accustomed to so far in the disk recorder evolution. We will look at two fundamentally different principles in storage delivery and structure: file transfers and streaming.

We will focus at first on why we've had to move into these domains, why striping is necessary, and how we've moved through bridging the gap between video and computer data.

The disk storage area is the other main component of the video server, and this is where the details get a little more complicated. Before we look into putting storage systems into place, we'll cover some of the reasons why drive arrays are used the way they are.

Standard hard disk drives are not adequate to support full-bandwidth video. Recording a single stream onto a drive is relatively easy. The video data is sequential, the heads can step routinely and predictably across the drive surface, and minor hiccups such as switching tracks of the drive surface are buffered adequately to make recording and play back straightforward.

This would be perfect if all you wanted to do was record or playback linearly. The video server brings many other challenges. First, in order to record multiple streams, the internal bandwidth of the server must be adequate to input, process, compress, and distribute to the drive array. Non-linear and random access to material, once recorded on the drive, posses another complication. Add in the desire for editing plus playback of multiple channels from the same common storage and the tasks becomes monumental. Therefore, it is obvious that the effective data rate of a system can be increased by using multiple hard drives.

Even though you increase storage time by adding more drives, another set of complications comes into play. The ability of the drive to seek out the correct track, to position the head over that track, to wait for the drive to spin around so the track you need is under the head, all takes time. The disk would need to be aligned exactly as needed, every time, before a continuous stream could be produced, and that just won't happen.

The effects of access time, made up of disk seek times and rotational latency, can only be managed when the data (video, audio, and timecode) is statistically spread over a wide number of drive surfaces. In order to provide a continuous video playback, you must spread out the challenge over several drives in an array.

If you consider that the size of data to be transferred is between 50 and 120 KB, the inherent delays of the system would prohibit any more than two streams for any single disk drive. Mathematically, for a 30 frame (60 field) per second, full motion video stream to be reproduced, the maximum permissible delay for any given disk is 17 milliseconds. Consider the newest 9.1 GB Ultra SCSI drives from Seagate, which have an average seek time of 8 to 9 milliseconds, even at 10,000 rpm spindle speed. The average latency is 2.99 milliseconds, but it still can take an average of 17 milliseconds to transfer 100 KB or more of video.

REASONABLE STRIPES

Mechanical performance is one of the key reasons for considering striping. As such, striping data across several drives is a common practice in nearly all video server storage systems. We also know that different manufacturers have made their own efforts toward balancing compression rates, recording time, number of disk drives, and the number of available additional features, including audio channels and ancillary data.

Some drive systems will use only RAID 0, striping across all the drives. This process works up to a point. At some time during the operations, timing errors or a drive hiccup will interrupt the stream of data and some of all of a field or element will be gone. If one drive

The Disk Dilemma 103

develops additional problems, or fails entirely, the entire data structure is lost and the system crashes.

Many servers have gone away from RAID 0 in favor of RAID 3. There are at least two reasons for this departure. First, RAID 3 has a built-in redundancy factor, that protects data against a failed drive. Second, data is inherently striped across all the drives (except the parity drive), and this also improves the ability for the video server to extract multiple streams of data simultaneously.

RAID 5, which is another commonly used form of RAIDing, is not generally used for video servers. RAID 5 spread parity striping is found mostly in transaction processing and information systems, such as a Windows NT server or larger mainframe. Much more detail on RAID can be found in Chapters 26 through 28.

Some media servers have a tendency to combine more than one RAID structure to benefit their own internal architectures. Tektronix's Profile used the benefits of RAID 0 and RAID 3, combined, for their multichannel server implementation.

Others use RAID 3, but in a different mode from conventional, SCSI interfaces. Fibre Channel drives have shown that higher bandwidth storage can be achieved. ASC Video Corporation has developed a concept that they are building their video server line around. ASC is using their FibreDrive and RAIDSoft, which is a software implementation of RAID 3, to speed up bandwidth. The result is that up to 24 streams of video (at 24 Mb/s) can play back simultaneously from a common drive array. This is made possible because the 9 GB class of 3-½" drives will produce sustained transfer rates of around 70 MB/s—from a single drive array. RAID in an FC-AL (Fibre Channel-Arbitrated Loop) system, using software striping and parity generation, goes beyond simple XOR parity RAID. This system also incorporates advanced error detection and recovery techniques that not only detect, but correct, dual data errors.

TRANSFER RATES AND DATA STRUCTURE

To understand the magnitude of data that must be considered in full-bandwidth digital video coding, let us consider NTSC video, sampled 640 pixels H × 480 pixels V, for a total matrix of 307,200 pixels. To generate what it takes to depict the entire image in terms of color for each pixel, we need to define what the depth of the pixel is.

When we say an image is 24-bit color, that means that each of the three channels, for RGB coding, needs 24/3 or 8 bits per channel. This results in a single frame requiring 7,372,800 bits (=24 bits/pixel × 307,200 pixels). Divide this by 8 bits/byte to obtain bytes, and you have just slightly under a megabyte (921,600 bytes). When you add the moving image factor into the equation, at 30 frames per second (30 fps), the number becomes 27 megabytes per second (27 MB/s).

In order to move this amount of data to and from disk drives, you must sustain a data transfer rate that is based on the following formula:

$$\text{DATA TRANSFER RATE} = \left[\frac{\text{HORIZ pixel/fr} \times \text{VERTICAL pixel/fr} \times \text{COLOR pixel/fr}}{8 \text{ (bits per pixel)}} \right] \times \text{fps}$$

When considering HDTV or other higher resolutions, these numbers reach astronomical proportions. This is also one of the reasons why bridging high-resolution video equipment into computer equipment has taken such a long time.

BRIDGING VIDEO EQUIPMENT AND THE WORKSTATION

Facilities of today will use disk recorders and servers for a variety of applications. The traditional post production environment is tasked with creating video content, multimedia content, and network delivery content (i.e., Web pages). Bridging the world of video and workstations has been a hit or miss technology, with many facilities creating their own independent approaches to solving similar problems.

High-bandwidth, data connectivity within the video house and post facility has had a long history. As Silicon Graphics workstations and servers became prominent, so did a variety of SCSI flavored interconnection schemes. Using both SCSI and Ethernet, the Abekas A60 disk recorder was one of the first bridge devices that connected workstations designed to render individual files to disk tracks and could in turn play back a continuous stream of linear video at 60 fields per second (NTSC).

The Disk Dilemma 105

Abekas was able to obtain high-bandwidth and capacity by omitting most of the ECC (error correction coding) and error detection on normal Winchester disks. While the A60 product was acceptable for storing images, there was no guarantee that the data was reproduced faithfully byte for byte.

Before there was a need to address streaming protocols, importing and exporting of data files became popular. In the late 1980s, this became routine at higher-end facilities, whether it was from a workstation to a video disk recorder or from videotape to disk to file. The capabilities became apparent once the bidirectional exchange of data and video became part of the work process. Digital media and video were about to become one.

The methods used to exchange video data by and between devices were fairly limited at best. Some facilities actually transported large, bulky, and expensive DDRs between facilities because there was (a) no other alternative, or (b) no faster way to render the files to data tape and back, given the short turnaround time on many projects.

For the post production and graphics industries, the introduction of the Exabyte 8500 series of data tape drives changed this capability considerably. Once it became possible to connect these SCSI tape drive devices directly to the DDR, getting video frames to PCs and workstations became a snap (see Figure 10-1). Using linear tape, full-bandwidth individual frames were transferred from the DDR to the tape, then the tape was taken to another computer (usually at another facility) where it was transferred to popular platforms such as Macintosh and SGI. This actually became a very successful backroom business for many postproduction facilities in the infancy of digital video facilities.

COLOR SPACE CONVERSION

Color space control (the conversion of RGB to YUV) and frame/field interpolation were early issues with file transfers to and from disk recorders, and again Abekas led the way in setting industry's unwritten standards on how scaled or weighted video files should be conveyed. The file formats are recognized as Abekas/Accom image file format with 8 bit and 10 bit formats supported. The data is in the YC_BC_R color space.

Gamma correction is the process of compensating for the nonlinearity of a television monitor's cathode-ray-tube (CRT). Since gamma correction is normally done in the TV camera, but in computer graphics these corrections should not be made unless you know in advance what the final output is intended for (print, video, or files). Video production devices (such as disk recorders) expect inputs that already have gamma correction applied, which should be accounted for before the conversion from RGB to Y, U and V components. The difficulties arise when because any RGB color can be encoded in YUV components, but not every YUV combination is a valid RGB color.

The elaborate details of scaling the various component formats to ITU-R BT.601 (formerly CCIR 601) were carefully considered in the first Abekas disk recorder products. Still, understanding the differences between C_B and C_R, rather than B–Y and R–Y (or U and V), is a complex topic, and we will leave that discussion to the reader.

Most of the modern transcoding devices, whether in a server or external in "glue products", and manufactured since the early 1990s have sufficiently addressed the conversion processes so that transcoding between formats meets the ITU and ANSI/SMPTE standards as they apply to component digital video.

ETHERNET FILE TRANSFER

Abekas further promoted the concept of Ethernet TCP/IP file transfers by offering a method to connect to the physical DDR and make it appear as a network node. Although it was not nearly as sophisticated as most of today's network systems, this approach did offer some viable means to compress, render, and transfer common file formats (such as Photoshop and Illustrator) to video via a gateway computer.

The Abekas A60 also provided remote control of the disk recorder over Ethernet, making it, in 1991, one of the only methods of moving and converting data between a computer in pure digital form.

WORKSTATION DISKS

Capitalizing further on what Abekas had created, the spin-off company, Accom, took the next steps in creating true functionality in digital disk recorders connected to computers and workstations. At first, the 30 second nominal length devices were basically extensions of the Abekas concept, done correctly. Interestingly, this may have been

The Disk Dilemma 107

because the promoters and developers of both Accom DDR product lines (the WSD and the RTD-4224) came from the ranks of Abekas.

The Workstation Disk (WSD) from Accom led the way to the next generation DDR for computer generated imagery. Much of the core file transfer technology was also applied to the Real Time Disk Recorder (RTD 4224), which was a dual channel DDR with Ethernet and networking built in. Audio was later offered as an add-on, but not before other products were beginning to take hold.

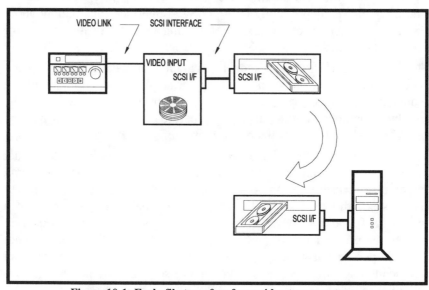

Figure 10-1: Early file transfers from video to computers.

The principles of these animation disk recorders started to extend to other products in the PC industry. Before long, several animation recorders, built around board set add-ons, were seen from a variety of PC peripheral manufacturers.

Once the SCSI and Ethernet interfaces on the WSD were stable, Accom went to work on developing a direct GIO bus interface to Silicon Graphics machines. The intent was to place disk drives as actual extensions of the workstation machines, bridging the benefits of video processing (in full 4:2:2 bandwidth) and the capabilities of 3-D and accelerated paint and video manipulating on an even plane.

MISSING ELEMENTS IN THE DDR

Disk recorders suffered from short record times, but that wasn't a significant factor until much later when entire commercials were being created in computer graphics systems. What was occurring, however, was a need for sharing a common disk recorder among more than one user throughout the work process. Also, there was a need for synchronized audio in different broadcast formats, which was still missing during this early period of disk recording development.

Disk recorder interfaces began to develop that would address these two needs by providing multiple SCSI interfaces and adding audio in AES and analog to the mix. Networking was still in its infancy for DDRs but connectivity using SCSI was progressing.

The next step was a short-lived attempt at making a file, clip and video store based upon networking principles but built around a conventional video backbone. The concept was extended and the products matured—but it was not long before compression, higher-speed networking backbones, and video servers came into being.

In general, there have been only a few successful products that adequately linked all the capabilities of a networking file structure and video disk recording technologies. Once the barrier of full connectivity is crossed, the next capability required is the ability to transfer a complete frame between a video disk and any connected workstation (given the proper network protocol, such as Ethernet). The desire is not to rely upon an existing application, such as FTP or NFS, but to create an application-specific protocol that addresses the frame boundaries inherent in video.

FRAME AND MULTIFRAME TRANSFERS

There is a strong suspicion among video server architects that streaming technology has a limited lifetime. If you think long and hard about where our technology is taking us, being careful to look far outside the box, it will become apparent that the synchronous world of video delivery we live in will just not be the only way to do business. Considering all the combinations and permutations of compression and the issues of system latencies, attempting to build a future digital plant based solely upon real time video transactions will be history in just a few short years.

The Disk Dilemma 109

Designing the digital facility of tomorrow will be a complicated task. Most television engineers fully understand the individual component concepts of integrating analog and even component digital video equipment into a facility. The tasks of compensating for system timing are not overly difficult because we typically will deal with delays from only a few milliseconds or at most 3 or 4 frames.

When we begin to marry the realities of requiring several seconds to process real time video into or from compressed digitally multiplexed streams, we will find that system timing issues take on an entirely new perspective.

Keep these thoughts deeply entrenched as you continue reading this book. It may very well be that the video (or media) server becomes the single common denominator in the broadcast facility of tomorrow that compensates for system timing latencies as signals move from one end of the system to the other.

Most of us understand the principles of delay incurred by satellite delivery. Few have had the opportunity to experience firsthand what encoding, multiplexing, bit stream splicing, and synchronization really involve. The future of compressed digital video delivery and air-chain processing for the master control on-air operations will be quite different. What we will really have, once you add in all the various time delays required for processing the signal, is an asynchronous facility. The digital broadcast facility of tomorrow will be tasked with delivering a signal, at a predetermined time, using bit streams that we won't know how long it will take to process until the processing begins.

We've digressed for a few moments to share an extremely important concept in the future of digital television technical operations. We should keep at least parts of this concept in mind as we return to the principle of file transfers to and from video workstations.

REAL TIME FILE TRANSFERS

Once we grasp frame transfers, that is, finding the needed frame and using FTP, RCP, or some other transfer protocol, and moving it from one directory to another, we can take the next step, which is discussing multiple frame transfers. Pluto Technologies International has been promoting these concepts as it develops and markets a unique product that can only be described as a hybrid DDR and network file-based workstation storage system. Pluto's SPACE product makes the DDR appear to have VTR functionality yet look seamless to other existing applications. Post production houses and graphics boutiques have been wrestling with this requirement for years and have only a few select products that address and meet the needs of their facilities.

Copying clips, now properly identified as multiple frames, can be accomplished in near–real time, and the future will bring them in faster-than–real time through interconnection schemes using Fibre Channel technology. The extension of the frame-based and multiframe transfer process is termed "real time delivery." When an arbitrary amount of video must be transferred, the concept of FTP or RCP fails because it is impossible to predict how much data will be transferred until the transfer is complete.

This is the dilemma being faced by tape archive devices sending streaming or partitioned files from an 8mm tape to a server. First, the protocol demands you reserve space on the file system to accept the entire file. The information is disclosed either directly from the header or via a lookup table in the system database. Second, the transfer between devices begins but the file is unusable until it has been entirely transferred to the destination folder or directory[23].

In Pluto's concept for real time delivery, the transfer time will be equivalent to the record time. Synchronization is accomplished by the transmission of clock sync messages from the transmitter to the receiver, much the same as with MPEG-2 transport stream protocol. In addition, both the receiver and the transmitter can obtain sync information from a common reference, such as a house reference.

Pluto uses its own native format to store data on the disk drives, following the ITU-R BT.601 specification but with frames (fields) being non-interlaced. The image contains the original 487 lines (525 format) and there is a 64-byte header attached to the beginning of each field. Pluto is also developing a file system for storage of its data in a protocol that permits exchanging that data over a network in a variety of formats.

Making the switch to real time delivery requires stabilization of the system so that underflow or overflow of the buffers does not occur. Minor clock adjustments can be made by the receiver to eliminate the potential for drift or slippage. Simultaneous exchange of Pluto File

---

[23] Research continues and products are expected to be available in Q2-1998 that will minimize the impact of latency in file playout. Refer to Chapter 35 for more details.

The Disk Dilemma 111

System files is limited to either play or record, but not both. At the time of this writing a resolution to this constraint was said to be in development.

STREAMING VIDEO, AUDIO, AND DATA

Industry is debating the merits, needs, requirements, and the futures of the two different structures of data delivery. We have touched on file transfers and have stated that this has been a developing and defining part of the industry for some time. We have explained the real time nature of video delivery from servers and how this relates to synchronous operations within a facility.

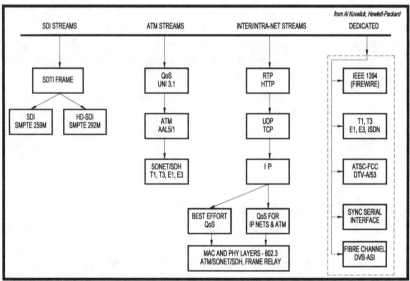

Figure 10-2: Mapping of video, audio, and data for streaming.[24]

Shifting our discussion to the streaming of content that is video-, audio- and data-bound, we define "streaming" as the delivery of a signal form from a source to one or more destinations. For streaming to take

---

[24] From "A Reference Architecture for Digital Video/Audio Transfer and Streaming," Al Kovalick, Hewlett-Packard Co., 11/26/97, presented in participation with the SMPTE/EBU Task Force (used by permission).

place, there must be at least a real time relationship between the delivery rate of the signal form and the presentation of the signal form.

The characteristics of streaming include some specific criteria. One characteristic is the pushing of data across channels or networks (refer to Chapter 21 for details on connection terminology). A stream is usually a push, but in some point-to-point streams, the receiver pulls the data. Point-to-multipoint streaming is also common.

Video servers of today are essentially streaming devices. They are representative of only a portion of the left-hand side of Figure 10-2. Streaming will be adapted to other forms of transports, as described below, and as such will need to follow basic mapping rules.

Content streaming, including connections and links, will function differently from file transfers. In a stream, there is usually no return path to request a retransmission of a missed signal or data packet, and the receiver must therefore make the best of any corrupted data. Isochronous and synchronous links are often required to support streaming, taking their performance quality directly from the QoS of the link, and using any error correction available.

Once a stream has started, receivers may join that stream without having to start at the beginning. The delivery rate is generally related to the presentation rate of the content. Streaming at other than the presentation rate is allowed.

GENERAL-PURPOSE AND DEDICATED-PURPOSE TRANSPORTS

Streaming data may happen over a variety of physical transport mechanisms, divided between general-purpose and dedicated-purpose categories. The general-purpose transports include the familiar SDI, ATM, and IP streaming, plus SDTI and IP-TV. More specific dedicated purposes have been classed into IEEE 1394, AES/EBU, conventional communications (T1, T3, E1, E3, ISDN), and Fibre Channel. These are extended into sync serial interfaces (for studio transmitter links), DTV, DVB, and other satellite links.

Streaming demands timing synchronization between the source and the destinations. This timing information may be derived from timing references embedded in the streams or a system-wide clock such as reference or timecode. On their way from source to destination, the streams may be routed through a variety of components. Should streams need to be switched, a method for the coordination of the streams, through the various components within the studio, is required.

The streaming model requires a format for mapping between the stream signal form and the physical transport (Figure 10-3). This model, presented by Al Kovalick, Hewlett-Packard Company, was part of a presentation and paper called "A Reference Architecture for Digital Video/Audio Transfer and Streaming" (used by permission).

When we look at how this mapping should occur, we see two factors to be concerned with. On a vertical axis, we see quality. Quality will vary from Internet, intranet, and browse level up through distribution of SDTV and HDTV. Production quality is at the highest end of the quality axis. On the horizontal axis, we see the relative distances that the streaming signal must traverse. The distances will range from the local studio, where SDTI/SDI for production lies, to wide area networks, where the range of mappings covers dedicated purpose (for DTV, DVB, T1), to IP-streaming over the Internet or private networks.

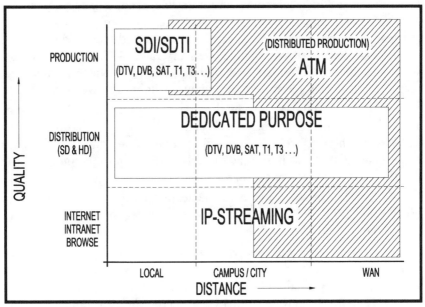

Figure 10-3: Streaming technology application spaces.

SMPTE 259M (270 Mb/s and 360 Mb/s) will make valuable sense as a streaming map in the studio, yet it is not suitable for file transfers between remote users. However, the inclusion of the optional SDTI framing layer would allow for the flexible carriage of compressed data instead of the normal 4:2:2 payload.

SDI (SMPTE 259M) at 270/360 Mb/s and HD-SDI (SMPTE 292M) at ~1.5 Gb/s are both considered isochronous links. These links both allow for precise timing of transmission. Although SDI switching is possible, it lacks the self-routing capabilities found in ATM framing. The quality of service (QoS) in SDI is quite high because jitter, delay, rate, and losses are strictly specified. As we mentioned, SDTI allows flexibility in data payloads, permitting compressed and uncompressed video, audio, and data—where SDI is inflexible and would not be used to transfer a file to a remote user. SDI presentation rates are also based upon video rates, where as SDTI is not time sensitive—that is faster or slower than real time video data may be transported.

ISOCHRONOUS, ASYNCHRONOUS AND SYNCHRONOUS DELIVERY

The differences in isochronous versus asynchronous and synchronous permit mapping into at least four types of streams. We've just described one mapping above for SDI/SDTI. Next we will discuss, which is not isochronous compared with a video delivery rate. ATM requires timing references to be embedded within the stream for synchronization to occur. ATM further permits asynchronous switching between streams.

Internet, intranet, and audio-video browsing use IP-streaming, with a QoS being set if needed. Usually, the QoS for most IP-based solutions is "best effort". Web-based off-line editing and IP-TV (Web-based channels), as well as unicast and multicast programming, all fit into the IP-streaming application spaces.

The fourth stream type is "dedicated purpose", which includes IEEE 1394, DTV (A/53 spec), and a host of other interfaces, including Fibre Channel and DVB-ASI.

The application for the various stream technologies as related to servers and storage devices is still being defined. We have to this point, understood and worked within the SDI interface as the primary form of interface and transport to or from a video server. We certainly expect the mapping of other stream types into servers as digital studios begin to support the matrices of uncompressed, nontemporal and temporal compressed digital video formats. The multiplexing, where applicable, of the various transport modes will also affect the way we distribute and store digital video.

Much of the work on these protocols and standards is just beginning to make its way out of committees and into standards. The SMPTE/EBU Task Force for Harmonized Standards for the Exchange of Program Material as Bit Streams has for over a year between November 1996 and April 1998, sought to develop a blueprint for new technologies that looks forward ahead the next decade. The first document Task Force produced, which appeared in the *SMPTE Journal* in June 1997, began to identify structures that will be necessary to support television production using compressed digital bit streams.

The SMPTE/EBU Task Force has since built on the results of the first document and is devising strategies that will permit development, testing and standards to be written. Some of the foundation principles have only been touched upon in this chapter. Many of the concepts and presentations described are making their way into living documents that will be refined as time goes on.

The influence of digital recording, storing, and delivery will make living with the disk recording and server technologies all the more crucial. As the television facility of tomorrow moves away from the isochronous foundation that it has lived with for decades, new challenges will force us all to deal with video in a different way.

It is through the disk recorder or the server that many of the concerns over delivery, recording and latency in systems will be addressed. The disk device may very well become the buffer and interface for the various transports and streams that we will need to deal with over time.

CHAPTER 11

WHERE SERVERS FIT

Determining what type of video server to place into a facility can be a challenging task. Deciding how the server fits into the enterprise is another element in this task. These are two of the many questions that should be answered before choosing one server over another. Even after selecting a particular manufacturer or server, you must review many options and combinations that make up the inner workings and the system integration.

Relying solely on the server manufacturer to provide all the answers or the total solution may not be in the best interest of the facility. Many potential purchasers aren't sure what kind of compression format they should use. Most don't truly understand how much or how little storage will be required. The question of redundancy keeps coming up, mainly out of a lack of understanding about how a server functions and what to do if it fails. The uncertainty of operational changes and media management add to the confusion over which manufacturer can provide the type of server best suited for an operation.

Today, choosing a server requires an understanding of technology and of application. The things we looked for in a server two years ago aren't necessarily what we'd want in a server today. This chapter will give some examples of applications and opportunities that servers have in the media delivery industry of today, yesterday, and tomorrow. We will take an introductory look at some of the concepts in file and video servers and at some existing applications that users may wish to explore themselves.

WHY HAVE A SERVER IN THE FIRST PLACE?

There are many reasons why a video server might be installed in a facility. Some believe the single most advantageous reason for a server is the replacement or reduction in dependence on the videotape recorder in as many operations applications as is possible. Others see the server as the answer to replacing worn-out videotape machines. Others find that the reliability and interface options made available by employing a server will make their operations easier and their revenue streams more dependable.

Others recognize that with DTV on its way, they will need to deliver more and different forms of programming. Since the server is by nature a digital device, there is a natural affinity to placing the video server into the facility, visualizing the server as a stepping stone to the delivery of digital video.

It might be easy to understand why a server is important in the facility if we first compare the video server for television applications to the computer file server for data storage. In a computer system, file servers are used to store large amounts of data in a central location with accessibility by many users, or clients, over a network.

The server provides for a number of services to be administered from a central location. These services include e-mail, printer management, accounting and data entry, scheduling, and applications that can be run remotely on desktop workstations or even dumb terminals. The server also provides a gateway to other elements on the network, including cross platform interchange between Macs, PCs, and UNIX platforms. It often is the connection point to the public network or acts as a node on a much larger LAN or MAN.

With the overwhelming shift to the Internet and the World Wide Web, the facility server is being used for more activities than ever before. The server is now becoming the single most important element in the office or network environment.

The modern network server must now be administered using a sophisticated set of tools that requires a knowledgeable and experienced systems administrator who can perform numerous functions. Some of these duties include software installs, updates, creating new client accounts, and controlling the amount of data that the storage devices contain.

Daily routines are generally straightforward until the system hangs up or slows to a crawl. At that point, decisions might need to be made that could affect data integrity, security, and the overall performance of the entire enterprise that relies on the server.

In the computer domain, before networking or file servers, the PC-based office computers kept individual user data on a local computer's hard drive (their C: or system drive). Every computer in the particular workgroup or even the entire office would have a unique data set, but in general, each computer had a similar set of applications to use. If one user created data, then needed to share that data with another user, the file was generally copied to a diskette and physically transferred to another user's computer.

Over time, computers started being connected via a peer to peer network (see Figure 11-1). Soon the simple network became an economical method to share data from one computer to another over a common connection scheme. Eventually, this elementary form of network grew and each user's individual computer would be slowed down as more and more users began to access each other's disk drives.

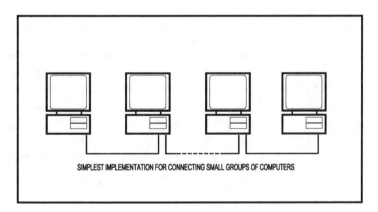

Figure 11-1: Peer-to-peer network.

The peer-to-peer network is only for the simplest of networks and smallest of workgroups. Peer-to-peers do allow for rudimentary sharing of data files and some applications, but the networking of today has grown by leaps and bounds compared to just a few years ago. We see far more sophisticated applications controlling the various

120 Video and Media Servers: Technology and Applications

connections and interconnections, hand-offs and conversions, and the management of system flow based upon prioritization of needs and necessities.

In much the same way, the television industry has grown larger and larger, needing to address many new methods of connectivity—some of it about to be dropped in our laps as the move DTV becomes more and more dependent upon compressed digital video and less on analog infrastructures.

TV FACILITIES: BEFORE SERVERS

Before the reliance on videocassette transports, which has developed over the past ten years or more, the physical facility consisted of a central library and a bank of videotape transports. At that time, a client would request a VTR operator to locate and thread up previously recorded material onto an available transport and play it back into a common routing system. The larger facilities at that time had some degree of point-to-multipoint router, and the signal would be delivered over coaxial cable to monitors, switchers, videotape transports, and the like. If the user desired a different program, a request was made to the operator, and some time later a new program would appear, either on the same VTR or on a different transport.

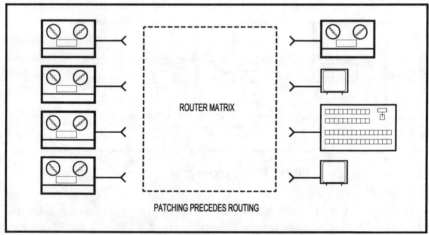

Figure 11-2: Point-to-point video connections.

A similar instance in the computer server model might hold for the video model. VTRs were first connected via the patch panel, then with video distribution amplifiers, and then eventually through a crosspoint–based video router matrix (see Figure 11-2). However, this solved only part of the problem of sharing video on a common basis. Prior to the video server, there was no physically or economically practical method to simultaneously make large amounts of content available to a substantial number of users. Adding more and more videotape transports and growing the router in proportion was not the answer.

Soon subgroups of VTRs and edit systems began to materialize connected by subrouters to main routers, and the cost equation continued to expand. The tape library grew exponentially and the format options, driven by technology and budget, expanded as well.

The concept of a central storage depot became more impractical because of the ever-growing number of clients that needed to be served. Some method had to be found that could address all these issues in a homogeneous way. Looking to the computer industry seemed to be logical, but before the video file server could happen, something else would have to become a reality first.

NEW CONCEPTS REQUIRED

For implementation to make practical sense, video compression codecs would need to become the essential element to storing video on hard drives. Concepts such as Quicktime and other similar software compression algorithms would work, but they depended more upon the availability of bus bandwidth and CPU utilization to function appropriately. At this early point in development, software driven compression schemes were not going to be the answer to streaming full-motion video from a disk drive—at least not yet.

The development of hardware compression codecs as motion-JPEG cards in PCs, suitable in quality to meet the needs of the non-linear editor, would be available in the early 1990s. It was then when digital would really begin to take off.

Other forms of video servers did begin to take form but met with about as much success as had broadband video-on-demand servers. At first, the concept of an optical video jukebox, albeit a primitive one, seemed like one of the possible video server solutions. The drawback was that laserdisc or magneto-optical systems required a material to be transferred to an expensive medium and were limited to only a few viable production resources.

Still, the jukebox concept was indeed the predecessor to the file server because it created a common library of files (clips) that could be accessed in a random nature. Under jukebox operation, the user could pull video baseband material, on demand, through a router and machine delegation subsystem to most any location in the same facility. It also provided an essentially nondigital approach to delivery that kept existing infrastructures in place.

At least one educational television facility actually used re-recordable laserdisc equipment for ongoing programs and interstitials, but found that upkeep was costly and the problems with ongoing misalignment proved the concept unreliable.

Once desktop non-linear editing concepts came on the scene, the optical jukebox became uneconomical. Non-linear editors paved the way to the video on disk drives before optical disc machines for video could become an installed reality.

SERVERS COMBINE ATTRIBUTES

The principle concept of the server would put a new dimension on the storage and retrieval of data and, eventually, video media. Once the computer driven office environment realized the benefits of networked, commonly accessible data, including delegated privileges, security and protection or backup, the server became the much-sought-after subsystem. Even though computer file servers were not without their own headaches, they did manage to address the expanding reality that more and more data would be collected and would need to be shared by more and more people.

The video server would strive to combine the attributes of file server technology with compression and mass centralized storage. It would be almost half a decade, though, between the introduction of video on a hard drive and true server implementation.

These first practical hard disk video storage devices came in the form of non-linear editing (NLE) on systems like Avid Technology's "Media Composer," introduced in 1989. Even though sharing of data between workstations was impractical, it set the tone for the ultimate need—a random access, video storage subsystem that could share media attributes between various users.

Non-linear editing soon became the craze as editors and videographers alike saw the benefits to compressed video storage and random access assembly on a computer desktop. Certainly, other disk-based video storage devices had been around for a few short years, but the price tags were enormous and the portability was almost nonexistent. It wasn't long before those with more than one NLE installed saw the real value in having the capability to share media (data) between one another. Fast-forward to 1995 or so, and we see that this story would continue to unfold.

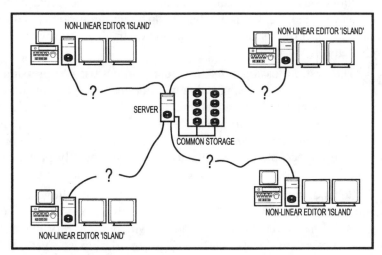

Figure 11-3: NLE "islands" lack common shared library solutions.

The lion's share of non-linear editing systems still remain "islands", that is, they stand by themselves, but the trend is moving toward shared, common, digital resources. Today, large organizations and even small news or production departments with multiple systems are able to share video information among many users.

NLE manufacturers have sufficient options in their product lines that allow managed storage on a single or distributed central library. The film special effects industry has used common file libraries on a server for many years, allowing groups to work on similar segments of a production without each having their own complete set of files or clips. News departments that are considering NLE for cutting stories are also beginning to recognize that a common library of digital video storage will save them time and frustration.

The video server, at any level, is now a reality. The concepts are in place, and now the applications become the key factor in implementation.

TO COMPRESS OR NOT TO COMPRESS

As we have discussed and will continue to discuss throughout this book, the development of compression for applications in servers and nonlinear editors was a critical element in making the servers real devices. Today, servers can employ a number of different applications and exist on a number of different platforms constructed in an ever-growing range of architectures.

The first professional disk recorders did not use compression for two reasons. First, video compression was not understood by the masses. Only a few commercial video compression codecs were available, and they were limited to desktop Macintosh computers targeted at off-line. In turn, video engineers refused to accept the quality of the image or the lack of industry accepted video signal structures generally provided by the codecs.

Second, storage technology had not matured to allow for much more than a few seconds of uncompressed digital video to be stored on a hard drive. The drives capable of storing video were very expensive, with the optimized A/V-drive[25] not even developed yet. The combination of these two factors prevented any accelerated growth in digital video servers for some time.

Video servers for broadcast applications would wait until the concept of video-on-demand was born, around 1991–1992. It is here that the principle of multiple channels of streaming video would begin to reshape the concept of a video server into a viable product.

---

[25] A/V-drive: Audio/Video drives are optimized for the streaming of large blocks and files. A/V-drives eliminate traditional drive thermal calibrations, which inadvertently interrupt the streaming of files, resulting in dropped frames or chopped audio.

Broadband video delivery, based upon streaming video channels, would not be a successful business venture for the masses. It did, however, open the door to a multitude of successful products that refined and massaged the video server concept into reality.

Once the parade began, everyone would join in. To answer the question "To compress or not to compress," the answer will be, "COMPRESS!" Even if video storage costs were as low as the cost of dirt, the bandwidth required to deliver higher-definition images is just not going to be there. It is also impractical and wasteful to carry all the redundant and unnecessary information along with every frame of the video image.

Efficiency is important and compression is the answer. All the pieces are not yet in place; this is a pure fact. Yet, as technology, manufacturers and standards all mature, we will see that the compressed video image will be the future of the video industry. Converters and codecs are all getting better. The heavy arm of the DTV commitment will accelerate this effort.

EARLY SERVER IMPLEMENTATIONS

In the early phases of video servers, there were only a few high-level broadcast video players in the arena. However, many more video servers have been delivering video and audio content to a more diversified set of end users worldwide. Before we look into the workings of servers, we'll look at some of the applications where video servers were in place and operational as of the middle of 1996.

We will look at 1996 because it was about this time when the first-generation broadcast class of video servers, shown for the first time around 1993–1994, were truly making a name for themselves in a variety of organizations. It is also about this time that broadband delivery was declared almost dead and that the real market was shifting away from exclusively unique system configurations.

One of the larger projects undertaken in 1996 was Sweden's TV4. This facility undertook the challenge of a complete system design and installation that uses server technology for day to day broadcast operations. The system was built with compressed digital servers employing motion-JPEG on Tektronix Profile platforms. A large-scale serial digital network employing a wide range of digital video production, distribution, and routing equipment was also part of this highly sophisticated system. The new station has over 150 hours of video server storage available, which could be more than doubled once the still-under-development (in 1996) enhanced MPEG-2 4:2:2P@ML equipment became available.

Tucson, Arizona, station KOLD was an early adopter when in April 1995 the station began networked digital video usage in its daily on-air operations. In its first use by a commercial station in the U.S., the HP Broadcast Video Server was a replacement for KOLD's robotics tape system and has been in use for commercials, promotional spots and PSA recording and playback. Eventually this MPEG-2 MP@ML server has the potential to grow into other broadcast applications such as time-delay for programming, news story play back, and other general station operations.

WOFL-TV in Orlando, Florida, placed Odetics' hybrid cache/tape library system into their operation. The system is based upon a combination of the Odetics TCS2000 cart machine, the CacheMachine and the Prophet cart subsystems. The system can maintain a full day's run of program material and provides on-line storage of thousands of commercials or other program material. Odetics, which for many years has manufactured a robotics systems that delivers videocassettes into conventional videotape transports, took the approach of integrating other manufacturers' video servers into a software and hardware package that become essentially a video cache for short form commercials and interstitials. One of the benefits to this concept is that an existing Odetics system can be upgraded with a server, in turn extending transport life and lowering maintenance costs. Odetics integrates its robotics products with leading server manufacturers, including Hewlett-Packard MediaStream, Tektronix Profile, and the Leitch/ASC Virtual Recorder.

CABLE, DB SATELLITE, AND NETWORK OPERATIONS

Cable companies have long dealt with the demands for low-cost, high-reliability ad insertion for their many channels of programming. Typically, these operations have had to install dozens of video tape recorders, usually in VHS format, just to address and fulfill the requirements of directed sales and spot insert advertising. Video servers from non-broadcast equipment manufacturers and control systems from other companies have teamed up to produce integrated solutions that could virtually eliminate a major share of VHS, or other types of transports, from the cable head end altogether. For the short term, video servers that employ MPEG-2 ML@MP, operating in the sub-10 Mb/s bit rate, will at least answer some of the needs for a completely computer controlled high-capacity ad insertion system. Ultimately, the ideal solution would be to provide a delegated video-on-demand or pay-per-view selection that is remote controlled by the end user.

Large cable and satellite-only program delivery services have been realizing the versatility of using servers since they first started shipping. Faced with the problems of time-zone delay, Hawaii's Oceanic Cable employs two of Leitch/ASC Audio Video Corporation's Virtual Recorders to delay East Coast feeds so they can be played back some five or six hours later (see Figure 11-4). Without using a single video tape recorder, these cache systems, sometimes called "store and forward" services, have been assembled to continuously record and delay programming that would otherwise require several machines, many personnel, and a lot of videotape.

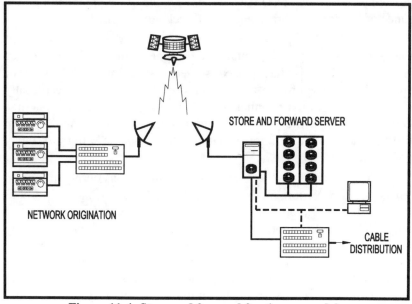

Figure 11-4: Store and forward for time-zone delay.

In order to reduce costs and streamline operations, Fox Broadcasting's Latin America Channel (FLAC) uses a combination of automation and server technologies for their ongoing daily operations. Through the use of high-volume storage and addressable video servers, Fox has reduced the daily preparation of five and a half hours of material for two sites with two edit rooms—running three full-time shifts, seven days a week—to one editor working a 40-hour week and a single ASC Virtual Recorder system. The initial amount of on-line storage was 14 hours and Sundance's "FASTBREAK" automation controller and software manages the system.

Some companies, including Philips-BTS, are convinced that disk drives are a more reliable, durable means of video storage than tape. The MediaPool video server from Philips-BTS is unique in that it can operate as a user selected full-bandwidth and a variably compressed system. This concept is beneficial to the facility that wants a high-end, high-quality server and a method to store finished product in a compressed format—all on the same storage system. This server features an intelligent video storage engine and efficient software management tools that offer the operator more flexibility and higher reliability. This philosophy sees disk-based solutions gradually supplementing the balance of linear tape storage and live, direct network feeds.

Avid Technology, which has extensive experience in a variety of server and non-linear implementations, recognized the need for a host of configurations. Avid has placed server equipment in hundreds of facilities that are used for on-air, broadcast and cable news and especially in production. For on-air applications, Avid makes the AirPlay system, a Macintosh platform file server, incorporating applications ranging from spot playback to full library routines that include program delay capabilities.

Storage sizes are a relative issue when considering a server. Cornerstone TeleVision, in November 1995, began using a five-hour storage system to simplify and maintain its one-operator master control room. CTV decided upon the Avid AirPlay system without previously owning any type of NLE system. This station, which also feeds a network of other broadcasters, plans to integrate additional file server technology and add increased storage for program delay capabilities and satellite network origination in the future. It visualizes the slow but steady demise of videotape for much of its operations in the long term.

Video server development extends beyond just one or two program streams fed from a common storage array. Hewlett-Packard's Video Communications Division is providing solutions using MPEG in their MediaStream server product line. HP's hardware architecture, based upon its own Video Transfer Engine architecture, addresses the need to stream large quantities of video data across networks and is not built exclusively from existing computer hardware architectures. HP realized that some of the new or expected applications of tomorrow will emerge from the various interactive-video home trials that HP has been involved with in the early 1990s.

While the marketing potentials for interactive and broadband video delivery have not been statistically impressive, the concepts developed from the trials are being extended to applications beyond just the broadcast or broadband realm. The reality of broadband communications using large scale VOD servers for interactive cable or other non-satellite delivery has all but vanished—forcing HP, and others, to shift from broadband servers to either broadcast quality servers or scaled back multiple stream servers for hotels, closed-circuit, and other limited pay-per-view type services.

AIRLINES, SPORTS AND CACHES

In areas not as obvious for the server market, the airlines are becoming real users of digital audio/video-on-demand (AVOD).

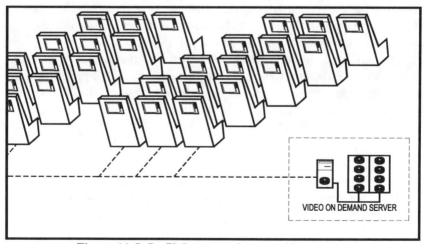

Figure 11-5: In-flight entertainment video server.

130 Video and Media Servers: Technology and Applications

Matsushita Avionics Systems Corporation (MASC) and the original Micropolis Corporation developed a system that can store over 40 hours of MPEG-2 or MPEG-1, video making it available on from 32 to 128 video channels.

Besides video movies and news content, including in-seat entertainment selections, some 96 hours of CD-quality audio in up to 12 stereo channels can be delivered from the same server. For in-flight entertainment on long flights, this effectively gives passengers a personal VCR and CD player at every seat. The initial implementation for this audio/video-on-demand server installation used 6.5" liquid crystal displays set into the seat back on 747 and 777 jumbo jets (see Figure 11-5).

The sports industry, once the near-exclusive user of video disk technology for replays, also saw the handwriting for video servers and was heading in that direction at full speed. TBS, ESPN, and the Sports Channel were all using video servers to access and air game clips and animation.

In a sporting event of any kind, using a server to store all the athletes' bios and stats, all the game highlights, and all the slow motion replays is a real plus. With only the important material kept available nearly instantly, it becomes a snap to reassemble highlights without waiting for the rewind and recue latency of a VTR. There is, however, one drawback expressed by several mobile unit operators: the hostile environment of a mobile unit makes reliability a critical issue. Servers used in this application must endure a lot of jostling and vibration, and there is still a lot of room for improvement in this area. This is one example where the application and the user interface are probably more important than the video server architecture itself. Reliability and flexibility are exemplified in this application.

QUESTIONS STILL LOOM

Those who remember the birthing of RCA's TCR-100 and Ampex's ACR-25 might still recall the impact those giant beasts had on broadcast operations. They changed the face of station operations forever. Now the decision takes another giant step forward as the universal acceptance of partial or exclusive disk storage for broadcast operations becomes reality.

Where Servers Fit 131

Despite the high praises gained so far by employing video servers into a variety of applications, some very real questions have developed that have yet to be fully answered, but answers are coming.

Will broadcasters turn to disk caches and video storage servers as replacements or just supplements to their aging robotics controlled library tape systems? The answer is both operational and financial. The video server offers many features that cannot be achieved in any other way except by disk storage. The server changes the face of operations by insisting on some level of storage management that requires a long-term commitment from the facility and from the vendor. In the past, even if a robotics operated video cart machine completely died, the station had a series of alternatives to keeping the revenue stream going. Server systems, when designed and implemented properly, can provide a level of security that will either meet or exceed the capabilities of a robotics cart only machine hands down.

We recognize that extensibility and scalability are important elements in the advance toward digital television. The scalability question is being addressed rather convincingly now, but the extensibility issue is one that doesn't have a clear direction yet. We appear to have solved the "need more storage, add more disks" dilemma just by opening the checkbook and adding more drives.

That scenario is about to change. We may still need to open the checkbook but this time we'll be adding MPEG-2 or some other form of native compressed digital storage onto an existing server. This will probably not be immediately deliverable, but the addition of MPEG-2 codecs will mean an instant increase in storage capacity for those that already have certain models of video servers. The transition from motion-JPEG to MPEG-2 in a 4:2:2 coded studio profile (4:2:2P@ML) is slated right around this book's publication date.

In 1995, as many operators still had the big DTV (then ATV and HDTV) question hitting them from all directions, they heard about the miracles coming from video compression and server technologies. Most weren't ready and couldn't see the reason for a quantum shift in comfort levels. Although in 1995 there appeared to be a final direction for ATV, the concept of having to convert the video plant to an all-digital environment was also a distant reality. It was obvious then that a video server was not a necessity, but probably more a desirability. History has already shown that delaying the decision in 1995, until HDTV or DTV came, had no real impact on whether a server should or should not be implemented.

Today, now that the DTV standard is finalized and the FCC timeline is set, the reasons for server implementation are even more profound. The broadcast facility will be able to program more than its heretofore single broadcast program stream. A server is a must for handling multiple, simultaneous, and varied program streams. Unless you have extremely deep pockets and you want to continue the exponential dependence upon videotape transports, a server seems to be becoming a necessity for even the broadcast facility of today.

Some operators ask if they should make the jump to disk storage since the application may have only a five-year life span. Operators should review the evolution of the videotape transport formats before answering that question. It is rather eye-opening to realize that every single significant tape format has survived at least its initial five-year life span. The non-linear editing system purchased in 1992 is most likely still in use today. We may find that the server of yesterday will find another use in the facility the day after tomorrow. When all the factors are added up, it makes good sense to invest in some level of disk storage if for no other reason than to relieve the dependence (and long term maintenance costs) of videotape delivery.

For the nonbroadcast sections of the content delivery industry, the answers fortunately appear to be quite clear. Airlines, prisons, hotels, cable systems, multimedia networks, training and even kiosk delivery systems have all begun to make the migration to video server or video file server technologies. Broadband delivery for interactive TV and wireless cable (quite an oxymoron in itself) has gone into hibernation for who knows how long.

For the video server purist, one big challenge lies ahead. That is the Internet, where great strides are being made on a weekly basis. Streaming video over a low data rate line will force the pipe to get larger, not because of quality but because of demand for bandwidth over the network. In the end, this may reopen the broadband delivery market all over again.

CHAPTER 12

ISSUES WITH SERVERS AND DIGITAL SYSTEMS

There are myriad other issues to consider when figuring out what to do with this digital television dilemma. The next two chapters are a hodgepodge of those issues and couldn't be broken down into chapters that would stand on their own. The writing is a collection of excerpts from the past three years (1994–1997), when we were all pondering the various implications of becoming digital.

When it comes to image quality, many decisions go well beyond just the server itself. The effects on overall image quality of various compression schemes, concatenation, or even the potential requirements for splicing different or variable bit rates become more important. With DTV on the schedule, new issues involving upconversion, downconversion, and even interlace to progressive conversion are all on the table.

How you handle the 16:9 aspect ratio will create a new agenda for production. Do we record everything in two formats? That would be nice, except it would halve the available storage on the server. The new domain of wide screen and compressed digital video brings with it many challenges. For example, videotape transports as well as video servers will have no knowledge of whether they are recording or playing back a 16:9 or a 4:3 aspect ratio. This could change once metadata is embedded in the bit stream and the servers contain the appropriate processing equipment that would be triggered by that data. Yet even if the data were there, is has not been determined what you'd do with it in the mixed-format facilities of the future.

Distribution equipment will be another factor in system design. If the server is made to record 360 Mb/s streams transported on SDTI, what will the processing equipment do to the signal? With the understanding of 360 Mb/s systems for 18 MHz sample widescreen came the notion that future distribution equipment had better pass 400 Mb/s signals. Now that we have a good five years or so behind us in building that kind of equipment (routers for one), that concern is minimal.

Graphics equipment will be another challenge to deal with. Character generators, still stores (i.e. disk recorders), and paint systems will need to deal with the two formats equally. There will be problems running the keyed image from a 4:3 image into a 16:9 converter. The characters will now be $1/3$ wider. To match equivalent fonts, they will need to be generated to $1/3$ fewer horizontal pixels. Building a graphic on a paint system and sending it off to the server could cause similar problems. So do we have two sets of servers? Certainly economics and practicality will answer that question.

OF FORMATS AND IMAGE QUALITY

Image quality and presentation will present multiple issues that we, as engineers and users, will need to deal with. With the predominant mix of servers using motion-JPEG compression, many changes will need to take place if we plan to move toward MPEG-2 as a choice.

It is quite likely that the server will become the prime medium for storing the various flavors of compressed digital signals we expect to see in the next two to five years. The reasoning behind recording full-bandwidth, ~1.5 Gb/s, 1920 × 1080I, HDTV will become quite obvious when you start looking at the alternatives to compressed digital production equipment. With that statement, we need to look at the expectations of various bit rates and formats we may see.

There will be more than one use for the video server of tomorrow. The most prominent use today is the storage of commercials and interstitials. In the future, as product is delivered to us in a variety of media and signals, the server will prove to be quite valuable as a buffer for holding the material until we decide what to do with it.

Issues with Servers and Digital Systems 135

It is highly unlikely that all the facilities of tomorrow will be fortunate enough to spend their capital on five or six different formats of videotape transports. So choices will have to be made, and the video *file* server is a likely candidate for becoming a universal bit bucket repository. This concept is one that is being explored by several manufacturers, as no one knows quite what to expect in bit rates or delivery formats of the future.

Already the networks are considering different approaches to the distribution of their HDTV programming. We expected they would continue to deliver their SDTV programming in much the same fashion as they have for years. For the future, as they make the transition to digital delivery and their existing equipment ages, it is a sure bet that they will change to a compressed format for delivery of both SDTV and HDTV.

If the network or any programming entity begins to deliver compressed digital that ranges anywhere from 19.39 Mb/s to 45 Mb/s, we should not expect to have to upconvert it; that is decode it to baseband just to store it on our servers or data transports. A noticeable reduction in image quality will result when it is moved to a server where baseband might be stored in JPEG (or MPEG) and then converted once again to a Grand Alliance DTV bit stream.

So, why not just store the bits? If we keep the compressed digital at the highest bit rate for the greatest length of time and minimize the compression/decompression cycles, then the final images—even when converted once to SDI for keys, dissolves or effects—will still look nearly as good as they did when originally delivered[26].

THE COST OF BANDWIDTH

Another growing concern is the availability of satellite transponder space-segment as the volume of programs, sports, news segments, and services continues to grow. With the new technologies in compressed digital, including DVCPro, DVCam, HDCam, and BetacamSX, the

---

[26] Alternatives for the handling and storage of compressed digital signals are continuing to unfold at the time this book went to publication. These suggested scenarios are intended to provide some of the then current 1997 thinking on the distribution and processing of compressed digital bit streams.

ability to pack more programs into less space becomes a viable and cost-effective alternative.

The converse is also true. If we can now deliver comparable quality images in considerably less bandwidth, then we can certainly deliver them in wider bandwidth at several times the normal play speed. The only things we need to do then is to take all those bits and store them until we can process them back to a real time baseline.

The server and even some of the newest tape transports make these options completely digestible. We just need the products and the processing to handle it.

CODING STRUCTURES

As we hinted earlier, more videotape formats are emerging that use some level of video compression. Digital-S, DVCPro-25 and soon DVCPro-50, DVCam, Digital Betacam, BetacamSX and now HDCam are all using different sets of bit rates and coding structures. Nearly all the manufacturers also currently rely on a common denominator for interchange between these formats. That common thread, for today, is the SDI interface.

The coding structure of component SDI (SMPTE 259M) is a serial 4:2:2 data stream defined by ITU-R BT.601-5. This coding is recognized as providing the best standardized solution for the 525/625 line, 59.94/50 field rate with 2:1 interlace. However, when we look to compressed digital, we find there are alternatives to coding that we will need to deal with right now.

Already DVCPro is quite prominent and, as of late October 1997, the number of units sold amounted to over 25,000. The expectation is that DVCPro will be profoundly accepted as a digital recording format because of the abundance of features that it possesses and its low cost.

The DVCPro-25 format uses 4:1:1 coding, which, when compared to the sampling rates for 4:2:2, has only ¼ the chrominance resolution but equal luminance (see Figure 12-1). From a production value standpoint this format will suffer from repeated processing, but not nearly to the extent of 4:2:0 coding, which has less chroma resolution due to the larger gaps in its co-siting of the chroma samples. When looking toward upconversion, today's thinking is that 4:1:1 is the winner over 4:2:0.

Issues with Servers and Digital Systems **137**

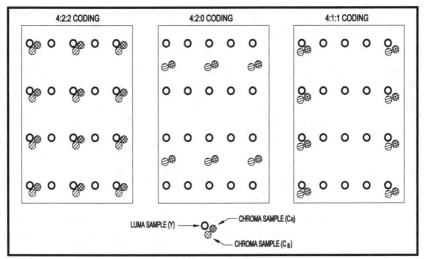

Figure 12-1: Coding grids for various Y/C_B/C_R sampling.

However, there is good news for putting DVCPro native onto a video server. At the point where we are able to record just the bits and not the reconverted SDI signal that appears at the output of the VCR, the 25 Mb/s data rate will yield additional space on the disks with a comparable first-generation quality.

Once the 50 Mb/s version of DVCPro is available, the process becomes much better. DVCPro-50 will use the 4:2:2 coding structure. This brings the performance back into a good perspective and should be a long-term solution for many of the resolution and upconversion issues we're discussing.

Server manufacturers are reopening the book on recording the native bit streams direct to disk. With the recent proposals before SMPTE, we think the mechanism for getting the digital video bit stream into the server, or across any other routing or signaling system, will become SDTI—where the native signal could be recorded to the server at from two (for DVCPro-50) to four times real time in DVCPro at 25 Mb/s.

In discussing compression and servers, we need to know more than whether it is MPEG. Already we're noticing that MPEG-2 must be qualified down to the profile and the level, or we could find ourselves trapped into a coding structure that might not hold up after three or four encode/decode passes. There will be other MPEG-2 profiles that we will

138 Video and Media Servers: Technology and Applications

need to implement, especially as production systems start using MPEG-2 and store the streams on disk recorders and video servers.

There are two significant MPEG-2 profiles that will have impacts on production-quality video when presented in compressed form. Servers that are considering, and in most cases implementing, MPEG in their codecs will be using either MPEG-2 MP@ML or MPEG-2 4:2:2P@ML (MPEG-2, 4:2:2 Profile at Main Level). The preferred profile is MPEG-2 4:2:2P@ML, with reasons that include the additional line samples (608 versus 576), 4:2:2 coding (versus 4:2:0 in MP@ML) and the capability to extend the bit rate upward to a maximum of 50 Mb/s (MP@ML is limited to 15 Mb/s).

	SIMPLE PROFILE	MAIN PROFILE	4:2:2 PROFILE	SNR PROFILE	SPATIAL PROFILE	HIGH PROFILE
HIGH LEVEL		4:2:0 1920 × 1152 80 Mb/s I, P, B	4:2:2 1920 × 1080 300 Mb/s I, P, B			4:2:0 4:2:2 1920 × 1152 100 Mb/s I, P, B
HIGH-1440 LEVEL		4:2:0 1440 × 1152 60 Mb/s I, P, B			4:2:0 1440 × 1152 60 Mb/s I, P, B	4:2:0 4:2:2 1440 × 1152 80 Mb/s I, P, B
MAIN LEVEL	4:2:0 720 × 576 15 Mb/s I, P	4:2:0 720 × 576 15 Mb/s I, P, B	4:2:2 720 × 608 50 Mb/s I, P, B	4:2:0 720 × 576 15 Mb/s I, P, B		4:2:0 4:2:2 720 × 576 20 Mb/s I, P, B
LOW LEVEL		4:2:0 352 × 288 4 Mb/s I, P, B		4:2:0 352 × 288 4 Mb/s I, P, B		

Table 12-1: MPEG-2 profiles and levels.

SMPTE is in the process of recommending a High Level version of the 4:2:2 profile for use in 1920 × 1080 HD production. The selection of this bit rate and coding is not trivial. With work ongoing in standardizing the SDTI transport, this 5:1 compression ratio for HDTV signals (from ~1.5 Gb/s to ~300 Mb/s) can be accommodated inside the SDTI 360 Mb/s transport. This reasoning will add future proofing to routing and terminal equipment in the serial digital facility.

ADJUSTING GROUPS OF PICTURES

Advanced codecs for MPEG should feature the ability to set or even change the group of pictures (GOP) length. This will be beneficial if we find that coded pictures develop unwanted artifacts or noise during the upconversion process or when transformed to another coding structure. If the sample upconversion or change in compression at the next level of the physical transport induces artifacts or visual degradation, we will at least have a chance of correcting the anomalies by re-encoding the original material at a different bit rate or GOP length. Of course, there is nothing we can do for the image once we've recorded it to MPEG and the settings are wrong. In that case we will have to live with the problem once it gets back to baseband.

There are certain coding structures and bit rates we may very well wish to avoid. Even though these pictures may display without undue problems, any future transformation or re-encoding might significantly decay the image, to a point where upconversion will not be practical or visually appealing. Furthermore, the possibility of getting a further degraded image when we eventually code for DTV at 19.39 Mb/s and then view it on a home display may also exist.

DEALING WITH FILM

One of the benefits of the MPEG-2 coding process is elimination of the 3/2 pull-down necessity for presenting 24 fps (frames per second) film at 30 fps. MPEG-2 does not code anything more than the 24 fps and therefore saves the bits. If we can receive motion pictures in MPEG-2 and take it straight into a video server, we will find storage capacity is increased once again. MPEG-2 has instructions to indicate that the data was in a 24 fps coding and that a decoder will understand it must replay those images at 30 fps for a proper video display.

All this assumes, however, that the data can get through the switching processes and on into the multiplexer in this state. The drawback is that if we need to take the MPEG-2, 24 fps images back to baseband, we'll loose the benefits of the MPEG-2 framing structure and may find it necessary to re-process the 3/2 pull-down for 30 fps all over again. Since there is no equipment commercially available at this time, this becomes another issue that will demand a "wait and see" attitude.

CHAPTER 13

LIVING IN THE DIGITAL DOMAIN

This chapter addresses other issues surrounding servers and integrating them into the facility. It covers subjects associated with mixing different servers and compression rates, networking, and measuring the quality of compressed video, and offers some words about compliance.

LIVING WITH VARIABLE COMPRESSION RATES

Some manufacturers say they are offering variable compression capabilities as a feature. What this essentially means is that the user can decide whether to support full-bandwidth storage or some level of compression from 2:1 through 40:1.

For a video server, variable compression may be used as a marketing tool to enable you to store more video data. The claim is that you can mix compression levels within the storage array and obtain a lower resolution, which takes physically less drive space. Only a select few video servers offer this feature, and its use should be approached with caution.

Some servers state that you can select different compression rates, but if you look closely you find that you can only store one bit rate on the same drive without stripping the entire disk again for an alternative bit rate. This is generally not the case with an MPEG coding, but JPEG codecs generally stay at a fixed bit rate once the recording process begins.

On systems that allow different compression rates to be stored on the same drive, ask if you can instantly and seamlessly play out one bit rate, switch to another, and then switch back. Be certain that a sequence of clips will play without pauses, glitches, or hiccups. Chances are there will be some latency while the software is reloaded into the codec to switch from, say, 24 Mb/s to 48 Mb/s.

The effect of different bit rates may cause different problems at other than a systems level. For example, what if your entire production library is compressed at 24 Mb/s and later you find it just isn't upconverting with the quality you expected. If the material is not available in its original form, or if it was video captured (or edited) at a 4:2:0 coding at 18 Mb/s, the library you've accumulated may only be of partial value.

We must plan in advance for options if the original material has the potential of no longer being available. One of the options could be compression at 50 Mb/s in MPEG-2 4:2:2P@ML. This would be a better preservation of the material, particularly if you plan to do post processing or post production at some time in the future. In this case, it might be wise to encode the material at this higher bit rate and store it to digital linear tape (DLT), essentially preserving the images for the future. This will be another good implementation for a video file server or bit bucket recorder.

Since we are generally not transporting bits around the facility yet, we still have time to watch hardware and systems develop for the future.

Finally, the term "variable compression" on a video server is quite different from the MPEG terminology "variable bit rate" (VBR). If you're using a codec that does do VBR recording, you could suffer consequences further down the pike. If you are managing bits at the input to the Grand Alliance coder or multiplexer, and it does not handle variable bit rates, your coding upstream may have been of no value. Again, since equipment is not readily available in the field, this issue may be a nonissue in the future.

PLAYBACK LATENCY DELAYS

When looking into servers for purposes other than spot playback, you need to know how quickly you can access the next clip on the disk. Some manufacturers will claim "instant-access," but that is not the full answer. Servers that are doing MPEG need buffer time to get the next clip in place and through the codecs. These servers should state a minimum clip or spot length that must be met in order to guarantee a seamless transition from spot to spot.

[Note: This issue is *not* the same as seamless or non-seamless bit stream splicing in MPEG. This form of splicing is where an MPEG bit stream is going to be appended or inserted into another bit stream. Those issues are being resolved in SMPTE so guidelines can be established for manufacturers to build to.]

The issue we're discussing has to do with playing out, from an MPEG-2 server, a sequence of short clips (such as a series of back-to-back two- or three-second station IDs. In this example, the server must be certain there is sufficient time for all the bits in the next spot to line up in the proper sequence and get pushed through the final pipe. Of concern is the displacement of the audio bits and video with long GOP lengths. It is entirely possible to take as much as 250 milliseconds to get all the bits aligned before the stream is ready to be decoded. If the disk access time and the decompressors aren't all lined up, the system could hang or actually delay getting the next clip started.

This problem might be resolved by having two decompressors on the same board set and switching between them after the signal is coded back to baseband.

RESOLVING INPUT RECORDING RESOLUTION

The input recording resolution, or sampling, may be another factor in your choice of a disk recording subsystem or video server. This is becoming more of an issue as new tape and digital formats, such as DVCPro and BetacamSX, come to market.

There are two issues to understand here. First is the matter of 8-bit versus 10-bit recording. For example, in some full-bandwidth video disk recorders, changing from an 8-bit to 10-bit resolution would require a complete restripe of the drive array. This requires either a complete dump of the video on the drive or an off-load to videotape, before the drives can be set to record at the higher bit level. Changing from 8-bits to 10-bits can take away precious disk storage space.

The other issue has to do with the front end of the server's codec, before compression. Server codecs are generally designed only for 8-bit video coding. So you must now ask the question and, "What happens to the other two bits?"

Some SDI input cards, especially those built with low-cost in mind or the intention of using only a PC as a platform, may simply truncate the last two least significant bits (LSB). If the image is laden with color ramps or low-luminance, high-chroma fields—visual stepping of the image will be obvious when dropping from 10 bits to 8 bits in depth. Once this occurs, how the codec handles the once-subtle gradations could be anyone's guess. You will know the results only from looking back at the coded image.

If your facility intends to do a lot of graphics or special effects work using a server, you'd be better off buying a device that will record full-bandwidth ITU-R BT.601 video without compression.

COMPRESSION IMAGE QUALITY, AGAIN

Today, and in the future, compressed digital video image quality will begin to take on a different definition than it did before. There is a plethora of digital editing systems now in place. The outputs from non-linear editing systems are still being transported on Betacam-SP tape to another non-linear system and redigitized, usually with reasonably good results. So then, to matters things worse, someone bumps that video up to D-1 or Digital Betacam tape and thinks they've got the highest quality that $100,000 in non-linear editing equipment can buy!

Unfortunately, degradation to the image is at its worst after the first pass through the compressor. The good news is that the images generally won't get worse—unless you begin to manipulate the image and continually take it back into another form of compressor. Being unaware of what is happening to your product is like trusting your only film of your baby to the local amateur film geek. Continually reprocessing the image might get you to the point where the image will drop down so far in resolution that portions become unusable.

In cases where MPEG-2 IPB with long GOP frames are continually decoded back to baseband, shifted slightly in effects units or taken back into compressors with a different GOP length, certain portions of the images are actually completely lost. While this is unusual in a facility that has similar video codecs at the various servers or production edit stations, this phenomenon is quite possible and leads us to understand more about the equipment we will be installing.

MEASURING QUALITY IN COMPRESSED DIGITAL

Measuring and determining the quality of compressed digital is quite subjective. This is not about the quality difference between a SMPTE Type C tape and VHS—these are factors that are completely different. Facility operators are now asking a daring set of questions including "How do I define quality?" and "Are there set standards for determining quality in compression?" and "How do I test for compliance and resolution?"

These are issues that SMPTE standards working groups are dealing with. Various research entities, including Sarnoff Corporation, are tackling the issues and have presented technical papers on the subject. A new instrument, manufactured by Tektronix and designed to measure image quality, was shown for the first time in January 1998 at the International Teleproductions Society's *Technology Retreat* in Monterey, California. As of this writing, static and moving image test materials, in a variety of higher-resolution formats, are being collected in order to determine what kinds of standard references we should be using and what we should test for.

BIT STREAM INTERCHANGE (STANDARDS)

Most off-line editing and the majority of video server systems use motion-JPEG for compressing moving pictures. The JPEG tool set was intended and designed for still images. When JPEG is used in non-linear edit systems the system is actually compressing multiple stills that will appear as moving video. It is entirely up to the manufacturer of the board set just how they will manipulate and integrate the JPEG tool set for the best image resolution.

Motion-JPEG has no true interchange standard. By standard, we mean that there is no universal decoder algorithm that certifies it will decode motion-JPEG clips from one manufacturer's implementation to another. Therefore, it is probably fair to say that each manufacturer most likely uses a different set of algorithms and possibly even different chip sets to accomplish what is now generically referred to as motion-JPEG.

These implementations can even vary from board set to board set, revision to revision, and upgrade to upgrade. What this means is that raw data files, as is the case of motion-JPEG, will most likely produce different results on different systems. Everything from resolution differences (essentially the compression ratios) to the filter sets used for compression or decompression will probably vary greatly from one type of server platform to another.

Some algorithms trade off black level noise for high-frequency resolution. Others change the overall video gain so they don't have as many bits to compress. Some of these artifacts become evident when changing and comparing the capture resolution differences in the same capture board sets with different software revisions.

The net effects are varied and unpredictable. That is one of the reasons you won't find a direct data transfer port that will transfer raw bits to another server from a different manufacturer. Anyone who is acquainted with multimedia applications, or desktop graphics, is aware of the complexities and effects of compressed image format conversions.

Unfortunately, transporting image data files is not like exporting a Microsoft Word 6.0 file to a WordPerfect 5.1 document over a LAN. In word processing, it took years to get all the various control codes and translators to be placed in the major word processor applications. Even after that, converting between word processors is essentially an easy task. In video imaging, transporting previously compressed video in its native compression format between different severs will most likely require bringing these file formats down to some common denominator.

THE COMMON DENOMINATOR

Most manufacturers of servers and video disk recorders consider that denominator to be the input/output interface at either SDI or component baseband video. This is not a wise thing to continually rely upon, as it means that the images have gone through a compression/decompression and then a recompression scheme. The effects will become obvious over time.

The common denominator is usually ITU-R BT.601 and ANSI/SMPTE 259M format. If you are dealing with HDTV, there are other standards (see Table 13-1) that need to be considered.

Living in the Digital Domain 147

Interface	Line Format	Type of Signal
SMPTE 240M	1035I	ANALOG
SMPTE 260M	1035I	DIGITAL
SMPTE 274M	1080I	DIGITAL
SMPTE 292M	1080I	DIGITAL
SMPTE 296M	720P	DIGITAL

Table 13-1: HDTV standards references.

Looking out somewhere into the future, what will really matter is what is received and displayed on the home theater screen. For this, we have the *display* formats, sometimes confused with the 18 various transmission formats. For today, the now infamous FCC Table 3 tries to state what we expect to be a universal set of display parameters that must (or should) be met in order to comply with the directives of DTV (see Table 13-2).

Vertical Lines	Pixels	Aspect Ratio	Picture Rate
1080	1920	16:9	60I 30P 24P
720	1280	16:9	60P 30P 24P
480	704	16:9 and 4:3	60P 60I 30P 24P
480	640	4:3	60P 60I 30P 24P

Table 13-2: The FCC's famous "Table 3."

If the facility of the future, servers and all, is going to meet this criteria, we had better find ways to verify the quality of the images we receive, generate, produce, and manipulate directly inside the walls of our broadcast plant. In reality, this becomes the common denominator (or so we think) for tomorrow.

MIXING OF SERVERS AND DISK RECORDERS

We can expect to be moving digital video on coax, in real time, to another server that will then compress the video again using, probably, a different motion-JPEG algorithm or possibly an MPEG-2 profile for storage on a different set of drive arrays. We can almost certainly expect this to be the case in a broadcast plant, and here lies the problem. Over time, as the newsroom buys one type of server for its applications and the production department gets a different one, the potential for different levels of image quality and compression resolution grows. There are now multiple types of systems in the same plant, even if there is only a software or board set revision. In some cases, it's a different manufacturer's server altogether.

Another system feature to consider relates to the number of channels or streams available from the server. Some server architectures state that they are capable of multiple simultaneous output streams. This is all well and good, provided the prospective user understands exactly what happens when a full-tilt test of random inputs (recordings) and a completely random, continuous output (playback) occur simultaneously from all the spigots.

If you're considering a multistream server, be sure it is put through its paces without fail. Look for any hesitations, overloads or pauses in either record or playback functions. Make certain the application interface software will handle this and see firsthand how elegantly it performs. Determine if the API can be simultaneously addressed when some channels act as spot playback while others act as editing source clips. Find out what happens when variable compression rates are scattered around the same array. Know how your automation system will actually address the server, for input, output, program delay and control. Demonstrating this is not only complicated; it may require specific interfaces using different control schemes in order to insure fail-safe performance under all the circumstances you expect to use the server in.

On a large system purchase, make sure the vendor or integrator understands your concerns. Look carefully for single points of failure and avoid them. Test the real transfer speeds from robotics data tapes and the timing between transfers on Fibre Channel when the server is at its highest activity. Look for the stress points and try them out.

NETWORKED CONNECTIONS

In addition to knowing the consequences of operating with multiple streams or channels, it would be wise for the prospective purchaser to understand the full scope of any networking that is implemented by the manufacturer. For example, some use ATM or Fibre Channel to achieve a high transfer rate. They may also use another topology, such as

Ethernet, for lower-level transport of image and non-image–related information between off-line workstations and central libraries.

This requires additional network switching or fabrics to handle various other network structures. The potential for complications or for failure or network crashes goes up as more indifferent systems get attached to the network. When evaluating a server system, spend time in an actual system-level demonstration, preferably at an operating facility, before committing your entire enterprise to any one manufacturer's technologies.

Server response time varies from manufacturer to manufacturer. System latency; that is, recording and playing back the compressed digital streams, demands another consideration. By the time all the time variations from original material are appended, including a satellite path, compression of the image, storing it or transporting it, reaccessing it, decompressing it and playing it back once more, delays might be generated that you can't tolerate. Your video server end-to-end solution may incur as much as two to three seconds of cumulative delay, depending upon how each subsystem is married to another.

While not all of these factors are server- or even codec-related, it is important to know what happens when latency in the server subsystem occurs. The time is coming when the real time nature of television will be applicable only to live programming. Even then, we can expect some unusual delays that will far outweigh what we're used to seeing on conventional analog satellite links.

To Comply or Not to Comply

A last point to ponder: If you're considering a system that states it is MPEG-2 compliant, regardless of the profile, check it against different encoders and decoders. You will eventually want to take advantage of bit stream transfers without converting back to SMPTE 259M baseband digital video. Full compliance is somewhat not a straight forward task, and it is possible that the full MPEG-2 format may not be observed—making model-A's MPEG-2 fall apart on model-B's system.

Just remember that it wasn't long ago when you had to be very specific and even more cautious about what you installed on, or in, your home or office PC. Ironically, as we watch the PC install base increase, the legacy hardware begins to slowly evaporate—more due to performance limitations than the cost of the new equipment. It is now far

150 Media and Video Servers: Technology and Applications

easier to install and operate good (that is, well designed and well orchestrated) software and expect reasonable results. As time goes on and the marketplace begins to shake out media server technologies, we can be certain that this new technology will mature and grow into greater acceptance.

CHAPTER 14

DIGITAL STORAGE SOLUTIONS FOR THE LONG RUN

Data backup and data protection are different and critical issues to any operating environment. Whether the facility is a computer IS department or a broadcast on-air master control using video servers, preparing today for the inevitable digital transition includes having a plan for the long-range management of the station's assets.

Developing a plan for long-term storage, whether on data tape, videotape or any other form of media, should be considered part of the economic and operational equation of any server implementation. The long-term storage process, referred to as archiving, involves finding answers to questions that really have multiple parts. Since there is generally no getting around the issue of asset storage, once a facility enters the digital age, it has no choice but to consider the questions.

We will assume the term archive describes the media in digital form. The digital conversion format and process will also be a part of the question discussed.

One of the first questions is, "Do you archive some or all of the existing assets in the facility?" The next logical questions become "How do you store it?" and "In what form(s)?" This can be challenging if you've never dealt with digital storage before. The deliberation should begin long before the choice of a server is made.

First, the facility managers must consider all the present and past uses of the video assets. This includes looking far forward into how those assets will be used, reused, and preserved in conjunction with (not separate from) the decision to build the "all digital" facility.

Unmanageable History

Videotape libraries have a tendency to grow to unmanageable proportions. Physical media continues to consume storage space, closets, and desktops. Tracking the content with any reasonable certainty and usefulness is seldom possible.

Over the years, we watched this never-ending process expand disproportionately to the amount of content created, often because everyone seems to want to keep their own original material. Ever since the advent of the videocassette, we've watched storage expand and media costs increase.

We've observed 1" video libraries reach a peak then slowly taper off as either the media deteriorates, the Type C transports diminish, or a new format comes along to take its place. Many facilities have invested in large areas of high-density storage and have considered expanding the library's physical spaces without truly analyzing the overall economic impacts. We just simply assume the tape library must get bigger over time.

Adding a video server creates a new set of criteria for the storage equation. How and what to store is changing rapidly as servers, compressed digital, and alternate delivery systems mature. We are starting to realize that digital is going to take the place of analog at an accelerated rate. With this impending reality, we should thoroughly understand what impact digital would have on all areas of the operation. Solutions will include adding video servers, new videotape transport formats, incorporating video compression, implementing automation, quantifying longevity of the product, and determining the needs for future repurposing.

Digital, while seemingly the answer for some things, poses new considerations for others.

Trust in Digital

We've learned throughout the first part of the '90s that we can trust digital to be pure, clean, and generally free of imperfections. On one hand, we've been told that digital allows for data to be copied and recopied without serious generational losses. On the other hand, we've

Digital Storage Solutions for the Long Run 153

received mixed signals about the problems caused by repeated alterations between one form of digital coding and another.

We've heard there can be detrimental changes to image quality, especially if a mix of compression formats is part of the repetitive process. We've been told about the problems of concatenation and we've been told editing of MPEG is just not possible.

Most have accepted that going digital is a step in the right direction. Despite the up-front costs, we understand that removing the ambient factors of analog video in favor of digital brings new definitions to the issues of purity, clarity and storage. Then, once we are convinced of the benefits, somebody turns around and warns, "You'd better be cognizant of what you do to the digital signal!" It is no wonder there is so much confusion about digital video and DTV.

When it comes to selecting a method to handle digital media, more questions arise: Can you or will you potentially place your facility and your data in jeopardy? Should you keep analog backup tapes of your assets as another form of protection? Should you transfer all your assets to an "uncompressed" digital form for further protection?

The general answer, in most cases, is "Probably not." One question that doesn't seem to come up very often is "Will the bit rate selected today offer reasonable preservation of the image for tomorrow?" The broad stroke consideration to remember is this: If you've carefully thought out your plans and are conscious of what can and cannot happen in the overall scheme of digital implementation, then you can implement choices that won't come back to haunt you later.

CHOICES TODAY STAY WITH YOU TOMORROW

There seems to be one thing that is still certain, at least in today's and probably tomorrow's known technology. The effects of your choice in digital processing and storage will stay with you forever. As an example, the selection of the initial coding process is a fundamental step to determining what the long-term effects are on image quality and durability for the life of the image. This is analogous to capturing an image on VHS, with only 240-some lines of resolution, and then dubbing and performing all the subsequent production processes on D-1 digital videotape. The D-1 tape may offer the highest image level, and the image quality won't degrade any more than it already has, but the original quality will never get much better.

One concept to understand is that there is more than one way to get an image to fit into a given pipe. One method is to filter the image so that its bandwidth is reduced, such as by bandpass or bandcut. The other method is to compress the image. Of the two choices, provided that the image is of high quality in the first place, compressing that image is the better choice.

You probably will ask, "What do you mean by "filtering"?" The simplest example is to consider the difference between recording an image in a component video format to a 76-minute cassette D-1, then transferring that D-1 videotape to an S-VHS tape playing in extended play mode. The physical amount of media it just took to contain the same running length of images is considerably less, both in length of tape and in cost. But you have now "filtered" that image to such a degree that you will never recover the original quality of the image.

However, if you were to have compressed that same image from the 270 Mb/s D-1 format to 48 Mb/s motion-JPEG, the image quality is effectively the same and the space (or bandwidth) required to store that image has been reduced by nearly a factor of four.

Let's take that concept a bit further. If you captured good-quality images in a component format at a reduced bit rate of 25 Mb/s (as in DVCPro), it is possible to preserve that image quality with little noticeable degradation, but you must know what can and cannot be done during any post production processes.

What this means is that storing material at a much lower bit rate does not necessarily compromise your ability to use the archived image in high levels of production in the future. This is really good news. Demonstrations have recently shown that after subsequent upconverting for use in ITU-R BT.601 production formats, manipulating the image, and then converting back to a bit-rate–reduced signal, a noticeable reduction in image quality is seen. The demonstrations actually showed that after eight generations, which included pixel shifting and macroblock boundary crossing, the resultant image looked worse than eighth-generation BetacamSP.

However, there is a good method to preserve that image integrity. You upconvert to a higher bit rate, in the range of 48 Mb/s, and then keep that image at the 48 Mb/s rate throughout the production process and finally take the completed works back to 25 Mb/s for air or archive (of finished product). It is the previously described, repeated conversion to and from a much lower bit rate to a higher bit rate, and

Digital Storage Solutions for the Long Run 155

back, each with a different sampling structure, that will destroy the image.

HIGHER-DEFINITION UPCONVERSION

We can mentally extrapolate where all this is going without a great deal of headache. We expect that devices not yet engineered or conceived will be incapable of making enhancements or other improvements to legacy video formats at a future date. There are astounding devices already available for the upconversion of 525/625 to 1080I.

Demonstrations have shown that a component recording at 25 Mb/s in DVCPro, transferred to 48 Mb/s motion-JPEG, then output in SMPTE 259M and fed into an upconverter for 1080I, can actually make the image look better than its original NTSC displayed counterpart when shown side by side on a high-end 525I monitors and a 1920 × 1080I, High Definition professional monitor.

Manufacturers tell us that these kinds of devices will be only some of the options facilities will have in the DTV transition for HDTV broadcast.

In short, once you shift to digital and you're considering how to archive for the future, you need to follow some important steps in planning:

First, keep away from the NTSC footprint at all costs. Record your future material to a format that will preserve the component nature of the video image. This could be component analog, but the better choice is component digital. The choice might be a bit rate–reduced component digital of any number of standardized compressed digital formats.

Second, keep the production processes on bit rate–reduced images completely away from NTSC. This goes back to the filtering issue discussed earlier. If you need to add a simple key or effect, do it as a first choice in digital but as a last resort in component analog.

Third, when storing material to a video server, try to keep the bit rate as high as your budget will tolerate. Never use an NTSC input format, as that will destroy the integrity of the image, especially once compressed.

Fourth, archive the material from the server to digital data tape in order to avoid any conversion, even a conversion to a component serial digital format. A digital streaming tape or linear tape (either Exabyte or DLT) will keep the data structure in a native digital state rather than a video state.

Once stored on a data tape, be aware of all future processes that you will take the image through. If you intend to do subsequent production of the archived material, make sure you have started with the highest possible bit rate–reduced format you can.

WHERE ARE WE HEADED?

By the end of the decade, most moving images will be stored using some form of compression. While motion-JPEG is the preferred method in mid-1997, most likely MPEG-2 will be the method of choice by the end of 1998.

There are already a variety of flavors for the storage of MPEG, including MPEG-2 4:2:2P@ML at between 24 and 50 Mb/s. This professional level maintains the 4:2:2 coding structure we've been used to working with for ten years. It converts relatively painlessly for inputs and outputs of ITU-R BT.601 production equipment without noticeable artifacts. Converting back to MPEG-2 4:2:2P@ML for servers or other long-term storage or air playback is straightforward as well. Furthermore, the 4:2:2 standard will allow for bit rates up through 50 Mb/s, allowing the manipulation of the image by production equipment with little compromise in image quality.

Once we begin to deal with higher-definition images, we will most likely see a mezzanine level of compression in the 45- to 60-Mb/s areas. While production equipment is not readily available to anyone, the planning for this level is underway at many manufacturers and research entities. There is also work in the area of partial decoding to facilitate keys or wipes on compressed images without having to return to baseband.

We will expect to see some digital servers that will store the GA bit stream at 19.39 Mb/s. This will then permit the facility to store, on a server, the complete bit stream for program delay and pass-through. While this process is not meeting with complete acceptance, it will be a probable alternative for some.

Finally, for serious production, the storage of HDTV images at ~1.5 Gb/s on servers will certainly be possible. However, this data rate is 62.5 times more storage than is currently used for storing 24 Mb/s motion-JPEG, the defacto benchmark for BetacamSP quality imaging on most video servers. Storing HD on servers will be expensive. There are many debates in varied corners of the industry asking whether the production or the broadcast communities will support this notion. Right now, 1.5 Gb/s is reserved for the elite—but with storage and silicon technology advancing at the rates we've all witnessed, the jury is still out.

CHAPTER 15

STORAGE AND INFORMATION RETRIEVAL MANAGEMENT

Another of the questions that surround the storage and archiving of digital video data involves the way we will manage the data once it has been stored. It's no secret that the video industry continues to store an overwhelming multitude of information. Even though video compression and physical size of the media will make the individual storage footprint smaller, we can expect the pure volume to expand to fill the space available.

As was stated previously, preserving the quality of the image at a reasonable resolution will require careful and thorough planning right now. The discussion should not wait until some months or years after you've started the conversion process to servers and digital. To ignore or postpone addressing the issues could result in a substantial loss of quality and integrity in the original material forever.

Broadcasters have been witnesses to a shift in storage and retrieval for playback that continues to mature as technology provides alternatives and improvements. The concerns for the processes, storage, and retrieval of broadcast inventories began with the first implementations of broadcast video cart machines.

The Ampex ACR-25 and RCA TCR-100 products were introduced nearly 25 years ago. At the time, few seriously considered the long-term needs for archiving commercial spots, station IDs and promos at the level with which we are investigating the issues today. Commercials and interstitials had relatively short lives and there just wasn't any purpose in keeping them for an extended period of time. The decision to store these today generally rests with management, who decide the value to the station in preserving this form of asset.

The daily, weekly and monthly management of the early spot or interstitial cassette library was a part of routine operations. Most facilities at that time had at best some form of database, albeit most likely a manual one, for tracking what was in the library, what would or could be purged, and what was active. Operationally, any degree of automation at that time involved the sequencing of spots in a break, usually by a closed loop, entered and managed only at the cart machine's console. There was generally no feedback to any device about as-runs, cart number played, and the like. After all, computers were relatively new devices, with only a few in use by relatively fewer broadcasters.

The station's programming assets were stored on an as-needed basis. In the mid-1970s, there were principally two formats of media— 2" quad tape and 16mm film. Larger group stations kept or owned their own complete program libraries, and most of them were on film. Programs generally stayed in the broadcaster's facility only for an abbreviated period, then were sent on to the next broadcast facility for their turn.

COMPUTER ASSISTED TRAFFIC MANAGEMENT

Computer assisted management for the purpose of traffic began to appear in the late 1960s. The earliest forms of computer assisted services were simple punch card batch management of schedules and billings for radio stations. It would be some time in the later part of the 1960s before a traffic system for television would become a consideration in the broadcast facility.

The early traffic systems of the 1970s would have little or no direct connection to the on-air master control operations. The principles of commercial contract management and billings were mainly for accounting purposes, and subsequently there was no real active concern with station automation or serious library management.

The management of most stations' assets in the mid-'70s was primarily a manual process, usually hand entered into a card catalog and manually executed by the staff. Videotape, still mostly 2" for broadcast and some ¾" for other purposes, was cumbersome. Most stations did not have a large library of programming on videotape. Film was still very much a part of the everyday routine in the TV station, both for news and

Storage and Information Retrieval Management 161

for longer-form motion pictures. Film was continually cut and recut for air. Timely programs on 2" quad tape would be bicycled from station to station and there were no satellites, only microwave and conventional land lines for live network distribution.

Toward the end of the 1970s videotape formats increased, the number of original programs produced grew, electronic news gathering (ENG) was added—and within a very few years the amount of media grew and the space necessary to store it expanded proportionately. Business also began to change dramatically as new technology provided new avenues for production—while at the same time labor costs began to rise.

New forms of delivery in the early 1980s brought another shift in how libraries would operate. The networks developed satellite delivery shortly after the syndication industry showed it could be done effectively. Following this, the program and commercial agency distributors recognized they should take advantage of the satellite medium and cut out the overnight delivery business for programs. More and more programs became time sensitive, and the syndicated rerun business changed perspective once again.

Over a ten-year time frame, all these changes forced facilities to find new ways to manage short-term and long-term storage. Each library in turn developed its own, generally unique, database. Satellite recording for program delay had to get linked into a record schedule and coordinated with the traffic system. Slowly, a basic station automation system began to find its place and become accepted for some operations. Yet the media and the amount of storage required to contain it just didn't seem to diminish.

In searching to account for this additional media, it might be desirable to take a look at those who have a long track record at library management and asset continuity. One of those entities that understands archives is the motion picture industry. It recognizes that storage and preservation of its works can be monumental tasks. Its policies, procedures, and methods have found their way into the broadcast domain in both commercial television networks and in the new era of satellite delivery networks.

As hundred-plus-channel direct broadcast services expand, they too are facing another kind of issue, one that the Hollywood studios haven't had to face in the same vein. Instead of programming a single program stream, direct to home (DTH), once routinely referred to as "direct broadcast satellite" (DBS), these systems are faced with programming hundreds of channels that are fed from a variety of sources. Some sources are merely direct rebroadcasts of satellite feeds from suppliers such as HBO, Showtime and the like. Others are stored libraries, at first stored on videotape, but now on data cartridges, that usually have repeat performances extending over months.

The DTH and cable origination facilities, such as Discovery and Home and Garden, have all dealt with new forms of automation, database management, and asset control in digital formats. An overwhelming amount of engineering and labor has been put into automation systems that can take a schedule and produce a program stream with almost no human intervention.

THE VOLUME ISN'T COMING DOWN

The pure volume of material to be stored, digitized, and preserved is not diminishing by any sense of the word. Fortunately, video compression is helping to control the physical space requirements necessary to store, yet preserve and conserve.

When you add the newest developing trends, such as connectivity of entire broadcast network systems, you can see that there are already several new facets that must be dealt with for today and tomorrow. Most facilities are dealing with the library and storage situations in their own independent ways. As we move forward in the digital era, this may become even more difficult to deal with as the demand for information and the various requirements for preservation continue to grow.

The needs for storage increase proportionately to the requirement for more services and more media. As we wait in the wings, DTV will come on-line and with it the programming requirements will expand proportionately to the number of channels the broadcaster transmits. Integrating automated control of media storage libraries on servers will become quite common as we shift from video on videotape to video data on data cassettes.

Even after some long-term storage methodology is selected, the management of that data may become another Herculean task altogether. The information services (IS) industry continues to develop more advanced processes for automated data storage, off-loading and recovery.

Hierarchical storage management (HSM) is one technology that offers promises in the management of data. We will spend the next portion of this chapter understanding HSM in an effort to put into perspective what the broadcaster will be facing in the future.

NEED IT NOW OR NEED IT LATER?

The archive of seldom-used information as a means to free space for newer, more often-needed information is, by definition, "hierarchical storage management." HSM, in the computer IS segments of industry, is being applied to the automation of migrating infrequently used data to less-expensive storage devices, such as optical and tape drive systems.

The HSM process, if adapted to a media server model, would intrinsically run in the background. Studies are being conducted to understand how this process might function in everyday operations of facilities such as cable head ends, broadcast plants, and post production facilities. In the early development of HSM for office environments, the IS industry set a benchmark of 50 workstations in any one working environment as the point where HSM becomes part of the storage equation. It found that at this point, conventional servers, jukeboxes, and other on-line storage subsystems became critically taxed and required some form of alternative access and control mechanism. Further, the 50-workstation number suggested above does not have the same value in a media server environment.

We are not suggesting that current server systems and databases are capable of HSM, only that if the model for IS could be extended to media servers, benefits could be realized. We have yet to see this HSM applied to the media server market because the definition of "seldom-used" versus "often-needed" was only being shaped during the broadband video-on-demand test. Since those tests have nearly vanished, the model is now being adjusted to a new implementation.

One might look at the theories presented in broadband VOD as being applied to multichannel operations in the future DTV broadcast plant. Of course in DTV, we will be looking at a lot more than just program video streams, so the data we will be concerned with will be a mixture of video, audio, conditional access commands, text and possibly historical or commercially pushed files controlled by outside sources.

The storage of and access to other associated data may be in different forms and reside on different subsystems. Some of the data

164   Video and Media Servers: Technology and Applications

may reside at the header of the associated data tape. This data may be off-loaded to another portion of the video server or another completely separate database server.

Other independent data streams could make calls to other subnetworks that in turn look up information, stored both locally and remotely, that would provide links to other services not necessarily associated with the video program information. This could be Web site data, biographies on cast members or details on similar programs being broadcast by the station at a later date. This supplemental data would be controlled by extensions to automation and control systems inside the station operations computers.

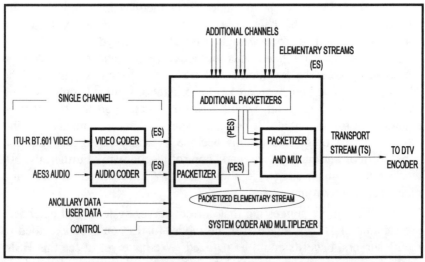

Figure 15-1: Simplified block for MPEG-2 coding system.

This type of data set is then integrated with the MPEG-2 coded audio and video elementary streams (ES) at the point they are multiplexed into packetized elementary streams (PES). From here, other similar channels are passed into the system coder to form the transport stream (TS), with its fixed-length packets of 188 bytes each. This model may be used for developing a subset of the electronic program guide (EPG), or for the generation of user specific data such as catalog print files for specific commercials aired on a particular DTV channel. Figure 15-1 depicts a basic implementation of the coding system from the point where individual audio, video and data are coded to the point where a transport stream for MPEG-2 would input to a 19.39 Mb/s DTV encoder.

The point being made is that some hierarchy will need to be established that will correlate the broadcast facilities' video servers, the data tape libraries, and the control systems so that seamless and timely delivery of the complete data package can occur.

Some make the argument that HSM may never become a part of the storage solution because it still limits accessibility. However, through the buffering of some segments of the "not-so-accessible" media, there is a possibility of putting HSM into the video server domain.

What is intriguing is that all the elements of data management used by the computer industry will somehow find their way into the central nervous system of the video media server. Eventually there will be only a fuzzy line between data and video media storage requirements anyway. Quite possibly, a long-term trend may follow that no system will be optimized to any one form of digital storage medium.

BROWSE SERVERS

As digital media servers become entrenched in the working environment, the need for access to program material by multiple users will also expand. The media or video server is not simply bound to the on-air operations or the program delay functions that, today, seem to top the list of desirable uses.

Once a central library begins to be constructed and the masters on conventional videotape begin to disappear, the data tape library will become the primary source for the materials necessary to create programs in the future. This is expected to be the case whether it is news, promotion, production, or traffic and sales.

Rather than managing a large image data base from one on-line server, supplemental browse servers may be networked to the on-line server providing secondary access to the same data at a faster and less bandwidth intensive data rate.

A browse server contains a low-resolution rendering of the content that is stored on either the on-line or the off-line library. The browser gives a user sitting at a desktop workstation the ability to search

166 Video and Media Servers: Technology and Applications

the database and call up a series of clips and view them without having to retrieve a higher-level rendition of the program material.

Browse servers usually store the images at a much higher compression rate than would be suitable for air or production. Browse servers also contain metadata that will link to other databases. These ancillary databases may contain audio-only clips, scripting, and other important data relative to the media.

The images on the browse server are generally a combination of thumbnail stills that identify some recognizable portion of the clip and a low resolution (¼ scale) video image, generally compressed to 352 × 240 (H×V) called source input format (SIF). Browse images can be scanned at high speed or VCR stunt mode for rapid clip or sound bite identification. The purpose here is to make a quantitative judgment as to whether this clip, segment or image is appropriate to the content the editor or producer needs.

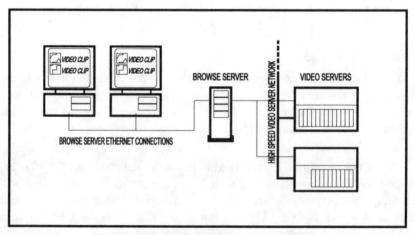

Figure 15-2: Basic browse server network.

Television news organizations seem to be the furthest along in the implementation of servers with browser subsystems, probably because they need more access to more types of files than the on-air operations. Figure 15-2 shows the basic browse server network which can be expanded to several workstations and several primary on-line servers.

The browse server can be quite valuable, particularly if you have multiple clips to review and none of them are presently loaded on the on-line disk storage arrays. These servers are usually independent of the larger video servers used for on-air or production. They generally operate on a separate LAN or occasionally, as in early large-scale library systems at colleges or universities, on a real time coaxial video system.

Once the user decides that the clip is indeed the one they need, a request is made from the browse server's engine to the main library server database for retrieval. It is then that the material, if stored off-line on a data tape library, is transferred to the on-line or near-line server disk drive arrays.

SPEED OF OPERATIONS

As mentioned, the speed of operations is a critical element in the functionality of any storage system, with or without HSM implementation. In the case of video system management where spontaneous decisions, selections, retrieval and storage are routine, the ability to decimate and resolve massive primary starting points (cue pointers) for video media will require prediction theories far greater than those used in many of the storage management systems available today. In most businesses, adding labor to the operations will not be an option. The entire process will be managed by computer and controlled by automation. We are finding this to be one of the primary selling points of both servers and automation systems.

To meet the definitions of "total random access" or "completely non-linear," there will have to be some finite rate recovery specification for the amount of data (video) that will be achievable without taxing the compute or retrieval capabilities of any given system. Those definitions need to become a line item in system specifications for this emerging technology.

The time required for the media server to transfer data from linear data tapes to air is a factor in overall system throughput. Suppliers of near-line or off-line storage solutions are facing these questions from facility operators who want to know just how long it takes to move a spot from DLT or Exabyte tape to air.

When a server manufacturer provides an off-line storage solution in the form of a data tape transport (either streaming data tape or digital linear tape), the user needs to know what control software or subsystems

168 Video and Media Servers: Technology and Applications

will best be suited to managing those libraries. How the library system works, under what protocols, what happens when you must update or append new material to tapes, etc., are all need-to-know questions that hinge greatly upon the automation and operating system software and interfaces themselves.

Making practical use of the data tape library, the on-line, near-line and archive, as well as that material still on legacy videotape, is a complex issue. Information retrieval is one area of applications software that will begin to a take on a new dimension as the digital media servers of tomorrow take form.

INFORMATION RETRIEVAL

Information retrieval systems work by building an inverted database index that is primed and ready for action. This type of system has been developed as an adjunct to HSM. We have taken these principles and extended them to a theoretical application for the media server.

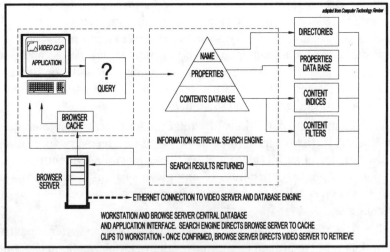

Figure 15-3: Browse server search engine.

Figure 15-3 shows one of the browser elements, the search engine, of what might be many browse servers located throughout the facility. Within the search engine application, a database index is built in advance. The engine becomes highly optimized for speed and space so that searches for media can occur based upon a continually updated process. Some of the elements that need to be accounted for in a high speed search engine would take into account the tape drives, the locations of the media, the transfer speeds of the network and a multitude of other subsystem resources.

Since the information to be retrieved may contain links to and from other databases, and may be carried in the metadata portions of the compressed data files, it is important that the index record position information for every occurrence of the media. This essentially requires management of the content on the front end, rather than sometime after the images and content have been transferred to the library system.

The integration of HSM and information retrieval systems gives the user the ability to search and browse data without ever recalling data from the primary video server. Once the user finds the correct clip or file to be reviewed, the index to that file is passed on in two parts: the index to the location of the files contained in the primary video server store, and the index to the location of the files on the browse server.

As soon as the index for the secondary browse server is received, the browser caches those files to the local desktop or workstation. The local workstation has appropriate applications to view the cached clips, the audio sound bites, the textural information and other elements of metadata retrieved from the local browse server.

Multiple browse servers could be dispersed to various departments within a facility, each with their own clips and databases appropriate to their individual specialty areas (e.g., promotion, production or news).

News and information libraries would have a much deeper database than the commercial or interstitial libraries for on-air operations. The degree of detail in this database will be the key to successful use of the entire system. When HSM is employed, most of the data must be robotically loaded before a user can gain access to it. The process includes coding original material to a resolution sufficient for production uses or air, plus a coding of browse-level quality with associated indexing, links and metadata attached.

It is a common assumption that anything short of instant access will be unsatisfactory, which is all the more reason that the browser concept will be support by operations much faster than those systems that are either manual or at full resolution with isolated access.

One of the goals behind HSM is a much deeper understanding of the uses that the facility will have for the digital content once it is categorized in the total system. For example, when a news department has a breaking story, the news director, assignment editor or producer enters a few key words into the database. Immediately, based upon previously entered data, the browse clips that match the key word links would be moved from near-line to on-line storage. They may even be transferred to a local drive on a workstation for the person to whom the story has been assigned.

Once the producer or writer begins searching the clips via the browser, a decision is passed to the main database that triggers another system that might begin loading the archive data tape to the primary on-line server. Since this process is not instantaneous, it can generally be accomplished in the background without the user even being aware of the operation. Should the clip not be needed, the transfer would be halted and the robotics would prepare for the next action.

HSM is a unique and developing technology. Most of the actions are background tasks and the user needs no training in their operation because they deal with the model only at the application level. Contrast this to current operations where the library tape must be located, loaded into an available machine, scanned for the material, then dubbed to another tape during the edit process. If more than one user wants to view the same tape, they must get in line.

Newsroom editing and server systems will value this technology greatly once it is moved into action. The same might be said about programming and promotion, where magazine shows need interstitials built from movie libraries and the like.

Cost of Storage

Another advantage to HSM is the reduced cost of storage. Cost of storage includes maintaining the existing legacy videotape library—including the real estate and physical space, the disk drive storage costs for the server, the data tape library for the robotics, and so forth. Once an interactive database system is constructed and the library is transferred, the need for retaining and maintaining legacy materials will diminish.

HSM should not be confused with backup, in the same way that archiving and off-site disaster recovery are different from backup. It is

Storage and Information Retrieval Management 171

suggested by HSM developers that HSM could provide relief to what will probably become an overburdened backup system in the future.

Backup and data protection can be accomplished by management software in some interesting ways. One method is to maintain a tally of how often the data tapes are accessed. When the tally reaches some predetermined number, based upon known specifications for passes on the tape, then an automatic replication is started. This is where the entire tape is high-level copied to another transport, creating a new data backup master, and the old tape is marked for removal.

Another method is a continual monitoring of the errors detected and either writing the data to the header on the tape or storing it in a secondary database. In either case, the backup and protection masters can be managed without human intervention in the same way that a protection break tape might be automatically built from the master control automation system.

Once a clear understanding of the operations is in place, HSM or similar backup, protection and information retrieval technology can be implemented. Assets can be easily and securely stored, disasters can be planned for, and labor and media costs can be saved and managed.

Only digital video and media servers will allow this to be possible. Data will remain secure and accessible no matter what other changes happen elsewhere in the system, provided planning and proper implementation are part of the enterprise solution for the digital era.

CHAPTER 16

MEDIA MANAGEMENT AND AUTOMATION

The broadcast facility of tomorrow will depend upon closely coupled implementation of video servers for broadcast on-air operations. As the complexities in storing, accessing and managing information expand, the use of software assisted automation systems, will become more prevalent. No longer will individual videotapes or cassettes remain the primary means for the playback of commercials, promos, and interstitial material. Traditional videotape libraries will be expunged from the shelf and replaced with a digital library. The video server will add a new dimension to the meaning of asset management in the broadcast facility.

This chapter, portions of which were first published in November 1997, will explore integrating the digital library into the broadcast facility using servers as the medium and automation as the mechanism.

Determining the amount of storage necessary for a digital library integrated with both servers and tape transports is a difficult task. The first question usually asked by video server manufacturers and their sales associates is "How much storage do you need?" In reality, if you've never put a server into a facility, there's really no way to know what is the correct amount of storage.

Disks, controllers, and associated storage array subsystems can be the single highest-cost elements in implementing a video server. If

174 Video and Media Servers: Technology and Applications

you have only a single-ended video server engine[27]—that is one in and one out—then storage becomes the only other principle element in the video server itself.

The physical amount of on-line available storage required for a particular operation can easily be misconstrued by not understanding how the server will be used and what will actually control it. Since most of the prominent server manufacturers (not systems integrators) do not provide an in-depth asset management and automation system interface (i.e., automation software), it now becomes the job of the end user to decide what best fits their operation. A complete server system will require some form of interface to be functional and practical. The interface may be as simple as a drag-and-drop spot list organizer (see Figure 16-1) or as complex as a complete station automation system.

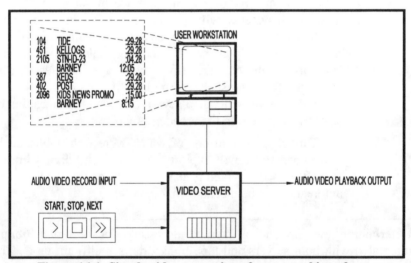

Figure 16-1: Simple video server interface control interface.

The process of selecting a video server system for your facility involves two major principles: understanding of station operations today and more importantly, an open vision of operations for the future. Segregating the vast amounts of video media, partitioning between on-line, near-line, off-line, and archive, entails knowing how large your

---

[27] The server engine, in this context, consists of the input and output, the codec, the operating system, and hardware other than storage drives.

present library is, how often it really turns over (not how often traffic gives you the okay to purge spots), what your average and your peak load for commercial volume might be, and many other items pertinent to areas the broadcast engineer generally doesn't get too involved in.

TOMORROW'S LIBRARY WON'T BE THE SAME

Using today's broadcast facility as an example, it is not unusual to find many different libraries controlled by several departments with rules that are typically made to be broken at all levels of operation. Traffic has its own needs for near-term storage–typically composed of the spot reels poised for dubbing to cassette or into the server. Each sales account exec keeps their own private library of their particular client's protection master—anticipating that the BetaCart's dub will vanish and the spot would have to be remade.

The ever-rotating video-cassette library and the on-air program library consisting of syndicated programming, movies, and a variety of specials aired by the station, is usually housed in a large room that typically gets visited many times a day.

When you add in the complexities of a news department, a magazine program and commercial production, plus promotional segments, you can amass a huge volume of different types of video content that is managed under different sets of guidelines and constraints for each area of operation.

Once the process of integrating a server into a broadcast facility commences, consolidating these libraries into one or two manageable entities is a requirement. While consolidation has always been a long-term goal yet only a dream for most station managers, the possibility of achieving this target becomes closer once the video media server is integrated into regular station operations.

The news department's library is, and should remain, separate from the commercial and program library. This is reasonable because eventually the news divisions will operate under its own server systems focused at managing a revolving inventory of material much larger than the spot or interstitial library in air operations. In addition, most news departments already have their own newsroom computer system and most likely won't change that unless moving toward non-linear news server systems like those from NewStar and Avid Technology.

176 Video and Media Servers: Technology and Applications

But for air operations, alas! The dawn of DTV is now squarely upon us. Over the next ten years, the face of every facility in this industry will change. Media servers and the management of the data stored upon them will be as commonplace as the linear tape transport.

When, not if, the time arrives when the facility makes its move to digital, several new media handling elements, including servers, networks, data tape transports and a level of automated air operations and library management will become a part of the overall equation. The accessibility of individual videotapes will change. The methods for viewing, editing, assembling and playing back to air will take on new dimensions.

Handling the transition to digital will mean that media storage and library consolidation is a process that will begin the moment the medium (whether program content or spot) enters the facility. No longer can the private hoarding of tapes and the unstructured storage of material be tolerated. New thinking and potentially cruel enforcement of rules that will be aimed at providing better and more accurate media management must be enacted and adhered to.

AUTOMATION CREATES CONSISTENCY AND ACCURACY

One of the elements involved in the transition to video servers will entail the organization of the existing and future inventories for the station. The storage methods and access to the content in native and on-air forms will be best handled under the control of management software, combined with traffic and automation systems, so as to insure consistency, accuracy, and protection.

Servers can be complex animals, but properly tamed, their benefits grow in proportion to their usage. Multiple channel inputs and outputs, coupled with simultaneous access, recording, and playback, allow a station flexibility in the growth curve. By segmenting an operation between multiple servers, connected by a high-speed backbone, many formerly manual tasks become background tasks, handled entirely by an overseer, or more properly an automation system.

The facility's media inventory might, for example, be broken into three categories. First there is program-length native media, usually libraried in Type C, U-Matic, VHS, MII, or Betacam format. This program material has a fairly well established and consistent air schedule. It need not be transferred to digital storage, at least not

Media Management and Automation 177

initially. As long as the station has an abundance of the playback transports, this media will most likely remain on the library shelves until needed for actual air.

The second type of media, commercials and public service messages plus rotating promotional spots, should be handled by the video server. The management and airing of this content is where the server will be used to its fullest potential.

At least for the foreseeable future, transferring new spots and PSA content into the server will remain a manual function. However, the backup process, even writing a protection copy to another server, is expected to be an automated process. Which material is protected, or backed up, will be determined by the life expectancy (contract term) of the individual content itself. In the new digital era, VTR personnel staffing master control operations will manage this process, even though much of the activity will be conducted with minimal hands-on intervention.

The third type of media content will be network and syndicated program delay inventory, which is usually delivered via satellite. Here is another application where time shifting from a few hours to several days might well be handled by a server. This is highly beneficial, provided disk storage space is available, because the life of the program is usually a few hours to a week or two at most. This is an ideal activity for a server and is preferential compared with tying up a conventional linear videotape machine, plus tape stock and the physical shelf space it occupies.

NATIVE OR DIGITAL?

Let's return for a moment to the handling of the commercial spot inventory. In order to maintain some semblance of order, both from a physical shelf space and a database perspective, station traffic or operations needs to determine just how much media remains in native form and how much is stored on the server, ready for air.

Some stations keep their agency-provided native 1[11] or BetacamSP masters throughout the air contract schedule, even after the spot has been dubbed into the server and then backed up in the server's disk or digital tape storage subsystem. Others elect to transfer each spot to either a single or multisegment Betacam or MII cassette so that they can use it in their cart machine. This tape may also be cached to a server, leaving the master as a station backup/protection tape. The latter method is more common when a digital streamer tape or DLT is not in the system.

Whatever the operational scenario, most likely the station automation system and the traffic computer will continue to keep track of where the spot is, whether on-line, off-line, or near-line. Should the need arise to use the protection, it is generally a seamless operation to engage the backup tape, provided the backup is not a discrete native format tape.

RESOLVING OVERLOAD

When content stored on the video server is not scheduled for air in the near future, something must be done to minimize the potential of overloading a server's disk capacity. Server manufacturers and automation vendors are working on hardware and operational solutions for what is termed archive and off-line storage. Deciding how far in time, the actual definition of "near-line" or "immediate" actually reaches becomes a pivotal point in determining just how much digital storage will be required as near-line storage.

Deciding which archive and near-line storage methods are best for the operation as an enterprise is a critical step in determining which products will best serve a facility's needs for today and tomorrow. Working through the proper solution for handling the nonactive media storage process will involve understanding both the entire server architecture on a facility-wide basis and the depth and choice of the automation system employed.

Both the automation vendor and the server manufacturer should become a part of the decision process, usually before any actual purchases are made. If you don't know the questions to ask or are unfamiliar with digital server architectures or automation, an independent consultant familiar with both servers and automation—or a group station technical team with real working experience—can be good, usually unbiased, resources.

HOW ACCESSIBLE IS THE CONTENT?

Efficient use of and access to all of the media storage and playback equipment becomes a true design goal in this systemization process. To extend the discussion, the facility must determine how fast the station will need access to any or all of the off-line, near-line, and/or archived media.

Gaining access to any or all of the data in a time frame that meets the demands of the services required, while maintaining integrity and reliability, are compound issues. One choice most likely becomes a trade-off with another.

In the broadcast and video industries, access time is crucial to getting material on the air or into the edit booth. The access time, sometimes referred to as latency, is a serious issue should the server, the backup, or the automation system fail. In most discussions regarding automation, which should include access to nonactive media, this becomes one of the most sought-after answers, for which there is no definable right or wrong total solution.

Having ready access to all the media in the inventory, on-line and at a moment's notice, is impractical. As was stated earlier, trade-offs will be necessary in order to qualify the amount of space available, both in the drive capacity and the near-line or archive tape library. Defining what those trade-offs are in order to maximize operational efficiency will involve the selection of software, hardware and in all cases, human factors of operation.

GETTING EVERYONE INVOLVED

The planning process for server and automation implementation should involve input from several staff elements. Typically, traffic, sales, production, master control, engineering, and station management all will have different needs and perceptions of operations. To successfully implement a managed digital server and associated automation system will require many new operational policies that will meld the station together into one functional entity.

History has taught industry a great deal about how to manage the storage of large volumes of data. In the broadcast industry, the appearance of large-scale video libraries grew over time to a point when it was realized you couldn't possibly keep every piece of media. Now that there are options in storage methodologies, users are finding that you must understand the implications of on-line or even near-line storage capacities, and include archive and off-line in the mix of daily operational strategies.

Presumably human involvement will continue to be in the daily accessing and moving of physical media for some time to come. Managing that level of involvement brings a cost factor into play that stations designing server systems into their facilities must consider.

As we begin to fulfill the transition to digital television and introduce multichannel programming, more activities involving storing, sharing, and retrieving of media will arise. By the time the transition is complete, with near 100percent certainty, we expect all the media to be in the form of digital data. The increased demand for program content will mean more people will want simultaneous access to even larger amounts of data on a recurring basis. Simultaneous or shared access may not seem like a big issue today, but once multichannel programming or news multicasts (including Internet IP casting) begin, the demand for nearly instantaneous access will rise.

Today, news editing systems seem to be further along in this model than the operations or production departments. Once high-speed transfers and networked transporting of media at greater than real time evolves, the news model experience will most likely be extended into other areas of the broadcast plant.

AIR AND PROTECT

Protection of the on-air product and the media itself, while two distinct concerns, are intertwined at various levels in media management philosophies. First is the protection of the air program, which is rather straightforward. Many facilities are looking at some level of mirrored operations, either through parallel-discrete servers or through the assembly of a break tape that runs in sync with the server delivery system. The break tape is the easiest to manage and costs the least to implement.

The break tape is constructed after the log has been assembled and the air operations staff has all the new material dubbed into the server. The automation system then assembles a rundown for the next day's operation in its own internal database. If the server is a multichannel server, usually one of the outputs is routed to the input of a videotape transport (see Figure 16-2).

A tape is loaded and then the automation system instructs the VTR to begin recording. The server soon starts playing out the first break, then pauses a few seconds, then plays out the next break, and so

Media Management and Automation 181

forth until the entire day's spot load has been transferred to a single backup videotape. This becomes the backup or protection break tape that can either be shelved or left in the tape transport ready to roll.

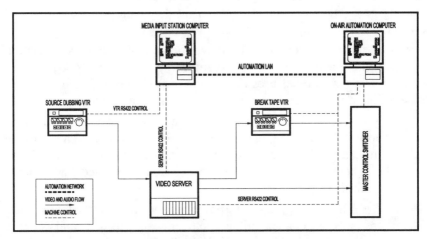

Figure 16-2: Elementary automation flow.

When the start of the broadcast day begins, the air operator can insert the break tape, and it will either be synchronized to run with the server playback or sit idle waiting for a failure to be detected by the automation system. Software will fast-forward the tape to each segment so that it is relatively close to the start point of the appropriate commercial break time in the schedule.

MIRRORED OR CACHED

In the mid-1990s, as early implementation of video cache servers were being installed, the conventional videotape library system (such as BetaCart, LMS, Marc or Odetics TCS-series) would run in parallel with the server. This configuration was generally dictated by the size of the broadcast station's market, which reflected the value of the spot revenue to the station. In addition, most facilities in the major markets were already running with two cart machines and this gave them the opportunity to have an operation that had backup and efficiency.

In the later part of 1997, operating with two complete servers, in some level of a mirrored server and RAIDed configuration, is a growing trend. However, due to the initial higher cost of implementation, this concept may not be in the first budget run for a station. If an LMS or some other form of tape transport assist system still exists, it is unlikely that mirrored and RAIDed operations would be implemented out of the starting gate, but many believe that in the long run, the budget should include this kind of upgrade configuration.

Over time, operational costs will begin to decrease as the station moves from linear VTRs to video server systems. As transports are either converted to other uses or outright retired, the station may find it wise to implement some form of near-line storage mechanism.

Many stations are choosing near-line storage devices made from small format, linear data tape, usually robot assisted. This storage subsystem lets semidormant spots reside in a near-line digital tape library until the schedule requires they be transferred or cached back to the server for airplay.

Near-line storage does not infer instant program content access. There are several considerations both in tape format and storage architecture that need to be considered. Today, only a couple of the near-line tape systems let content be played back direct from the digital linear tape in real time, and there are restrictions on that mode of operation as well.

One true benefit to near-line storage is that it will provide space to store digital media that relieves a crowded server from overload without adding excessive disk arrays.

CHAPTER 17

FACILITIES FOR THE INFORMATION AGE

In January 1996, when the original text of this chapter was published in *TV Technology*, broadcast facilities were just beginning to address the integration of the video server into some segments of operations. Servers had already taken their place in broadband test applications, video-on-demand, and in some high-profile applications for nonbroadcast systems.

The early adopters in broadcast facilities had already used servers as cache devices for robotics-assisted cart machines. In 1995, few had placed servers into actual full-time exclusive use, mainly because of control and software issues that made it less than easy to manage the systems on an enterprise basis. The road ahead was going to be challenging.

At the close of 1995, it became apparent that connecting an entire facility together or "networking," would bring some interesting challenges for both hardware and software vendors. The turning point was clearly ahead. The future of this industry would require a mixing and marrying of equipment, control architectures, network protocols, and file structures. The solutions would require a level of effort that was just beginning to be understood.

Even before the DTV challenge was solidified, it would become obvious that the facility of the future would need a lot more uniformity if it was going to remain interoperable. Networking and data interconnectivity were high on the list of must-haves. The development of systems would become more prevalent as major manufacturers sought VARs and partnerships that would expand the potential of their own respective products.

The Internet was just beginning its kick into high gear and the PC software manufacturers were starting to perk up and pay attention to the challenges about to be levied by the FCC. Broadcast hardware manufacturers were paying serious attention to this convergence between computer files and professional video, developing meaningful products that would bridge video and the PC.

The ascension of the digital video server continued in parallel with video's transition to the digital domain. As the broadcast industry continues to make progress toward an all-digital environment, multitudes of complications present themselves to owners and operators alike. With this transition comes the management of systems spanning a much broader horizon than broadcasters have had to deal with over the past 40 years of the television industry.

FACING THE NEW CONNECTIONS

The issues of connectivity were changing before our eyes. The coaxial cable of the broadcast facility had predominantly been used just to move single channel (or sometimes multiple component) analog video from one point to another. Twisted pairs of shielded cable were typically used to carry single channel audio information. Multiple pairs of cable carried control information from transport to machine and back.

Specialty equipment has been developed to convert user inputs and decisions into commands relegated to operating linear video- or audio tape transports. Other equipment manipulates either the audio or the video content in a fashion that assembles the medium into a program stream for conveyance to an audience.

Most of these devices used multiple cables to carry single or component streams of information and commands. Most of the equipment was designed for single application-specific functions and was built from the black-box dedicated-purpose concept.

Now, with the transition to digital, comes repurposing and reengineering of the facility, the hardware, and the operations. Hardware, now permanently married to software and generally attached to a multipurpose control architecture, has created an entirely new subset of conditions that must be dealt with. Linear storage, while still prominent in a majority of operations, has taken on a new dimension as random access and non-linear recovery become commonplace.

Users are now forced to address both the science and the magic of digital storage and transmission. New techniques and technologies are developing on parallel courses. Keeping pace with the advances in information technology demands making critical decisions that typically have long-lasting economic impacts.

Corporate computing environments are providing management and information processing methodologies with worthwhile models that broadcasters may wish to examine more closely as they transition to the all-digital environment. Networking technologies, matter-of-course in the corporate IS computing environment, are finding their way into the broadcast facility.

The television traffic systems, generally tasked with composing and reconciling the station log, are now an integral part of facility operations. When video servers become a part of the facility record and playback mix, the integration of the traffic system becomes essential— and with it comes an even stronger case for networked media management.

When comparing the overall dollar contribution of information technology (IT) hardware and software for industry versus broadcasting, the broadcaster and video production company seem to be only small potatoes. Ironically, whether video media is found on the corporate network, the Internet, or elsewhere, the foresight that broadcasting continues to exert in producing, presenting, and transmitting media still remains *the* bench mark that the IT industry strives to achieve.

ONCE ISOLATED, BUT NO MORE . . .

For decades, the broadcast facility was perceived as an isolated loop. In the U.S., any single broadcast facility had little direct connection to any other facilities. The broadcaster's primary aim was to deliver essentially the same material to an all-encompassing market audience. The broadcast model is similar to the "push" technology deployed on the Internet of modern day.

As microwave systems, satellites, and the Internet have evolved, the broadcast facility is becoming a connected entity no longer bound to exclusively over-the-air delivery. Cable television has offered one of the first variations on that theme, providing broader choices on a single transport. Direct to home (DTH) programming offers the first real paradigm shift in broadcast delivery in the past decade.

Yet, other than for wider selectivity, more channels and more specialized programming, television has remained essentially a one-way "push" model. The viewer, although offered some striking alternatives to program content, is essentially a receiver only, accepting what is delivered to them with little variation from source to source. This is what the model for interactive television was supposed to change, but it couldn't work over the existing infrastructure. It would take developing the Internet from both a business and a technology framepoint before the market would understand and accept the choices we'll eventually be offered with DTV.

As we mentioned, technology and content-based applications are changing the delivery model. Content has seen a marked change, primarily as result of the Internet, but also due to delivery selectivity and user choice. Both can be viewed as a teeter-totter hinged around technological change and market acceptance.

How the new digital model affects the future broadcast facility is being debated in all of media's professional circles—from manufacturers to politicians to end users. The benefits of having multiple disciplines in all forms of content and media delivery are enormous. We, as consumers and providers, will force faster-paced, technological development while simultaneously finding new opportunities for all facets of industry.

Tomorrow's video and audio interface will address many of the same rudimentary issues that corporate IS environments, deeply rooted in information technology, have been wrestling with for at least a decade. Developing a sophisticated network for any industry is an evolving technological nightmare.

Broadcasters, thanks in part to its acceptance of industry standards and generally similar methods for inter- and intrafacility connectivity, have for the most part enjoyed a neat, clean, and structured method of getting the moving image from point-to-point.

Technology is bringing more change in facility interconnections. Cross-implementation of video, audio, data and control are rapidly, and most assuredly, connecting more elements together than ever before.

Facilities for the Information Age 187

With these cross-implementations come more complications and less universal acceptance.

What the broadcaster is planning for, and in many cases now having to build, is a new kind of network that integrates a wide variety of information technologies. There is opportunity for a new courtship between industry and broadcast television. Broadcasters need to look forward much further in order to address the same issues that the corporate information technologists, designers and integrators are preparing for.

A COMMON HARMONY

We recognize that broadcasters and production companies will continue to use much of their existing infrastructures for as long as feasible. We expect that the physical hardware will remain as mixes of videotape transports, local video storage arrays, short-term and long-term storage libraries, multiformat cameras, sound processing equipment and a variety of remote transmission feeds.

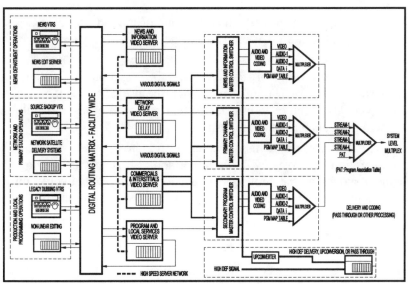

Figure 17-1: Functional multichannel model for DTV.

Broadcasters are expected to retain certain levels of their current operations, especially in editorial and news operations. We expect that news will become more entrenched in the daily operations as DTV and multicasting come on-line. Television news may eventually be given its own DTV stream (channel) and operate under its own control system (see Figure 17-1), feeding its output into the system multiplex with little concern for other programming streams being broadcast under more conventional scenarios. This may eventually provide the news departments with the variations in identities that they've always desired, providing more flexibility and counterprogramming opportunities than ever before.

Through the concept of digital islands, as shown in the functional model for DTV (Figure 17-1), various building blocks can be added to the digital broadcast facility so that the immediate impact of complex and expensive hardware is not so strong. The model depicted shows how a mix of news and information, network SDTV, local SDTV programs, and HDTV pass-through might be built. Each island adds its own video server, coding hardware and potential master control switcher, incrementally, as the market needs them or as the station can afford them.

It is certainly possible that portions of the existing NTSC plant can be adapted to the DTV model, but the long-term goal should be an all-digital plant nonetheless.

Multipurposing of content will be essential to filling the available time slots. Through this model, the use of the content can be extended enabling an even broader system model in order to handle the expected growth areas of the industry.

All of these systems, their respective program streams, and the new operational policies, will live in harmony. It is easy to see how the business model of the broadcast television facility will evolve into a multiplicity of subchannels—all vying for identity and profitability.

CONCURRENT AND SIMULTANEOUS TRANSITION

We must certainly accept that the integration of software-based, computer-structured media development—including its storage, management, and delivery—has become an essential part of our industry. With that acceptance comes denial. Previously the broadcast industry has relied upon a select group of manufacturers to provide most of its hardware solutions. Today, we are seeing those players change. The magnitude of hardware and software providers coming on-line creates an entirely new set of reliability and compatibility issues.

Tomorrow's facility transitions will happen concurrently and systematically. The days of upgrading equipment in the facility as a few proprietary boxes, added one at a time, are probably gone. Most organizations making purchasing decisions expect to integrate each new added subsystem homogeneously on a schedule that will get them to DTV at a future date. This fundamental concept brings a complication to systems engineering, especially when technology makes its rapid migration to new and independent media formats. Consequently, the once relatively easy equipment ordering process becomes more involved all around.

The decisions we must face in making the migration to a near-tapeless, all-digital facility are standing squarely before us. The choices for compressing the ever-expanding data banks of content into a manageable bandwidth for both storage and production continue to grow. Some of the solutions are available today, but in general, many of the long-term equipment choices have not been created. Furthermore, much of the equipment we'll be using is becoming more complex as technology expands to meet the desires of the information revolution.

INTERCONNECTING THE FACILITY

Today's manufacturers of video equipment are looking at some of the same models that the corporate IT people are looking into. Corporate computer infrastructures trying to determine if they should upgrade or completely replace existing LANs are faced with quarterly changes in system concepts as technology unfolds to provide faster, more efficient solutions.

Many are looking at intrafacility transport of video data at higher speed and at greater bandwidth. Some of the alternatives include ATM, Fibre Channel and Gigabit Ethernet. Expectations for studio-transport of compressed higher resolution formats, with data rates of 540 Mb/s (two times SDI's 270 Mb/s rate) and 742.5 Mb/s (ten times the HD clock rate) are in the proposal stage. Plans for mezzanine level compression of HD, permitting the storing of HD at one-half to one-third the bandwidth would reduce storage cost, as well as make practical sense.

Considerations for peripheral device interconnect and mass storage architectures are being given to topologies such as SSA, Fibre Channel, and various flavors of Ultra SCSI. Each of these solutions is viewed as a viable option for faster I/O and greater storage per unit cost. Not all of them will fit into the broadcast facility in the same manner as they might fit into information systems. We do, however, expect a hybrid of multiple network topologies to exist in the facility of the future. There are no alternatives, simply because the equipment and software necessary to address the future technical issues will no longer come exclusively from the broadcast equipment vendor.

The broadcast industry is no different now that a new storage medium besides linear videotape is firmly in place. Yet the broadcasters remain skeptical because they have observed, albeit in retrospect and from afar, what the computer and IT industries have already had to endure.

ARE THE LIGHTS COMING ON?

For the most part, the broadcaster has not yet seen the brilliant light of digital with the same saturation as the production and multimedia industries. True, broadcasters have dealt with non-linear editing; Mac, UNIX, SGI and Intel-PC graphics; even composite and component digital transports and islands within their existing plants. Still, with several digital standards already in place, broadcasters face some very real questions on how they should model their facilities.

Paramount to the decision is the unknown full impact of DTV—a prospect still filled with much confusion. Now that a timetable is in place, we see that significant alterations to existing facilities are necessary. Adding to the difficulties of making decisions, the sources for this needed equipment are also in a constant state of flux and portions of the entire system chain hardware have yet to be standardized, let alone designed. There are elements of the transmission system for DTV that will require upstream (production) considerations be identified and practices established if the downstream side (the viewer) is to get any benefit at all from the future of digital broadcasting.

For any number of reasons, the overall systemization of the facility is misunderstood and consequently most answers are taking a back seat to the questions that must be raised. Some of these questions include:

- Will a significant percentage of or even the majority of my future equipment, including its interfaces, come from "traditional" video manufacturers?
- Will I need to integrate more equipment from nontraditional sources such as computer workstations, desktop video systems and the like?
- What makeup of equipment will the digital facility of tomorrow require and when will that equipment be ready?

These are tough questions for both manufacturers and users. Many think that standards should be in place *now* or users will be forced to let marketing dominance create a de facto product without standardization's due process. Technological change presents a distinct possibility for alienating the interconnection of facilities, forcing the end user to purchase additional translation equipment, driving up costs and fragmenting the market even more.

We expect that all of the equipment necessary for DTV will not be ready on day one and that the first stations to go on the air will be the guinea pigs for the industry. A lot of opportunities exist to develop equipment and systems that will provide the foundation and the mortar to assemble the new generation of systems. Digital media servers are just one component in the puzzle. As engineers, if we apply what we already know are required elements in a media server (scalability, interoperability and extensibility) to the balance of the physical plant, there's a safe bet we won't need to rebuild again in just a few short years.

SCALABILITY AND INTEROPERABILITY

Many are asking, "Will the equipment in tomorrow's facility look more like networked switching systems—or will we see attempts to keep traditional facility structures as they are now with supplemental hardware added to bridge the road to digital?"

The chances for overnight radical changes in equipment as we know it today are small. A facility simply cannot discard all of its existing equipment in order to address the DTV transition. Stations have grown to expect reliability and interoperability of their technical and production systems—something we've seen only a few long-term examples of in the personal computer industry since IBM announced the first PC in 1981.

The broadcaster will demand the same level of performance it is used to in traditional legacy equipment. However, from a cost standpoint, the broadcaster may be forced to use less-expensive alternatives, bringing the computer industry back into play from a serious perspective.

We all recognize that technology is changing so rapidly that no one can afford to invest in products that lack scalability or that might actually impede future growth. In essence, we've already seen the turn in this direction.

PCs of the mid-1990s already possess as much horsepower as the dedicated video blackboxes did in the late 1980s. The difference today is that now the problems center on the applications and the continual upgrade requirements that become a circular problem with no resolution.

Seamlessly integrating manufacturer "x" with manufacturer "y" remains an unrealized wish. Perfecting the human interface remains a challenge. Buying a product that claims to use a connection over a Fibre Channel interface still won't guarantee uniform integration with a different manufacturer that uses the same protocol and transport.

How will the broadcast industry address the multitude of other questions that will surface? How will this industry that hinges on live, real time delivery of timely information employing mission-critical systems be operated by nontechnical personnel and supported by a continuing reduction in qualified manufacturer support staff?

Perhaps the answers will lie in looking at other information technology domains. The IT industry has continued to grow at a phenomenal pace—driving the technology in directions as new and broad as those the broadcast industry expects to face over the next decade.

CHAPTER 18

THE MODEL CHANGES

Just as the business model for corporations is changing, so may the business model for the broadcaster. If all goes as planned, the broadcaster will be able to offer information technology in a new perspective before the millennium. We may only see this model in the largest of the television audiences and only a few will be able to take advantage of the new services.

As soon as marketing and sales see new avenues for increased potential revenues, we can all expect the growth curve to shoot upward. Once the number of channels expands from essentially one to many, there will be an onslaught of hardware purchases to service those demands.

The first consumer DTV sets with 16 × 9 aspect ratios and HDTV capable displays were shown at the Consumer Electronics Show in Las Vegas in January 1998. These first sets, expected to be available in October 1998, will cost, on the high end, around $11,000. On the low end these high-definition capable sets will be around $3,000. The prediction is that second-generation sets at the high end are expected to drop by a factor of three and at the low end by a factor of two within the first year or so.

The launch of receivers and the availability of programming for both broadcasters and DTH satellite delivery systems will set the foundation for new growth. DirecTV made a profound statement by demonstrating live HDTV resolution programming at the January 1998 CES exhibition—and then stated that by the time the sets are available for the public that they'd already be transmitting HD programming to the home via satellite.

193

With more than just the forty required broadcasters delivering pictures in the top ten markets, the model now changes. In fact the entire country should be able to receive digitally broadcast HD before the end of 1998, creating a potential for DTV on a much greater scale than was possibly expected.

NEW REQUIREMENTS

With the new growth comes a new staffing requirement. In broadcasting the engineer may have a new job description that may include the functions of an information technologies (IT) manager. If that becomes the case, and there is little reason to think otherwise, the engineer of tomorrow will need to be tuned in to the same kinds of issues the corporate IT managers are addressing today.

There are advantages to being where we are in this industry today. Video, as the broadcaster knows it, remains a relatively virgin structure for computer IT people. The broadcast and IT industries will be comprised of video, computer, and software specialties. Both industries are in dire need of engineers and support personnel who have a systems foundation in both video and computers. It will take time to cross-train these specialists and their front-line workers.

Primary issues, such as interoperability and systems, can be solved only by understanding all the elements of the communications models. We need to understand operations, technology, relationships and the economics of the new business models. We must understand that hardware is no longer the only component in making the system function effectively.

Systems issues will include far more than just the distribution and routing of video, audio and control inside a production environment. While broadcasters understand these concepts fluently, they remain completely foreign to most corporate IT managers.

Image processing is understood from a document and static picture standpoint in the corporate arena. The broadcast professional has only a marginal understanding of this domain. Add the elements of moving images to the corporate IT genre and you're most likely going to lose them. However, moving images are well understood by the broadcast professional until you place them back into the multimedia or computer versions of digital video.

Although quality was initially compromised by the varying early flavors of desktop and non-linear systems, we are seeing greater acceptance in this domain with far fewer quality issues than five years ago. Image quality is being improved by the driving forces of the video server and the necessity for compression in advanced delivery systems for DTV.

EXTENSIBILITY FOR FUTURE BROADCAST SYSTEMS

What the broadcast industry does not know well is something the corporate IT manager excels in. The distribution and management of pure data from the mainframe down through the workstation and on to the desktop is the meat and potatoes of their jobs. In this model, the integration of video and audio to their environment is for the most part dependent upon the application programming interface (API).

The IT manager can do little about "quality" in imagery because that was already dictated by the hardware and fixed by the API. To look cleaner and perform better, you wait until a better software package comes out and a more powerful piece of hardware is brought to market. In turn, this brings the requirement for continual upgrading and retrofitting of both equipment and software.

Extensibility, the ability to extend the performance or features of a system without hardware replacement, will be an ever-increasing specification for the facility overall. To accomplish this requires a flexible design centered on software or sublevel circuit board–only upgrades. The computer hardware industry now seems focused on the issue whereby a processor upgrade does not necessitate a complete motherboard replacement.

This is good news, because if one uses the past decade as a yardstick, the average desktop processors should exceed 1,000 MIPS (million instructions per second) before the "nines" roll over to a "two plus all zeros". Ironically, it may come at just about the same time rollover date coding software mathematically catches up to handle the years 2000 and beyond!

At these speeds, the desktop will be capable of real time operations in much the same way on-line transaction processors are today. We fully expect digital video, as we have envisioned it in future DTV models, will be available on the computer in the same fashion we'd expect to see over the air. This may force the broadcast facility to require even greater performance with a heavy reliance on network input-output (I/O) operations.

THE I/O BOTTLENECK, AGAIN

The distributed model in which the video facility operates today may have some real problems trying to address the network issues of tomorrow. The present model of tape transports supplying recorded media to a switched point-to-multipoint routing system is becoming more and more expensive to expand. We will need to find more value in the conventional SDI routing switcher rather than continue to add crosspoints and expansion frames.

As multiple channel video servers are added to the facility, this input-output bottleneck will grow exponentially. This could be the case whether the I/O is baseband (digital) video or shared data—like that carried over a data network.

As users start routinely copying their media "files" to their local drives, access to the central library via the network (or the router) will become overtaxed. Reliability will diminish if this is done over a conventional network as well. This dilemma will place costly burdens upon departments, which will react by demanding additional local disk storage just to meet their ongoing production needs.

Even though the cost of drives is relatively low, the management of that data becomes increasingly more complex. If I/O bottlenecks cannot be controlled, the whole purpose of a networked central storage environment could be defeated.

Management of the central library will become essential to revision control, resource sharing and data backup. The network will become an integral part of the broadcast environment from the outset. This requires that the entire network perform better and more reliably, or the entire operation will suffer. From a technical workflow perspective, software upgrades and network system management will become a regular part of the facility technician's job description.

FUTURE PROOFING WITH NEW TRANSPORTS PIPES

Philosophies differ in how and where media storage should be located. For example, if the entire facility's library is stored on a common single server, the user will demand comparable I/O performance to that of a local storage drive. The user expects no outside influences that would interfere with getting work done.

Tomorrow will bring about a different case. For example, to handle the growing data demands in the corporate IT system, larger pipes (channels to carry data on the network) and better traffic managers (hardware and software) are being installed. A larger pipe might be Gigabit Ethernet or FDDI with bridges or routers acting as the traffic managers.

In the broadcast infrastructure, the larger pipe is just beginning to be defined. While some areas of the broadcast plant will not change, the information traveling down those pipes may be different.

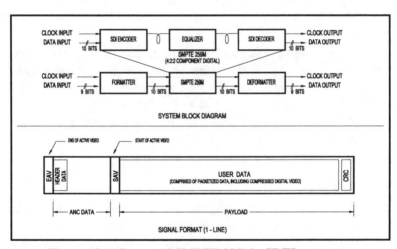

Figure 18-1: Proposed SMPTE 305M - SDTI structure.

To provide an example of extensibility to satisfy new and broadening data structures, we'll look quickly at what SMPTE has been developing over the course of the past couple of years. This and other standards are being developed to extend the use of existing interfaces, such as serial data transport interface (SDTI).

The proposed ANSI/SMPTE 305M SDTI[28] (see Figure 18-1) will allow the use of conventional ANSI/SMPTE 259M serial digital

---

[28] As of the publishing of this book SDTI was in the final stages of the standards process (if confirmed it will be ANSI/SMPTE 305M).

interface (SDI) routing switchers to carry and switch packetized data throughout the facility.

Since packetized data, such as MPEG-2 transport streams (TS), will become the future of advanced digital television, this concept is important to understand when looking at future proofing the digital facility.

The SDTI protocol parameters are compatible with the existing 4:2:2 component serial digital interface (SDI) format. The data stream is intended to transport any packetized data signal over the active lines that have a maximum data rate up to ~200 Mb/s (for 270 Mb/s systems), and ~270 Mb/s (for 360 Mb/s systems). This allows the transport of compressed digital video, or any other data stream, on the same interface (cable media) and through the same switching equipment, provided the distribution equipment can properly synchronize, as today's present component serial digital video facilities.

The maximum data rate maybe be increased through the use of extended data space. In order to increase the amount of data carried on a given line, beyond that which can be incorporated in the digital active line, it is possible to insert an additional ancillary data packet following the header data in the HANC (horizontal ancillary) so as to carry the extension data.

EXPANDING THE NETWORK

Other "faster than real time" digital distribution is already available, making it possible to couple user definable levels of video compression over Fibre Channel. These processes will extend to the movement of data files between servers and other near-line/off-line storage systems.

Although wideband digital video, AES audio, and machine control routers are already in place, when it comes to addressing data networks, most broadcast facilities are barely breaking ground. Some postproduction and graphics facilities have been bridging video and computer data for many years. Non-linear editing stations are also capable of being connected to each other on a baseline level. Few broadcast facilities have taken the steps to address network backbones for the transport of compressed digital media.

Except for those now implementing server technology on an enterprise scale, little thought is being placed on facility-wide data

The Model Changes 199

networking. This is not necessarily the broadcast engineer's fault in any way. The development of homogeneous efficient, functional, or practical hardware and software is still a long way off. The best that can be done now is to prepare for the inevitable by staying on top of current and advancing technology.

The other "bridges" in video facilities might include those devices that implement the migration from videotape archives to data archives. These devices may bridge network interfaces, facilities, remote servers, or simply converters for file format and video compression interchange. Many of these concepts will require the continual management of the data, the structure, the short-term storage, and the long-term storage of the station's most precious assets—its media content.

NETWORK MANAGEMENT FOR VIDEO SERVERS

It is possible that the video server of the future will need to function in a way similar to that of the corporate IT environment. In order to manage the flow of data, the archive, security, protection and backup, as well as managing the usability of the stored media—the video server may need to be managed from a perspective similar to that used to control a data network. Controlling future video server networks might also resemble data network administration techniques—where privileges, trusts, and file sharing allocations are common.

Already we notice that efforts to control local data storage must be pursued. Today, the network systems administrator places security restrictions on data access, sharing, and transfers in an effort to accommodate the various users and to retain control of assets. The facility of tomorrow will require a control (management) structure that will depend on high-end performance and intelligent asset management. Without a well thought out network, data flow restrictions will result in confusion, poor system performance and ultimately in rejection by the user.

Minimizing network I/O bottlenecks requires attention to many issues. Standards are being developed at all layers of the various interfaces and network topologies. Those that stay tuned in to industry technology recognize that for many years the development of networking systems had no relation to the development of video standards. Much of the systems for data networking were standardized without regard for their true potential in carrying video media information. That is also changing.

First came ATM, which developed haphazardly and remained at arm's length from true real time delivery of video. Next, Fibre Channel took its place along side the development of the video server. Gigabit Ethernet is the next great hope for extending data transfers over existing systems—with the prospect of integrating packetized video on a more conventional topology.

Several broadcast equipment manufacturers are already staking their ground in all these areas. Lest us be forewarned—interoperability, extensibility, and flexibility between hardware platforms and network transport layers is essential. It is simply not wise to subscribe to a single source solution for all of your facility's needs, locking yourself into one technology, possibly forever.

The next two chapters deal with integrating video servers into operating environments. Chapters 21 through 25 will look into the various network issues related to data, video servers, and transmission.

We will then explore the details of disk drive systems, connectivity and finally we will revisit some technologies that have had peaks and valleys in their success as storage devices or systems for the delivery of moving images on media servers.

CHAPTER 19

INTEGRATING THE SERVER IN THE FACILITY

The video server is an extremely flexible tool for the broadcast facility. The makeup of the server architecture allows it to be reconfigured, upgraded, expanded and used in a variety of ways. Once we escape the concept of using the server as a cache device for a video cart machine, we can see that the opportunities are abundant as the transition to digital television moves forward.

This chapter will deal with some fundamental configurations for implementing the video server in several operational scenarios. We will look at single station, multistation, multichannel and protection modes for the server and the station's operations. As we go forward in the quest for expanding the business of broadcasting, it will be easy to see how the server will help in that challenge.

PREPARING FOR MULTICHANNEL OPERATIONS

The video server is quite possibly one of the best assets the facility will use once multichannel broadcasting begins. The video server will provide access to like content (spots, IDs, promotions) on a continual basis without having to constantly dub videotape.

Another reason for using the server in a multichannel mode is that it provides a redundancy path for protection. The servers of today typically have at least two outputs available simultaneously, and in some models, many more are available as add-ons and options. Having more than one server channel means that one output can play material to one broadcast channel and the other output can feed the second channel. The second output can also be used for previewing, dubbing, or off-loading material to videotape without affecting the primary on-air output.

TWO-CHANNEL SERVER CONFIGURATIONS

We will start this chapter by diagramming some very basic operational systems and then move into more advanced solutions. The first example will be the simplest of systems, consisting of two video server channels, one set up as a record (input) channel, and the second as a playback (output) channel.

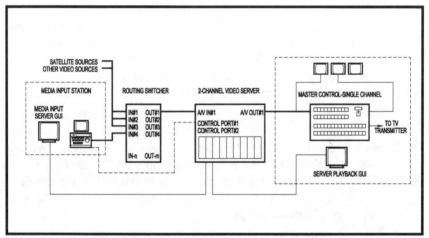

Figure 19-1: Two-channel server operations.

The media input station on the left side of Figure 19-1 is generic. There is at least one format of videotape transport and a GUI from the server. Since the server will be controlling the transport during the dubbing from tape into the server, this monitor and associated mouse and keyboard should be adjacent to the tape transport and monitors. The master control switcher is on the right side of Figure 19-1 and shows only the server's second channel (output) for simplicity. Obviously, other network feeds, cameras, tape machines, etc., would be connected.

Master control should have a second server GUI to control the playback (output) channel of the server. This position would assemble the spots to be played back using either the server's generic software or a third party resident to the server option. Since the server is essentially an

Integrating the Server in the Facility 203

instant start device, prerolls become a thing of the past. This makes timing easier and recuing a snap.

This solution does not show a traffic or automation interface. However, the user might find a software application that would reside on the server that could take a log output from traffic and dump it into the server's internal software. The type of software to do this would depend upon the server and the traffic system.

SERVERS FOR MULTICHANNEL BROADCAST OPERATIONS

The next example (see Figure 19-2) is a very simple extension of the one previously described. It involves using a four-channel server with two channels as inputs and two as outputs. Under this operation, we've added a second media input station and input it (via the routing switcher) to the second input channel on the video server.

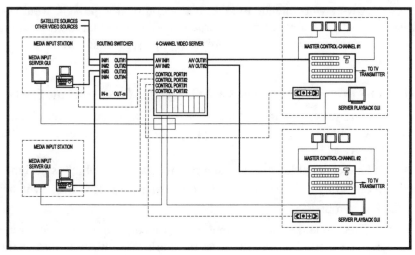

Figure 19-2: Four-channel server for two-channel broadcast operations.

Two output channels are available now from the server. Each channel would feed a separate master control switcher, one for each broadcast channel. Remote controls have been added to the server control ports. These would be optional and possibly unnecessary, depending upon the control system available from the video server.

In this type of operation, which is really two concurrent and different operations under two different schedules, the software within the server would need to allow for four separate operations (two dub stations and two air chains) to be handled from the same GUI system. The difficulties arise when there is only a single keyboard, mouse, and monitor output port from the server.

Under these conditions, a remote control system or a third party interface would be suggested. As a buyer, you should address these configurations with the server vendor when considering this option.

The operational configuration shown in Figure 19-2 is just about the limit of what a single server without external automation or control software can handle. At this point, management of the spot inventory, the simultaneous control of two channels, and the uncertainty of scheduling staff that needs to be conscious of not only their own station, but the other program feed as well, is probably beyond a simple implementation.

A user who intends to take this operational approach, would be wise to contact a few of the automation systems manufacturers or a systems integrator well versed in providing this type of solution.

SERVERS FOR SPLIT STATION AND REMOTE TRANSLATOR FEEDS

There is, however, another scenario that makes use of a four-channel server (see Figure 19-3). This concept uses two output channels from the same server.

One channel feeds the master control switcher, which is designed to be operated by one person. The second channel feeds a 2 × 1 switcher or the house router.

This model provides a supplemental feed from a common master control. The second channel inserts (or overlays) only spots that would feed the station's translator in a distant city. The station on the other end could be unattended from an on-air operations standpoint. The same approach could be used to send a news channel to a cable headend or even as a backup/protection operation for a single-channel station with one master control.

This simple set of three operational models provides some ideas on supplemental uses for servers that we have, until their arrival, had to handle with multiple tape transports and much higher personnel staffing.

Integrating the Server in the Facility 205

Next, we will begin to look at air and protect schemes followed by a case study on a multichannel installation destined for completion in the spring of 1998.

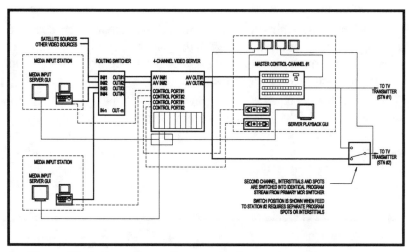

Figure 19-3: Four-channel server feeds two transmitter systems.

AIR AND PROTECT

In a situation where the station becomes conscious of spot revenues that would suffer greatly in the event of a failure, adding a second protection server is an option. Some ask "Why a second server?" The perception is that a server already has lots of backup and protection. Yet, that is not necessarily the case.

Let's consider a two-channel server in the arrangement shown in Figure 19-1. We are showing only one output from the system, leaving the other channel to be used either for input or for a second output. While it is unlikely to happen once a system is fully up and running, if the server does fail—and it will happen sometime—you have no alternative but to go completely to manual. That is, unless you plan in advance.

First, review the daily record and run schedules for master control. After you've entered the schedule of breaks for that day into the server, whether by hand entry via the keyboard, with drag-and-drop software or via a download from the traffic computer, you then prepare a

206 Video and Media Servers: Technology and Applications

backup or "break tape". The break tape is made by recording a separate tape that matches the schedule, leaving a short segment of black between each group of spots in each break. This can be assembled by playing back the schedule of breaks on output channel two even while channel one is used on the air.

After all the spots, promos, and interstitials are recorded to the break tape, in the order that they will play throughout the day, you rewind and eject the tape, set it next to your master control program tapes and hope you don't ever have to use it.

This break tape is your protection. The process is time consuming and labor intensive if not handled by some level of automation. The good news is that most master control automation packages provide this service as a background task. It is usually scheduled during long segments of network or syndicated programs fed via satellite.

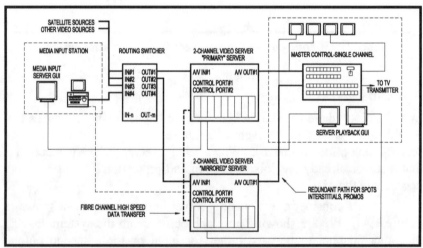

Figure 19-4: Mirrored server operations provide 100 percent protection.

The other alternative is to add a second server set up in a "mirrored" configuration (see Figure 19-4). Under this configuration, two identical servers are installed. The two servers are connected together at the data level using Fibre Channel (FC). The FC link between the servers is managed by a secondary program that keeps the identical material on each server in identical locations. All media inputs are shown as being dubbed into the "mirrored" server, which keeps the maximum amount of storage available for the day's schedule on the primary server.

Provided there is sufficient storage, the mirrored server could act as the secondary library for all the running inventory, plus the upcoming spots that need to be held for a few days or weeks before they're scheduled to air. A mirrored server provides several benefits to the facility. It first is a 100 percent backup of the spot inventory (subject to the timing between updates and transfers from mirror to primary). There is a completely separate system available in the case of a catastrophic failure.

One of the two servers is always available for maintenance. If additional protection is needed for scheduled maintenance periods, a break tape is made ready and shelved in case it is needed. This is valuable should software or hardware updates be required.

If additional channels are added to either of the servers, those channels could be used for other purposes. Having additional channels and storage space available would allow bumpers for news shows to be aired from that channel without sacrificing the on-air chain. The additional channels could be used for making that break tape or as a confidence monitor channel during media inputting.

Some areas may need some clarification when we discuss "mirroring." The concept of mirroring is similar to, but not the same as, RAID 1. In a data server, RAID 1 means there are two separate independent disks running exactly in sync. Two identical sets of data are stored on both of the drives. However, this does not offer complete system protection. In RAID 1 if the controller fails or a power supply quits, it doesn't matter that there were two sets of data, because none of it can be recovered. For this reason, the mirrored concept described previously, and shown in Figure 19-4, is a better solution than RAID 1 drives. (For more details on RAID, refer to Chapters 26 through 28.)

One other important concept: The physical disk drive compliment in each server can be RAIDed. In this example, RAID 3 or a combination of RAID 3 (parity) and RAID 0 (striping) is completely appropriate. When using an external chassis RAID storage configuration, this further provides the ability to grow the storage capacity on an as-needed basis.

Adding Archive and Backup

Now we have completed initial discussions on systems implementation, we need to discuss the archive and backup of the digital media. The first device that comes to mind is the conventional videotape transport. This is certainly an option, but not one that is recommended.

The linear tape transport could be the method for storing the station's assets for a couple of reasons. First, if the server acts as a cache for the LMS[29] (BetaCart, Marc, FlexiCart, or any other robotics assisted cartridge-based videocassette system), then the majority of the inventory is already on these devices. It may not make operational sense to immediately abandon that mode of operation.

If the server lacks sufficient capacity to hold the station's entire inventory plus an adequate amount of buffer space (usually not less than two weeks of rotations), then it makes sense to continue to use the cart machines. In addition, if there is only one server, with no backup protection in house other than the LMS, then it makes sense to operate this way.

Second, if the server lacks sufficient space to store the entire inventory, you might want to continue to use the cart machine as the archive method. The LMS most likely has the ability to continue to operate in a slave mode rather than a master mode. This would be acceptable, as it will certainly reduce the amount of exercising the transports and robotics mechanisms would normally endure. Further, most of the automation vendors have some form of control and archive management routines that would allow the station to continue using the LMS as a backup, archive, or support system.

The concept of using LMS devices as described falls apart once the size of the server and the number of available channels begin to grow. This becomes obvious whether you have multiple discrete servers tied together on Fibre Channel or several channels of video I/O engine accessing a common digital library on disk. Once the maintenance and parts costs, plus the continuing investment in videotape is removed from the equation, it becomes easy to see that investing in additional server

---

[29] We will use LMS, or library management system, in the generic sense of a video cart machine with robotics-assisted loading. Although the LMS is a trademarked name of Sony Broadcast, it has the same generic connotations as "Betacart" or "DVE".

Integrating the Server in the Facility 209

channels, storage or a robotics-assisted data library makes economic and operational sense.

DIGITAL TAPE ROBOTS

The concept of storing digital archives to a streaming tape system is nothing new to the computer data or transaction processing industry. They have been using them for years and have found that they are reliable and easy to manage and can interface sufficiently to make operations much simpler.

While we will not get into the details of the Exabyte, the Ampex DLT or the StorageTek MediaVault systems, these three are the primary offerings in storing digital media for near-line and archive uses. Each of the devices is different and each has its own benefits (and drawbacks). The purpose of the remainder of this chapter will be to describe how these types of devices fit into the overall system.

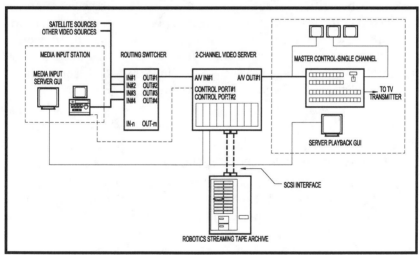

Figure 19-5: Interface for digital streaming tape backup.

Adding a streaming tape system to the server is not a terribly difficult problem from a physical sense. Managing the assets between the server and the library requires some insight. Figure 19-5 takes the earliest and simplest version of the server and attaches a robotics drive (such as an Exabyte tape streamer) directly to the server's SCSI expansion bus.

The configuring of the tape library is generally managed by the automation vendor's software subsystem. In the simplest of implementations, without automation assist, each cut or clip would need to be manually transferred as a file from the server database to the tape library database. Some server manufacturers have this software built into their own systems, but others, like the Tektronix PDR Profile and PLS-200 Exabyte, and Hewlett-Packard with StorageTek, rely on third-party VARs and integrators to develop that interface and/or supply the hardware.

Digital streaming tape systems, such as the Exabyte at ~3 MB/s, are not particularly fast. This format is not meant to be played from tape to air because the linear tape has insufficient buffering to allow a smooth continuous and error free stream at the required data rate. Therefore, in the scheme of utilizing the tape storage system, there must be sufficient space on the disk drives of the server to make the complete transfer, and there must be sufficient time to make the entire file transfer (at around 24 Mb/s) into the reserved directory before playback can begin. In some data tape systems, although it is called a streaming tape, the process is really a file transfer. The important difference here is that like a file on a computer, you cannot begin to use the data in that file (i.e., begin playing back the video) until the entire file is transferred.

The load and setup time prior to the start of the transfer can inhibit a last minute spot from making air as well. Hopefully, the operations department won't cut things that short, but it is a concern in some facilities that operate very short-staffed.

To get the tape ready for transfer we must include all the steps involved. The typical cycle time for a streaming tape is right around two minutes from the start command until it is transferring straight to disk. This means that once the location of the spot is determined in the database, the robot arm must extract and thread the tape into the transport. Then the tape must go through some preliminary shuttling to ascertain where it actually is in relation to the SOT (start of tape). The transport will rewind the tape to the header and read that information into the software database. Then, if the items match, it will shuttle the tape to the approximate location of the spot—which is identified by partitioning on the tape itself—and begin playing.

Integrating the Server in the Facility **211**

The transfer will be in near–real time if the compression rate equals the tape transfer speed, the error correction is low, the data integrity is high and the SCSI bus activity is minimal. For an Exabyte tape drive, this is about 24 Mb/s and at this data rate with motion-JPEG coding, the video is representative of BetacamSP quality. With some overhead and some servo interjection, the transfer will be complete and the disk can play the spot.

Recording to the tape is another matter. Since the tapes are partitioned and the only information about the location of those partitions is on the head of the tape, two actions must occur to get material onto the tape. One step is to read the header and determine that space is available to append onto the tape. If not, and this depends upon the software protocols controlling the subsystem, then the database must "free" that space and permit an over-write at the proper location. Some protocol interfaces allow only one insert over material per header read/write. This places a limit of one spot that can be added each time the tape is loaded into the tape drive transport.

Once a data segment is written, the tape rewinds and then re-writes the header information indicating what has changed to the library tape, as it understands it. Other protocols allow for multiple appends or inserts before rewriting the header and ejecting the tape.

Other larger-format tape drives, such as digital linear tape (DLT) are more sophisticated and have higher-speed transports. They are also integrated into servers differently because their transfer rates are high enough to allow a spot to begin playback (from disk) before the transfer is complete. The lead time for a system like this is on the order of a few seconds as opposed to a few minutes.

Even though a DLT can begin its cycle time a lot faster, the video server must still reserve all the space required for the transfer before the transfer can begin. This is a little like emptying the bucket before you begin to refill it. Failure to do so will cause overflow and loss of data. If there is a 30-minute program stored on DLT, then 30 minutes of disk space must be made available prior to ordering the DLT transfer.

USING MULTIPLE SERVERS

Finally, in this rapid-fire venture through server system implementation, we want to show you a real life model of a server system that was being

212 Video and Media Servers: Technology and Applications

installed at the time this book went to press (see Figure 16-6). The facility has the task of combining two network affiliates under a shared service arrangement, all in one building. The stations each use different robotics cart machines with different tape formats. The new facility will be an all-digital facility and is to remain operational during the construction and transition.

The video server systems will be the largest part of the physical plant investment and will be the most complicated of all the systems. There will be 24 hours of on-line storage, spread over three servers in three different configurations. An archive library will hold 4800 minutes (80 hours) of interstitials, commercials, and the like on a digital streaming tape library. Here are the important highlights of the server and automation systems:

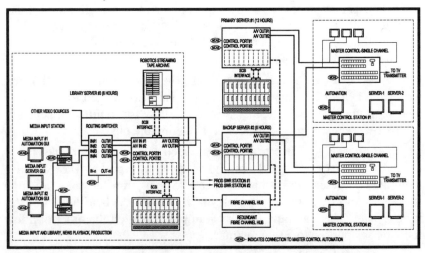

Figure 19-6: Multichannel automated servers with archive.

Server #1—Primary commercial insertion server, with two channels, each directly feeding the two stations' master control switchers. The RAID 3 chassis is fully populated with 12 hours of on-line storage. The server is under direct automation control, and no recording is expected.

Server #2—Secondary or backup for the primary on-air server. This server does not use an external RAID chassis and has the internal drive allotment filled. It will store approximately six hours and is also available directly on-line.

Server #3—Library and input media server. This server will also have the RAID expansion chassis, but it is short loaded to six hours of on-line storage. The server also has the Exabyte robotics archive/near-line library system that will hold 9600 30-seconds spots on 80 individual Exabyte tapes. The library server accepts all inputting of raw materials from agencies and production company commercials. Programs are timed and bar code IDs are assigned.

The library server is a four-channel server that is configured with two input (record) channels and two output (playback) channels. This will also serve as the playback medium for news bumpers, intros, and other interstitials. Finally, the library server's output channels are used for record confirmation during the media input.

Automation—Two distinct, but networked, automation systems will control the playback to air portions of the facility. Two were chosen for various reasons, including the protection of assets (the stations are competing network affiliates) and for backup in the event one fails or needs updating. A common media database will manage the movement of assets between library/archive and the two servers.

Mirrored Server Operations—The structure is essentially a two-server operation, where the backup server has somewhere between 30 and 50 percent of the entire day's program inserts stored on its interior drive array. The primary server will hold 130 to 150 percent of the station's one-day requirements, providing a buffer for long weekends or heavy workloads.

Fibre Channel Interconnect—The three servers are connected via Fibre Channel, which serves as a superhighway for the transfer of materials from library to server and library to backup server. The databases are all managed by the Clip Vault automation software (provided by Columbine JDS).

Shared Resources—After analyzing the stations spot inventories, their daily dub lists for selected periods, and various station logs, it was determined that somewhere around 70 percent of the spots being aired are common to both stations. This will lead to a significant reduction in dubbing and tape costs. This is also an important discovery in that it reduces the amount of disk storage the stations, combined, need to have on-line.

Protection—The primary reason for three servers was developed after thoroughly looking at failure modes and single points of disaster. Each server contains identical codecs that can be swapped between the other servers in the event of a failure. All the disk drives in the RAID arrays are interchangeable. If the robot fails or the library server fails, it is only a matter of minutes to physically disconnect the SCSI cables and move them to the backup server. Once the automation logic is reset, and devices are rebooted, work can continue. At no time will either station be without a primary and some level of secondary server.

All of the internal elements of the servers were studied to make sure each had at least one backup available somewhere in the system. There is a standby (off-line) Fibre Channel hub in the equipment racks and at least one additional power supply. All other items could be obtained via overnight shipment, limiting down time for any particular server to less than 48 hours.

CHAPTER 20

SHARED AND DISTRIBUTED SERVER ENVIRONMENTS

Video servers are powerful tools that, when placed in a collaborative working environment, can ease the traditional burdens of the media workplace. With the capabilities of video server technology, the routine functions of acquiring, editing, archiving and managing the thousands of volumes of stories and programming-related material collected annually can be electronically organized into controllable segments for concurrent use within an enterprise.

Video servers will become a part of a common pool of resources for the production of commercials, short- and long-form programming, and promotional interstitials. The repurposing of existing material can be targeted for direct and specific messages with greater ease and flexibility.

To accomplish these tasks effectively, the systems necessary to produce program content must be configured as a mixture of on-line, near-line, and data management servers that are interconnected at both the video and the data levels. Entire video server systems, such as Tektronix's NewStar/EditStar and Avid Technology's Media Recorder, Media Server and Newsroom systems, are examples of how integral and central parts of daily operations within the broadcast facility can and should function.

Digital news gathering (DNG) is just one of many areas where video and media servers will excel. DNG principles can also be applied to the program production process and the instructional development processes. The concepts can be extrapolated to research, law, special interest groups, and higher education.

The structures of the processes centered on digital media servers are not necessarily unique to any one industry. To better examine how servers will fit into enterprise organizations we need to review and possibly redefine the functions and workflow of the system as a whole. These processes may not apply to all organizations. They may be scaled back for some groups or expanded for others. Regardless of the physical size of the organization, most of the principles remain the same. So let's explore the concepts and the directions.

FUNCTIONALITY FOR A LARGE ORGANIZATION

The structure of daily operations in a content creation enterprise revolves around several areas. The organization must be able to record feeds directly into a central storage library. The database manager must process requests and manage access to the central storage library. Capabilities must be provided for the editing of locally produced pieces, the preparation and transmission of finished pieces, and the playback of edited or raw material directly from the central storage.

Individual workstations must be capable of scripting and managing the programs developed on the systems. Their work tasks extend into automated processes such as the preparation of the teleprompter from finished scripts and the forwarding of approval scripts to managers on a timely and unencumbered basis. Systems must be capable of matching video to sound bites that are timed to the speed of the anchor or on-air talent who will eventually read the introductions and voice-overs live on the air.

We expect to have searchable databases comprised of thumbnails that are keyed to important elements of previously produced stories. Data banks will be coupled with browse servers that can match words to images without the requirement for a human to manually review or select the content. Intelligent scanning and word recognition software already exists that can file a story according to a process that understands the linking between the scripted words, the spoken words, and the visual images attached to a segment or story. Live coding of video into compressed clips that can be retrieved at lower resolution and manipulated from a video workstation in both written and visual form will be constructed as background tasks.

This is the future of content development, research paths, and independent data organizing and warehousing. Not all the elements exist today, but many of them do in segments or subsystems that will eventually become a way of doing business on a widespread basis. In much the same way that search engines work on the Internet, the intranet (the internal network of the enterprise) will develop and mature to a workable, efficient, and cohesive system.

THE PAST IS FADING

News editorial departments and off-line editing environments have worked for a couple of decades on an independent, nonconnected set of systems ranging from discrete editing transports to compartmentalized non-linear editing systems. The edit bay was where you lived when it was time to finally put your efforts to a magnetic form. The creative process always seemed to be encumbered by the unfriendliness of the hardware we've grown to hate—but have to use.

Admittedly, the methods of composing a finished video piece for the five o'clock news are much simpler and considerably faster than they were in the past. Gone are the days of waiting for the film to come out of the soup, shuttling through the wet emulsion until the piece you needed can be located and "cut" from the filmstrip. Those who've worked in television news long enough should remember those days. The rush to complete a story is still there, but the means to the end have changed dramatically since the late '70s.

Yet we still are encumbered by the mechanics of tape transports that make us wait for the tape to spool forward, all the while stalling the creative process and disrupting our paths to completion. We have no real time archiving, no intelligent decision-making processes—only computer assisted mechanics that control equipment to make the edit process mechanically simpler.

Graphic artists and editors alike have been working "unconnected" since the inception of the desktop. These professionals have managed to circumnavigate the system using tools such as sneaker net and dubbing, both only forms of copying video graphics from one system to another when the capabilities of one system outweighed the capabilities of another.

Once computer networking began its acceleration into facilities, the model shifted. As video routers were added to newsrooms, productivity picked up. As microwave broadcasts or liveshots came, the immediacy picked up. When satellite news gathering emerged, the workflow changed again. The evolution of electronic journalism continues, driven by competition and technology. One underlying problem is that many of the processes that have developed over the years have remained bound by the fact that bandwidth is typically so low that efficiency suffers.

As digital video gains momentum, the shift is away from routing switchers for video to file transfers from workstation to workstation. Non-linear editorial and computer generated graphics have taken hold, providing a variety of new methods for content development.

Despite all this technology, there remains a long sought after missing element in all these new methods for doing business. A common database of images or storage that would serve all those connected on the network is becoming a highly desired goal. The video server will play an important part in the development of this database, and it isn't that far away.

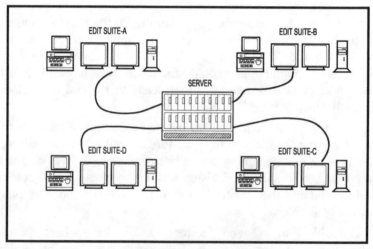

Figure 20-1: Island connectivity.

Community library systems use computer searchable databases as replacements for the traditional card catalog. While cataloging was once a paper-driven task, the computer screen is now an electronic crutch that, while it aids in locating available material (books), misses one of the most important possibilities available in using the computer as a tool.

A library system could use multimedia and video databases coupled with artificial intelligence to find a variety of subject matter that the reader couldn't locate easily by any other means. A cumulative set of queries could be assembled that would narrow the choices and expand the depth of available information. The concern many have is that it requires years of effort to develop the content and years more to implement it. If the ingest process were automatic, dynamic, and not so labor-intensive, then the databases could actually construct themselves and be self-generating over time.

This is just one example of how a video server implementation could solve some of this kind of dilemma over time. Instead of using cards with links to authors, general subject matter, or cross-references, video servers with common data formats could pole other database engines and reveal the highlights in a multimedia presentation.

FILE SERVER VS. VIDEO SERVER

Video servers, by themselves, do not stand as the total solution to the common sharing of all media assets within a facility. We will discuss next why the video server is different from the data file server.

A data file server does not demand a great deal of computational capability from the CPU because in a file server the concern is over network and bandwidth management. Files servers typically deal with file sizes of 1 MB or less, which are distributed over a relatively narrow pipe on a point-to-point basis. Performance of a file server can vary based upon the particular application it is handling, the volume of demands placed upon it by numerous users and the overall network traffic into and out of the server. Multiple CPUs in file servers can be delegated to handle multiple processes, and the speed at which they are handled will depend upon CPU speed, amount of memory and the amount of available resources required to perform the designated tasks.

However, the video server is an opposite case in many aspects. The increased handling of video data rates is not necessarily accomplished just by increasing CPU speed. Increasing bandwidth is a necessary requirement if you intend to delivery higher frame rate video with more channels or streams, and all in real time. At least for today, a video server must deliver streams, not files[30]. A file server is not bound by a set of constant streams that must meet a synchronous delivery of data to a wide variety of end users. Video servers differ from file servers because files servers are in their best domain when delivering bursty data. By "bursty," we mean the data comes in quick bursts, a file here, a file there. Because the uses for that data vary greatly, when it is delivered is of little importance compared to how it is delivered (including how accurately).

File servers are customarily random access devices. Data delivered from a file server is not generally needed or usable until the entire file is delivered to the host computer. Once the files are opened by the user and displayed using a local application on the desktop, the connection to the file server is irrelevant and it can go off and do something different.

A video server, as we said, must stream (video) data continuously and without interruption. The quality of that service must always be high. From a hardware perspective, the performance of the video server then becomes dependent upon hard disk access and interconnection speed.

Interconnection between other systems consists of video-based ports and one or more data ports. The video ports are conventional video, analog and timecode inputs or outputs. The formats for these signals are generally independent of the data structure within the server itself.

Data connections present the greatest variety and are the most likely to change in structure as digital television evolves. Data, or network, connections are already made up of both low-speed control and high-speed data oriented schemes. Internally, each will take a different path within the server, with the higher-speed connections providing for the routes to and from the disk storage and compression subsystems.

When two network systems are utilized, the response time for any particular subsystem can be tailored for the specific usage of that subsystem. This means that control structures and database management can ride on a 10BaseT network and the video transfer backbones can ride on much faster, better optimized networks.

---

[30] It is the opinion of many that video servers will eventually move toward a file-based structure once packetized video becomes more prevalent in the physical plant.

Later in this chapter we will see why manufacturers have opted for different LAN topologies for different needs.

ADEQUATE CAPABILITIES FOR THE TASKS

Configuring a server with enough input and output channels, plus adequate storage to hold several weeks of production video for multiple non-linear editing environments, would be cost prohibitive for all but the most lavish production houses.

This, however, is what news organizations would really like to see. They'd also like to have edit capabilities at the desk of every reporter and staff editor in the facility. Realistically, implementing full services on every desktop doesn't make economic sense. The manufacturers of video servers and newsroom editing systems recognize this and have alternatives in the making.

NETWORKED BANDWIDTH

We have already discussed bandwidth and speed issues to varying degrees and conditions in other chapters. Now, if we look deeper into what the real desires of these working environments are, we quickly realize that not everything needs to be the highest resolution or have the fastest access. To understand how we segregate what needs to be higher resolution and what can be lower "browse"-level imaging, we'll quickly review some digital video basics and network throughput issues.

Uncompressed NTSC (640 × 480, 30 fps, 24-bit color) requires a data rate of 27 MB/s. This, of course, is far wider than the best performance expected out of ordinary 100BaseT connections. If we still desire to keep images at full bandwidth, they would need to be stored on a very large server and accessed only when absolutely necessary. Therefore, we introduce compression into the equation, which brings a quality issue into play.

Compression brings opportunities and solutions. By using compressed video, we can divide the functions of the server into other tasks. A portion of the server could be used for full-bandwidth recording, preserving image quality and flexibility—but at the price of sacrificing transport bandwidth and storage.

Another portion of the server could be used for lower-resolution images, used for confirmation or scanning (browsing) without concern for image integrity or quality. This method preserves bandwidth and aids in accessibility. It also allows video to be presented over a more conventional network to more users on a shared and simultaneous basis.

Before we shift into the distributed and browse server models, lets quickly consider what we can ship around on the conventional network topology we have in place. To understand what kinds of bandwidths we must deal with and how they affect a network topology inside a broadcast or production facility, refer to Table 20-1 for a listing of common broadcast video bandwidths. You can see the data rates are all over the map, leading to the conclusion that transporting video data around a facility is a challenging task.

Format	Bandwidth
4:2:2 (10 bit resolution)	200 Mb/s
Motion-JPEG	12 to 48 Mb/s
MPEG-2 4:2:2P@ML	15 to 50 Mb/s
Compressed HTDV	150-300 Mb/s
HDTV	~1.5 Gb/s

Table 20-1: Common broadcast video bandwidths.

Many facilities have 10BaseT data networking already in place. At best, the data rate is about 1 MB/s throughput on a relatively inactive network. If we move up to 100BaseT, you would think that the data rate would increase to 10 times the 1 MB/s. In reality, the 100BaseT network throughput is on the order of 30 Mb/s, and that depends upon the processing power of the computer and the number of other tasks it is doing while it is accepting the data at the NIC.

Using conventional data networking architectures is therefore impractical and not recommended. For these and other reasons, the concept of different types of servers and different levels of networks for the collaborative workgroups makes sense. The idea is to have a central server that manages the higher data rates, the internal processing, and the movement of data between on-line and near-line storage systems. This server may be connected to several other workgroup servers (edit servers) that would pole and retrieve only the files they need, when they need them.

Shared and Distributed Server Envrionments 223

Here we have different networks specializing in different tasks. Provided there is sufficient bandwidth, control/communications and low resolution video clips could be transported on a more conventional 100BaseT Ethernet network. Part of this would be a subnetwork that would service a series of browse stations. Workstations would have their own browse server that would contain compressed video at a resolution much less than what is needed on the edit servers' workstations.

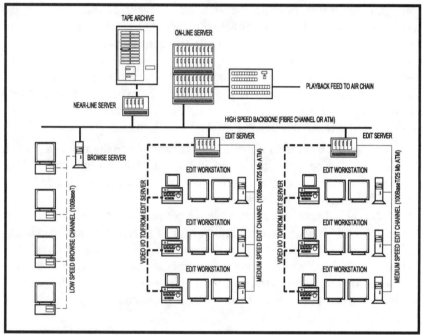

Figure 20-2: Collaborative server-based editing environment.

An entire collaborative server-based system, as depicted in Figure 20-2, would be composed of the following elements:

Edit Workstation—with a conventional videotape transport to handle field tapes and legacy products. At the start of a session, the editor would log on to the station, insert the field in the transport and the session would be identified by story, shooter, tape number, date, and a brief summary of the content. This would immediately be entered into a system database that would track all the elements of the story and archive that data as either metadata (embedded into the data stream) or as ancillary data attached via user bits or another tracking method.

With the newer in-camera logging systems, some of this work might be done before the session starts and might be already recorded as metadata. In any case, metadata would be "attached" to the central database so that it could be properly logged, tracked, and retrieved from other nonvideo databases.

Once the tape is placed into the transport, the local editing computer assigns an input to the local edit server. In this example, we show three edit workstations attached to one edit server – but this could be two or four stations, depending upon the configuration. Video is sent either at baseband (SDI) or as a bit stream to the edit server. The disks begin capturing the material the instant the transport rolls. The editor views the material on a local monitor and successively places marks into an edit decision scrub list. Every piece that is marked is cached to the edit server and those that are not marked are discarded.

The edit can begin immediately because once the mark is placed, the material has been digitized and it can instantly become part of the finished story. As a story is assembled, the edit server is also coding a thumbnail version of the story at a much-reduced bandwidth, as low as 1.5 Mb/s. The thumbnail version would be available to send as preview data to either a browse server or another for of signal transport (including a T1 data line to another location) as soon as the editing is completed. Using a lower bandwidth allows faster access for previewing and other functions, including the attachment of thumbnails adjacent to copy as scripts are reviewed or printed hard copy is produced.

Edit Server—where actual video clips are stored. This is also the device that performs the actual "editing", which is something of a misnomer. Video servers are true multitasking devices; cuts only and outboard effects editing can be accomplished on the same device.

Cuts only editing may be accomplished in one of two ways. The first is where the editor views either reduced-resolution versions of the segments (such as cuts stored on a browse server) or the actual video clips, as they are stored on the server. Decisions are made and entered into an edit decision list (EDL), but the video is never actually transferred or moved from the locations on the server where it first appeared. No external devices such as switchers or effects units are

Shared and Distributed Server Envrionments 225

used; the video never needs to leave the server as baseband. EDLs merely point to the in point and out point (which are registers where the data resides on the server) of each edit. During playback, the list is essentially autoassembled without the need for any recues or prerolls.

The other method is nearly the same, except after the story's EDL is constructed, only the clips pertinent to the story are collected and autoassembled to create another complete clip, stored on another segment of the server. The finished segments are transferred over the high-speed backbone to the primary library server, where they await further instructions. Further instructions could include playback to air or archive or output off to videotape, depending upon the facility's operational requirements.

At the time the transfer is made from the edit server to the library, a compressed clip of browser-scale video resolution is made available for viewing by anyone so designated on the network. If the compressed clip is going to be used only on browse servers, this may be formatted to a multimedia file structure (such as .AVI or Indeo) for ease of viewing on conventional PC or workstations.

Air Playback/Library Server—in the model we've described, all material is played back from the primary central server so that last-minute changes in format or order can be accomplished without having to move work from an edit server. This is the point where data is converted back into conventional video.

The air playback server does not require the capabilities of the edit servers or the browse servers. The playback server will have a much larger amount of storage capacity. It will also contain the decompressors and video and audio format converters, plus it will have the ability to intake other material (such as satellite feeds) as required for operations.

Near-line Server – this is where the archive material is buffered to and from the archive library. In instances where transfer from the library to server must be entirely completed before the material is available to others, the near-line server reduces the tasks on other servers and keeps the overall bandwidth higher over the balance of the network.

SINGLE SERVER VS. DISTRIBUTED SERVER

A single, large-scale server, with only one common shared-storage array, would eventually create bottlenecks that would be resolved only by a temporary reduction in productivity. Having a single server and a single operating system further requires a tremendous amount of computer power to service the large number of clients expected in the collaborative workplace.

Remember, today's news organizations are used to having only so many edit bays and usually only one copy of any given worktape or archive tape available at any one time. The distributed model makes near-line and on-line storage more practical. It also prevents a total corruption of all the data in the event something fails or the network crashes. In a failure situation, each distributed server could continue to operate, even if the larger on-line server experienced difficulties. Workflow could continue; it might just be hampered by an inability to access near-line stored data in an instantaneous fashion.

In a distributed server environment, there is no immediate need to synchronize all the data or to provide instant access to every item in a library. Only that information requested by the user, through the database, need be available.

Another feature in this concept is that of security. Through the use of passwords, privileges and trusts, only certain individuals would have access to certain materials. Security levels are set up by an administrator and can be changed at any time if need be. Privileges would be granted only to certain work groups, and sometimes – in the case of investigative reporting – only to a particular individual. Only a server can offer this level of security and media management.

By providing common access to all of the data, more than one person can be working on segments of the edit and still share the same material. An associate can review the flow of the edit composition before the project is too far down the path. More than one version of the same story can be cut, simultaneously. Scripts can be reworked to fit additional material or missing material.

All these concepts become real and easier to manage once the collaborative and distributed server concepts are put in place.

ONE SIZE DOES NOT FIT ALL

It is easy to see that servers can be of different sizes and different internal makeup. As bandwidth decreases, the number of users decreases. Bandwidth can be managed if the degrees and types of servers are varied and distributed.

The two structures we've described can be looked at in terms of a "broadcast" server versus a "video-on-demand" server. The broadcast servers in our examples were depicted as the library and edit station servers. They function as storage, input, output and editing.

The browsers would be the VOD servers, since their primary function is to order and review video. These are the desktop workstations that use databases to select material and use only the material on a temporary basis.

Keeping these concepts in mind, you can see how distributed servers–of varying classes and capabilities–make sense in the working environment of the future.

CHAPTER 21

NETWORK BASICS

In the fall of 1996, the uproar over Interactive Television had just about subsided, and that over the Internet had only begun. Video-on-demand delivered to the home had seen just about as much growth as it was going to see for quite some time. Direct to Home (DTH) satellite broadcasting was firmly planted and DTV was still only a promise.

The FCC's Christmas gift of December 24, 1996, would give the broadcast industry the impetus to truly start moving forward in its utilization of digital technology and video server applications. It was as though we knew our wake up call was just about to be heard!

All the while, we in the television industry were being forced to change the way we had been accustomed to thinking for so many years. There was now sufficient evidence to tell us that these new systems we were about to indulge in required understanding a relatively new form of connectivity and communications. For many broadcast professionals and technicians, networking was a strange new world. Less than a decade before, we had thought that networking applied only to computers. How quickly we forget that routing switcher control panels had been strung together; that is, networked, for many years prior—we just hadn't put together the concepts of connecting nodes, branches, switched topologies and the like. Now, with packetized television on the horizon, we must reflect on some basics and apply them to our future.

The convergence of digital video and computer networking technologies is forcing us to apply communications technologies to both datanetworking and video-multimedia. Connecting video servers, graphics workstations, disk recorders and even video routing and master control switchers demands a practical working knowledge base in the principles of networking.

We need to have a fundamental comprehension of how and why these systems can meld into one homogeneous system. Without a doubt, the facilities we build or renovate for the future will be a mixture of various digital video technologies and networked computer topologies. Chapters 21 through 23 will discuss networking as it applies to the video and the datacommunications industries. We plan to provide an introduction and an insight to networks—from inside the computer and out in the facility.

CHANNELS, NETWORKS AND CONNECTIONS

At the foundation, there are two basic types of data communications connections. The first occurs when connections are between a processor and a peripheral device. Such a connection is called a channel and is considered a direct or switched "point-to-point" connection. Channels are used to move data with minimal delay and highest speed. Examples of channel connections are graphics boards to PC buses, PC buses to video display accelerators, or graphics overlay boards to processor-based video effects subsystems.

The second connection type is when data is distributed between edit or graphics workstations, central file servers and other workstations. These are called "network" connections. Figure 21-1 graphically depicts the broad distinctions between the two forms of connection.

Although there are certainly other connection methods, network connections might also be made to outboard encoders or decoders, multiplexers, and even video servers. When the information is transferred between these devices, over a common transport medium, these data transfers are handled in the network connections environment.

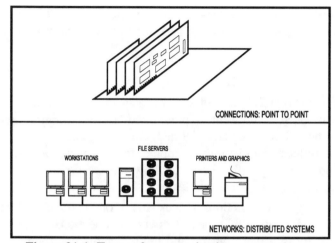

Figure 21-1: Types of communications connections.

Information transfers, whether over a channel or a network, generally require that ancillary information be transmitted along with the payload or data itself. This information is tasked with ensuring the accuracy of the data and the timing of its delivery. In computers, there is a great sensitivity to instructional errors. Depending upon what the error is, the magnitude of the problem might be slight or it might be devastating. In video and audio, errors are subjective and might have little or no impact on the overall quality of the image or sound.

Since we are concerned with moving video information, as data, over a network, we must now be concerned that the guidance of that information is relatively free from errors. The instructions must be accurate enough to ensure that the proper packets are routed to the proper destination. We must also provide some mechanism that tells the receiver that it actually got the correct information—even though it does not necessarily know what that information actually means.

When connections between devices are based mainly in hardware, as in a channel, the requirement for error correction is not as high as in a network connection. Overhead, that is, the additional information or error correction added to each byte or word of data, is considerably less in a channel connection than in a network connection.

In a channel connection, data moves much faster. Channel connections tend to work "in sync" (synchronously), so timing is important. Whereas in a network, these communications are not necessarily in sync (they may be asynchronous) and they may arrive at different times, at different rates, and sometimes out of order.

CLOSED OR OPEN SYSTEMS

Channels are typically closed systems. In the channel connection, the address of each component is predefined to the operating system and remains bound to a strictly defined domain. This is much like the interrupt requests or memory mapping found in computers. Specific commands tell specific systems how to react to specific data. Consequently, channels can run faster and more efficiently.

Network connections are usually more software intensive and handle a wider range of tasks. Network connections, by their nature, communicate in a more hostile and unpredictable environment, and are mixed with a variety of both high- and lower-speed communications applications. More overhead is needed because specific formats, addresses and procedures must be adhered to or the data will become lost, scrambled or misinterpreted.

Channel connections are more closely integrated because of their inherent clock or reference signal timing. In a computer bus, whether it is 8, 16 or 32 bits wide, timing of information transfer from point-to-point is controlled by a strict set of tasks (the operating system or the processor microcode). In a network, communications connections must deal with unpredictable traffic, mixed modes of signals, collision of those signals, and smaller packets of data with a myriad of additional information and constraints attached.

Comparing the channel connections to what occurs in digital video equipment, we find certain parallels. For example, digital effects systems for video can be either hardware- and software-based (as in the ADO) or entirely software-based (as in a graphics workstation). The processing power of a digital video effects device is awesome because the software and hardware are both required to perform calculations in real time on continuously moving picture information. The data speeds internal to an effects device rely heavily on dedicated hardware connections, or channels, that have been optimized for the specific tasks they perform.

Contrast this process to a software only effects product, even a non-linear editing workstation. When effects are processed strictly in software, a lot of overhead is consumed in processing elements on a

Network Basics   233

pixel by pixel basis. This is why software-only systems typically have latency or rendering time built into the equation. First, the system must capture and digitize each analog image before it can begin any number crunching. Then the system must store the digitized image to a format that software can operate upon, and then transform, render and finally reformat the digits into video. If the device requires an analog output, it must decode the digits to an appropriate analog video format.

Most real time digital effects units are predominantly hardware-based. Complex compute-intensive tasks are accelerated by directing signals to application-specific hardware components driven by a combination of firmware and programmable software elements.

Returning to the network model, when video data is stored on a server or is to be used in a collaborative network environment, data must be added that conveys information relative to the transport, which is the data carrier. This additional information is called "overhead" and consists of added bits such as headers, trailers, checksums, and error correction coding. All of these added bits are system constraints that must be dealt with in similar fashions, and in most cases, following the same rules or protocols found in networked data connection schemes.

For a network to be efficient, its communications systems must remain flexible, exhibit a high degree of performance, and provide for a low-cost interface that has a relatively high degree of future proofing. As higher-speed network schemes and connection methods are developed, several additional constraints continue to limit high-performance computing in both LAN and WAN environments.

The physical distance between computer processing units and storage peripherals is just one of the limiting factors for high-speed data communications. Even with the newest of 300 MHz Pentium-II processors, system speed is still constrained to the local server or desktop. For the most part, connectors between chassis plug-in boards and peripherals still remain large, and restrictions continue on the physical quantity of connections (nodes) that are needed as systems grow to meet enterprise needs. In recent years, development has occurred that takes the command sets of parallel SCSI protocols and extends them to the serial environment. As a result, we are well down the path to topologies and connection schemes that will include:

- Serial Storage Architecture (SSA)

- Fibre Channel-Arbitrated Loop (FC-AL)
- Universal Serial Bus (USB)
- FireWire or IEEE 1394

These serial connections increase data connectivity and do not rely on the principles of networking topologies found in standard Ethernet, the most widely accepted form of network topology.

COLLABORATIVE ENVIRONMENTS

In the production community, there is an expanding desire to work in a collaborative environment for the production of material using a variety of tasks, individuals, hardware platforms and software applications. The key to the success of this collaborative work effort lies in fast connectivity and reliable, secure performance on a global basis.

Many still ponder the questions "What is required to digitally connect my facility to another?" and "Should I be making that investment now?" Today, wide area digital connectivity still depends on those who developed the communications backbones when hand crank signaling and manual patching interfaces were king.

The telephone companies are keenly aware of what is happening in high-speed connectivity. The cable equipment manufacturing companies see the light, and satellite technology, albeit a more expensive investment, offers another set of options altogether. We are now beginning to see that the door is opening for terrestrial datacasting over the television broadcast channel as we approach the frontier of DTV. With change, we will see a paradigm shift in business and opportunity.

OBTAINING HIGH-SPEED COMMUNICATIONS SERVICES

When an organization needs more than an analog line for interconnection (i.e., something greater than a 28.8-k dial-up modem), it now turns to digital data services (DDS). These dedicated point-to-point synchronous connections operate at 2.4, 4.8, 9.6 or 56 Kb/s and guarantee full-duplex bandwidth by establishing a permanent link between each point.

DDS provides transmissions that are nearly 99 percent error free and do not require modems. This reduces some of the error correction overhead in a network connection and therefore improves efficiency. DDS lines are available in several forms as described in Table 21-1.

Data Rate	Carrier System	Signal Level	Voice Channels	T1 Channels
64 Kb/s	N/A	DS0	1	fractional
1.544 Mb/s	T1	DS1	24	1
3.152 Mb/s	T1C	DS1C	48	2
6.312 Mb/s	T2	DS2	96	4
44.736 Mb/s	T3	DS3	672	28
274.76 Mb/s	T4	DS4	4032	168

Table 21-1: Common data rates for communications.

Conversion equipment is always required when transporting data from one domain to another. This equipment reformats the signals into a transport that can match the local signal to the wide area signal and back. When using this type of service, DDS sends the data from either a bridge or router, through a channel service unit/data service unit (CSU/DSU). This unit converts the local area network's signals into bipolar digital synchronous signals consistent with wide area network communications standards.

T1 is probably the most widely recognized type of service. For a full-duplex service, T1 requires just two pairs of copper wires and the appropriate service equipment. Another service, called fractional T-1 (FT-1), is available for users who do not need that kind of service or do not want to incur the monthly fees it charges for occasional use.

High-bandwidth services such as T3 and T4 lines are used for transporting very large amounts of data over microwave or fiber on a point-to-point basis. DS3, for example, is a bandwidth that is appropriately sized for the common carrier and the satellite system. DS3 is being considered heavily by the CBS network for delivering HDTV signals (in a compressed ~40-Mb/s form) to their affiliates all over the country.

On the lower end of the scale, if a modest connection with occasional use is desired, rather than a completely dedicated line at 56 Kb/s for local loop or long distance data connections, a dial-up service called switched 56 is available. This service is merely a circuit-switched version of a dedicated DDS 56 Kb/s line. The computer must also be equipped with its own CSU/DSU equipment so that it can dial up another switched 56 site.

Communications systems are finding their way into more and more media environments. The audio recording industry uses Switched 56 to send digital data between audio workstations located either locally or remotely from the primary studio. Multimedia clips of audio, lower-resolution video or still images are also being transmitted between sites in an effort to help collaborative product development expand. We now find that advertising agencies are connecting to their clients' desktops so they can exchange publishing or print information; and even the postproduction facility is using this medium for the video-clip approval processes.

PLAYING CATCH-UP

Even before MPEG burst onto the scene, network connections have been on a catch-up race headed for a bandwidth restriction that no one really wants to experience. The solutions to improving performance lie in both expanding the pipes and lessening the restrictions on connection schemes.

Extrapolating for growth remains an exercise that calls for a crystal ball approach. Over the past five years, we've heard a lot of talk about how far the growth of the Internet will go. We have also heard similar talk about the development of interactive television services. Even when applying both Moore's Law[31] and Amdahl's Law[32], the quintessential determinants for computer technology, there exists a very real market factor that controls acceptance. This acceptance remains

---

[31] Moore's Law: Gordon Moore, cofounder of Intel, predicted in 1965 that the capacity of a computer chip would double every year. Ten years later, in 1975, that forecast changed to doubling every two years and now, some 30 years after the initial prediction, it turned out that the average computer chip capacity has doubled every eighteen months.

[32] Amdahl's Law: Eugene Amdahl, a senior engineer at IBM, who in 1970 formed Amdahl, a competing company that built IBM 360 software-compatible computers, stated that "a megabit per second of input/output capability is required for every million instructions per second (mips) of processor performance".

Network Basics 237

based in part upon the old adage, "How would you like that: cheap, fast or good?" with the answer restricted by: "Pick two."

New solutions to network connection methodologies continue to loom. Buzzwords and concepts flourish. For data communications, ISDN, FDDI, Fibre Channel, and ATM either are in place or sit on the not-too-distant horizon. For video applications, only one or two seem suitable at this time. In Chapter 31, we will look at why Fibre Channel seems to be the current leader in the quest for server and data connectivity.

NETWORK DESIGNS

In the simplest of definitions, a local area network (LAN) is a communications network used by an individual organization that connects resources so they can share information over a relatively short distance. Connecting computers and servers to the LAN, whether for office management of information systems (MIS) or for a group of collaborative graphics and editing stations, requires a good understanding of existing and future network systems.

In the simplest of schemes, a network can exist when two computers are connected such that they share data between them. The connection could be a twisted pair of copper wires, a coaxial cable, a fiber optic cable, a modem with dial-up networking (including the Internet), or a closed satellite connection (as in an intranet).

No matter how large or how small, the critical elements in developing any network design include efficiency, reliability, high performance, security, and low-cost interface. These elements become significant when the physical distance between computer processing units and storage peripherals lengthens.

TWO APPROACHES TO NETWORKING

There are two broad-based approaches to networking (see Figure 21-2). One of these is called a peer-to-peer network. In a peer-to-peer arrangement, there are no dedicated servers and there is little hierarchy among the computers or peripherals. Sometimes referred to as workgroups, they are typically found in situations where there are less than ten computers that need to share information.

238 Video and Media Servers: Technology and Applications

The other type of network is referred to as a "server-based" network. These are generally built around dedicated computers, or servers, that are optimized to service the network quickly and efficiently.

Early LANs employed something called a "disk server," which is essentially a hard disk in a peer-to-peer based network that could share information with other workstations. The hard disk appeared to others on the network as a designated drive. In Intel-based workstations this could be depicted as the E: drive, or some other named designation. File servers, which use software to build a shell around the rest of the disk operating system, are a more sophisticated and efficient flavor of the basic disk server principle.

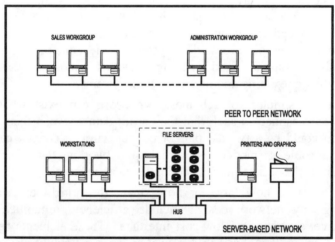

Figure 21-2: Peer-to-peer and server-based networks.

A network need not be restricted to just one server. An extension of the single file server concept is called a multiple or distributed file server scheme (see Figure 21-3). These additional file servers are typically parceled out to specific work groups or islands where there is no need to share all the same data resources between all the workstations in the environment. Specific tasks may be spread out between specialized servers designed and optimized for specific operations such as file servicing, printing, applications, mail, fax, and now even multimedia and video servers. Communications servers in turn handle data and e-mail between one network and other networks, including mainframes and WANs or MANs (wide area or metropolitan area networks).

Today's modern file servers are generally dedicated to the network and optimized for the sole purpose of handling data and file management operations. In the file server, the internal CPUs and RAM resources are optimized for specific server functions, such as managing input and output, handling network hierarchies and moving data between storage devices. They generally do not handle applications such as database queries, word processing or spread sheets.

In a computer graphics environment, a dedicated render engine might also be coupled with some degree of file serving that collects and manages the various elements of the graphics process. It is not uncommon to find a "render farm" where multiple processing engines are connected to a file server that contains the raw and rendered elements. Individual workstations are then connected via the LAN to the render farm, where the heavy number processing involved with 3-D imaging is performed. A network of this type keeps the individual members isolated from the mainframe rendering so that they may work more efficiently.

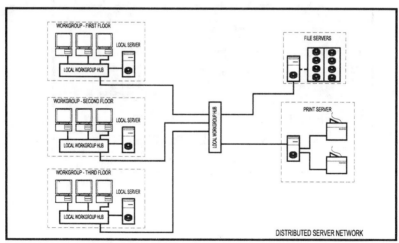

Figure 21-3: Distributed server model.

240 Video and Media Servers: Technology and Applications

In a cost-conscious operation, a non-dedicated server may function as both a workstation and a file server. This principle requires the workstation's CPU and RAM to be shared between data management and applications functions, resulting in less performance on both sides of the operation.

Implementing a hybrid combination of both peer-to-peer and server-based networks is also possible, depending upon the needs and level of services required. It would not be unusual to see this in a production or broadcast environment when "island" NLE systems are integrated with desktop graphics or multimedia settings.

Peer-to-peer networks may also allow computers to act as both a client and a server. This concept is simple to implement on a small scale, especially where costs must be contained and system management is not critical. By contrast, server-based networks allow for greater resource sharing and can be used to standardize applications, control security and broaden overall performance.

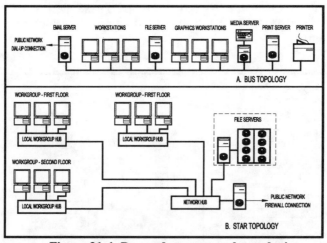

Figure 21-4: Bus and star network topologies.

NETWORK TOPOLOGIES

The physical design or layout of a network is called its topology. This is a map or diagram that takes into account the types and capabilities of networking equipment, their growth and their management. Considerations for cabling, distances, communications methods and operating systems place definite constraints on system design. Each method of implementation influences the network's performance and its capabilities; hence, understanding what exists today and what is required for tomorrow is extremely important in planning for growth.

Network designs are grouped into three basic topologies. The "bus," or linear, topology (Figure 21-4A) is the simplest. This consists of a trunk (sometimes referred to as the backbone or segment) made up of a single cable that connects each computer, printer, file server or device together in a string. No repeaters, amplifiers, boosters or hubs are used to manage or segregate the computers or their signals. All computers essentially communicate on the same wire, and therefore only one computer can "talk" on the network at a time.

The second type of topology is called the "star" (see Figure 21-4B). Here the computers are connected by individual cables to a hub in a star pattern. The hub is a centralized point in the system that can be either active or passive in nature. The passive hub can be as simple as a punchdown block. The active hub can receive and transmit signals (repeat) while amplifying, equalizing and isolating them to some degree. Active hubs prevent single points of failure by providing a barrier between a failed element in the network and the rest of the computers or subsystems.

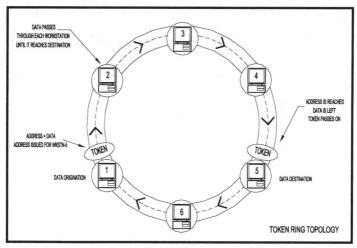

Figure 21-5: Token ring network topology.

An extension of the star topology is called the clustered star. When several stars are linked together, the failure of any one star does not bring down the entire network. This is a common implementation when stars are spaced out on multiple floors of a facility.

The "token ring" or simply the ring is the third type of topology (see Figure 21-5). Here a token, or carrier, moves from computer to computer, carrying both data and the transmitted address from the sending computer around the ring until it finds the receiving computer. The data is dropped off and the receiving computer sends another message around the rest of the ring via the token back to the sending computer indicating that the data was received. There are no terminal ends and data passes in one direction only, being amplified by each successive computer as it passes through one to the next. One drawback to this topology is that when any one computer fails it has a tendency to impact the balance of the network.

Topology	Advantages	Disadvantages
Bus Topology	Easy installation and expansion	Slows down in heavy traffic
	Simple and cost-effective	Media problems may affect many users
	Single workstation failure does not cripple network	Difficult to troubleshoot media problems
Ring Topology	Uniform network access	Single failure of workstation affects entire network
	Uniform performance independent of number of users	Difficult to isolate problems Media installation more critical
Star Topologies	Easy to expand	Network fails if centralized point (hub) fails
	Individual computer failure does not affect entire network	
	Monitoring and management from central point	

Table 21-2: Network topology advantages and disadvantages.

Table 21-2 provides a summary of the different topologies' advantages and disadvantages. Different types of topologies can coexist in the same network architecture. This is especially true in larger organizations where several smaller peer-to-peer groups may be connected in a bus type of topology and a new addition might be connected through a managed hub to a central media library. Some networks, out of necessity, have mixed star, bus and token ring topologies using hardware and software devices that merge and manage these topologies such that they do not adversely affect each other.

Considering the structure of network communications is the logical next step in understanding how the network functions as a whole. In the next chapter, we will begin a basic review of the major standards setting groups who have sought to develop the procedures or protocols that allow different hardware and software solutions to exist harmoniously in the network. Our next chapter will begin the discussion of network protocols, the rules of the highway.

CHAPTER 22

PRINCIPLES IN NETWORK PROTOCOL

For a network to function efficiently, uniformity and consistency are required. This is one of the underlying reasons why standards setting organizations exist. Standards for network communications, hardware, and protocols have been in place for several years. The setting of these standards dates to 1978, when the International Standards Organization (ISO) began its description of network architectures that eventually became the guidelines for modern networking and Internetworking. The Paris-based ISO functions as the umbrella for several worldwide standards setting organizations composed of government, business, educational, and research groups (see Table 21-1). It continues to work toward the establishment of standardization for all services and vendor products.

The ISO's original work defined a model that was further refined in 1984 to become the "OSI model" or Open Systems Interconnection. Today, the OSI model is the most widely used guide to describing networking environments. These guidelines are applied to the majority of network systems and can be found in most connection schemes, including some of those being developed for the digital video networking architectures of today and tomorrow.

For networks to communicate properly, such details as when the data word starts and ends and how long it is must be universally defined. These definitions are termed "protocols" of communication, the rules of behavior.

Organization		Charter and functions
ANSI	American National Standards Institute	Codes, alphabets, signaling schemes
CCITT	Comité Consultatif Internationale de Télégraphie et Téléphonie (International Telegraph and Telephone Consultative Committee)	Consists of some 15 various study groups related to preparing recommendations for protocols in modems, networks, and fax
EIA	Electronics Industries Association	Industry standards for interfaces between data and communications equipment
IEEE	Institute of Electrical and Electronics Engineers, Inc.	Publishes standards for data communications (among others)
ISO	International Standards Organization	International standards for services and vendor equipment
COSE	Common Open Software Environment	Multimedia, graphics and object technologies
COS	Corporation for Open Systems	Subscriber organization: interoperability in OSI, ISDN, and certification
OMG	Object Management Group	Promotes languages, interfaces and protocol standards for vendor applications
OSF	Open Software Group	Combines vendor technologies and distributes results to various other groups
SAG	SQL Access Group	Creates standards in interoperability of different database systems

Table 22-1: International Standards Organizations (ISO).

The OSI model divides network communications into layers that describe specific network functions, activities, and behaviors. The model spans seven layers and begins with the lower layers addressing physical media, with the higher layers addressing how applications access communications services. Each successively higher layer becomes more complex as it defines narrower tasks for specific applications.

While the OSI model broadly addresses data communications along the network, other committees have worked toward developing standards for topologies and access methods that interleave within the various layers of the model. The IEEE's "Project 802," comprised of twelve functional committees, was tasked to sort out and standardize the wide range of LAN incompatibilities that came out of various manufacturers' implementations of topologies (see Table 22-2).

IEEE Project 802 Committees	
802.1	Internetworking
802.2	LLC: Logical Link Control
802.3	Ethernet (CSMA/CD Network)
802.4	Token Bus Networks
802.5	Token Ring Networks
802.6	MAN: Metropolitan Area Network
802.7	Broadband Technical Advisory Group
802.8	Fiber-Optic Technical Advisory Group
802.9	Integrated Voice/Data Network
802.10	Network Security
802.11	Wireless Networks
802.12	Demand Priority Access

Table 22-2: IEEE Project 802 committees.

By the mid-1980s, vendors were individually approaching solutions to network LAN problems with their own methods. It became evident that for network software to work, standards would need to be followed. The committee created subgroups that eventually became the widely recognized 802.x specifications (where x indicates the standards committee's group number) describing the various network topologies. Today, the IEEE 802.3 or "Ethernet" specification is probably the most widely recognized and implemented specification in networking.

NETWORK TOPOLOGY

Network "topology," its physical design or layout, is a map that describes the types and capabilities of networking equipment, its growth and its management. The three primary structures for the LAN include bus, star, and ring topologies, but when you add the capabilities of a remote access service (RAS) or intranet and Internet, the network topology takes on an entirely new dimension. Before getting into the details of complex services and the hierarchy of the network model, we'll first take a few steps backward and get back to basics.

On the surface, the functions of sending data from one computer to another seems generally simple. In actuality, this is a rather complex task. To examine the activity within the network, we should break those functions into discrete tasks.

For each piece of information that moves on the network the following basic activities occur: First, data must be recognized. Next, the data is divided into smaller parts called packets. Information is added to each of those packets that determines the originating location of the data and the address of the receiver. Also included is timing information and error correction data. Finally, the physical transport method is specified and the entire data capsule, consisting of more than just the original information, starts its path through the network medium.

Earlier, we introduced the Open Systems Interconnection (OSI) model, a 1984 revision of a foundation or structure of network architecture that addresses dissimilar device connectivity. In the OSI model, network communications are divided into seven layers (see Table 22-3). The lower layer addresses physical media, with the highest layer addressing how applications access communications services.

LAYER 7	APPLICATION
LAYER 6	PRESENTATION
LAYER 5	SESSION
LAYER 4	TRANSPORT
LAYER 3	NETWORK
LAYER 2	DATA LINK
LAYER 1	PHYSICAL

Table 22-3: OSI model.

Each layer in the OSI model provides services to the next higher layer. As data migrates from one application to another, each layer is designed to address different network activities, hardware or protocols.

Each layer also acts as an insulator from the other layers—handling all the activities necessary to move data from one layer to the next—without concern for what the other layers beyond its immediate neighbor may have done to actual content data. The first three (lower) layers of the OSI model are sometimes grouped together and referred to as the network services.

An important concept to understand is that each layer is set up such that it could communicate with the same layer on another computer. These are called "peer" layers and they theoretically act as virtual communications channels. In actuality, they use software to communicate between neighboring layers on the same computer.

LAYER 1: PHYSICAL

When two computers are connected together, they require a physical connection, connectors, and a wiring or transmission method such that data can flow from one device to another. The bottommost is Physical Layer 1 and is where raw, unstructured bits move from the connector through some form of media (fiber, twisted pair, coax, even the air) to the other connector. Layer 1 also describes how the media is connected to the network interface card (NIC). It includes information about connector pinouts, electrical or optical levels, and functional characteristics about the media connection methods.

Physical Layer 1 is concerned only with raw bits that have no real meaning when compared to the original application. Layer 1, however, is also responsible for determining that if a bit was sent out as a "1," it is received as a "1." Further information is also represented, including data encoding, synchronization, length and electrical characteristics of the bit and its type or form (for example, negative going, balanced, differential).

To compare this in television technology, Physical Layer 1 might be equated to the BNC connector on the back of the distribution amplifier. It must, for example, be a true 75-ohm connector that meets a specification of particular size, electrical specificity and material. The main point here is that the physical connector isn't concerned with whether it is presenting serial digital data at 143 Mb/s (D-2) or 270 Mb/s (D-1), or plain simple analog video (PAL or NTSC) at 1 volt peak to peak. That is the job stated in another specification.

LAYER 2: DATA LINK

Layer 2, the Data Link layer, interfaces data from the Physical Layer 1 to the Network Layer 3. Layer 2 translates the raw bit stream into data frames or packets that make sense out of the bit stream before it is acted upon by the Network Layer 3. The simplest data frame consists of the elements shown Figure 22-1. The forwardmost portion of the data frame is called the header. Just like the information on the envelope of a letter, there is a sender ID and a destination ID. A control ID is added as a guide for functions such as synchronization, routing or frame type. The bulk of the data frame is the actual data (the meaningful information that needs to find its path to the application layer). Finally, a trailer or cyclic redundancy code (CRC) is attached for error correction and frame verification.

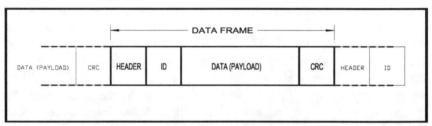

Figure 22-1: Basic data frame construction.

At each of the layers, this root process is ordered, defined and adhered to so that the interchange between layer x and layer y is consistent regardless of where the hardware or software resides. Enough information is contained in each data frame so there is certainty that data from a word processing document will end up as a word processing document, although many more functions occur before the data really "appears" as a word processing document.

LAYER 3: NETWORK

This is the Network Services' upper level. Layer 3 addresses packet or data frame switching and is where the virtual circuit is established, that is, the translating of logical names and addresses into physical addresses associated with the connection method. The Network layer determines

Principles in Network Protocol 251

the data path within the network based upon network traffic and congestion.

Another function of Layer 3 is to resize the data so that it fits within the confines of the router. Should data be too large to squeeze into the space defined by the Data Link Layer below it, the Network Layer will break data into smaller units and send it on. At the other end of the virtual connection, the information sent along from this layer enables reassembly of the packets into meaningful data as it passes up through the network layers and on to the application.

LAYER 4: TRANSPORT

The barrier between the Network Layer 3 and the Transport Layer 4 is a zone where many anomalies must be accounted for. Packet orders may be wrong, bit errors must be accounted for (and corrected); even the multiplexing or demultiplexing of several messages on the same bit stream must be accomplished.

Transport Layer 4 manages these functions. Here another piece of header information is added to the packet that identifies it to the transport layer on the other end. This layer will regulate data flow by controlling the message based upon a priority it received well ahead of its layer. The Transport Layer 4 also resolves receiver and transmitter problems associated with all the packets that pass through its layer. This is also the layer responsible for sending acknowledgments forward or backward through the network stating that the data was either okay or needs to be resent because errors it detected were unrecoverable.

Still left to discuss are the three remaining layers, Session, Presentation and Application. As you can see, the tasks become more complex and more information is added to (or stripped from) the overall data frame or packet as you move through the OSI model. We will only touch on the major functions of these layers, leaving the details to the future.

LAYER 5: SESSIONS

Up to this point, the emphasis has been on bits and messages within the bit stream. The user on the network has been of no concern to this point. The Session Layer 5, is that portion of the network model that establishes when the two computers begin, use, and end this virtual connection. As the name implies, this period of use is called a "session."

You can think of a session on a network connection like a video editing session. Similar types of events happen. For example, a client books a session with the editor, who allocates either directly or indirectly the edit suite and associated videotape machines, and then requests the use of pooled resources for that particular edit session. Then the editor initiates the edit session by creating a name for the session. Starting with "edit-1," then "edit-2," and so on, a number of active elements in each "edit," or event, are arranged to make an overall flow to the program. The edit listing might not be in sequential order in the EDL or on the screen and that is left to other tasks, but the process where these tasks are grouped is called a session. The session therefore becomes the binder of the processes.

Network Session Layer 5 also places markers that act as checkpoints in the data stream. These markers provide synchronization that in turn allows the session to resume, from that point forward, should a failure occur that temporarily interrupts that session. A network failure may be catastrophic or completely transparent to the user. Something as simple as a continuous collision of data or a hiccup on a router might signify a millisecond pause in the session and may not even be visible without a network analyzer. However, these are things that slow down performance and become part of the software tuning that network LAN engineers and specialists look for in peaking up a network's overall throughput.

The Session Layer 5 goes a few steps further. Its traffic control function determines when transmissions can occur, who goes first, and how long they'll have to transmit. These sessions are virtual in nature and occur continuously with numerous other activities on the network overall.

LAYER 6: PRESENTATION

The Presentation Layer 6 is the network's "translator." It formats the data to and from the Session Layer 5 so that it can be interpreted and utilized properly at the final stop, the Application Layer 7.

In Layer 6, protocols are converted (for example from "terminal protocol" or "transmission protocol" into another protocol), encryption/decryption occurs, and character sets (ASCII or EBCDIC) are

Principles in Network Protocol 253

managed. For some of the older applications that were machine to machine–related, such as ancient word processing, the boldface, underline, or extended graphics codes are inserted.

Data compression is also managed here and a function called redirection, which relates to server I/O operations, also occurs. Some of these activities are more related to data communications than video, so we will leave the explanation at this point.

LAYER 7: APPLICATION

Finally, the topmost layer, the Applications Layer 7 acts as the gateway to the actual user-specified applications such as higher-level applications, e-mail, and multimedia. Layer 7 is also responsible for remote log-ons and the statistics of network management.

This is the layer that can be adjusted to fit specific industry needs. As such, and because different user programs have different requirements, it is difficult to describe, in general terms, the details of the protocols found at this level.

DIFFERENT RULES FOR DIFFERENT PURPOSES

Other organizations have established different names for similar activities or layers. One is the CCITT X.25 standard that developed many of the early telecommunications standards. This is where confusion adds to frustration. In the CCITT X.25 standard, the names "frame" and "packets" have different meanings than what we've explained within this overview (see Figure 22-2).

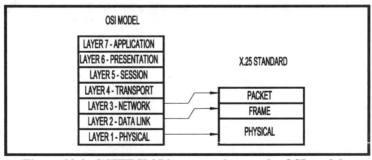

Figure 22-2: CCITT X.25 in comparison to the OSI model.

The lower three OSI layers (Physical, Data Link, Network) have different names in the X.25 recommendations (Physical, Frame, Packet respectively). Be conscious that a telecommunications explanation of the layers may vary from a data network explanation.

X.25's Physical Layer, corresponding to the OSI Physical Layer, is where the recommendation in CCITT X.21 defines full-duplex, point-to-point synchronous and RS-232 asynchronous data transmissions. These are the connection points between DTE (data termination equipment) and the public switched network. The X.25 Frame Layer is where data is exchanged between a DTE and a public network.

The switched public network requires data to be presented in packet form, so X.25 states that at the Packet Layer, information sent from DTE is assembled such that it can be understood when received on the public network.

HOLDING ON TO THE BASICS

Understanding the basics of the OSI model is helpful when trying to comprehend the changes occurring in compressed digital video and packetized television transmission. We are finding that video servers and other networked digital video equipment are all moving toward Ethernet LAN connections for the control of their equipment. Control panels, switcher frames, even connections to automation equipment seem to have standardized on Ethernet as a transport mechanism.

This should not be confused with moving compressed digital video between servers. There are alternate methods, such as Fibre Channel, for accomplishing this. We will look into this in later chapters.

CHAPTER 23

CONNECTIONS AT THE NETWORK LEVEL

Interconnection, data transfer modes, networking protocols and the like continue to be hot topics in today's ever expanding digital media technologies. The technical methodology associated with moving data around in a system begins at the transport level (sometimes called transport layer). Even though it is the physical layer that makes the actual contact with the media (the wire or optical fiber), it is further up the network model, at OSI Layer 4-Transport, that all the details get put together.

Standards aimed at making the interoperability of data communications possible are constantly evolving. The establishment of the MPEG compression standards, the transport protocols such as SDTI, and the multiplexing of the packetized elementary streams, are critical elements in the system. Part of the standards process includes making certain that data can be efficiently and effectively exchanged, stored, and forwarded—on a variety of cabling, networking, and transmission topologies.

The most prevalent of network media is Ethernet. The name is nothing new to most of us. What it does and how it works still raises eyebrows even in broadcasting technical circles. In the past couple of years, those getting into multimedia and digital video on the computer desktop arena have dealt with Ethernet as the method of choice. We venture to bet, though, Ethernet is still one of the more overly confused of the various network topologies and connection interfaces.

As data, voice and media technologies grow ever closer together, some difficult and expensive decisions will need to be made in the network interconnection domain. How the facility is retrofitted for the digital age is a matter of concern for all domains. The considerations will go far beyond the purchase of a new 360 Mb/s serial digital router or embedded AES audio in a SDI video transport. We expect to have a number of different network topologies and wiring infrastructures spread through the physical plant—and it would be nice to know, now, what impact that will have on the future.

BANDWIDTH AND TRANSMISSION RATES

Today, when connecting one computer (host) to another, we are faced with a variety of interconnection solutions. Two critical areas to be concerned with when selecting a network topology are *bandwidth* and *transmission rate*.

"Bandwidth" is the range over which signals can be transmitted without undue loss. The "transmission rate" is an instantaneous measurement of how quickly information is transferred over a transmission facility (network).

ETHERNET TOPOLOGY

Ethernet, as it was once generalized in the 10 Mb/s transport flavor, used a large-diameter coaxial cable connector (with RG-6 or RG-8 cable). Today, Ethernet has grown as the need to address more efficient and economical interconnections between computers expands. The more familiar 10BaseT is common in most business LANs and in many television video production environments. Nevertheless, many don't know what the nomenclature really means.

Descriptively, the "10" still refers to the 10 Mb/s speed that the data would run at if it were not hindered by things like network interrupts or data collisions. The suffix, usually the "T", "2", and sometimes the unfamiliar "5", refers to the physical size of the cable and hence its overall capacity or transmission rate to transport data.

The "2" in 10Base2 refers to two meters and the media is a 50-ohm coaxial cable with 50-ohm BNC connections. The "T" in 10BaseT indicates "twisted," refers to two pairs of unshielded twisted pairs of telephone gauge wire (UTP), one pair for transmit and one pair for receive. The connector is an 8-pin, RJ45 phone-type modular jack.

The "5" in 10Base5 refers to the five-meter variety. This big, heavy RG-8 (or RG-6) size cable requires in-line taps, called transceivers

Connections at the Network Level 257

and an adapter called an Attachment Unit Interface (AUI) that provides the bridge between the Ethernet cable and the computer host.

For data networking, Ethernet, in 10Base"x" configurations (where "x" is any of the above suffix designators), was once considered the best for moving data around in a network. Discounting the actual network topology, such as token ring (in 4-Mb/s or 16-Mb/s offerings) or basic star topologies, there just didn't seem to be many reasons to expand data transmission capacities until the first half of the 1990s. Suddenly, the explosion in desktop workstations, hard disk storage capacity, digital media, and high-resolution imaging hit. Digital video technology demanded that systems run much faster than in the previous decade.

Facilities began adding more hosts to their networks, files became much larger, and then all of a sudden, everyone realized things just weren't moving very fast anymore. A change seemed to be immanent but the direction was sketchy.

FASTER, BETTER, BUT NOT CHEAPER

In the heyday of the early network expansions, before the Internet was the hottest topic on the planet, many data networking companies began planning hardware for the information superhighway on ramps. At this same time, development was continuing in areas related to moving data much faster through LANs.

The rapid success of fiber optics resulted in a new technology called fiber distributed data interface (FDDI). FDDI promised data rates in excess of 100 Mb/s, ten times that of conventional Ethernet. However, the problems included the expense of connecting host devices into the FDDI system and the relatively complex task of fiber cable installations. Cost estimates for connecting each point in the network grew to over $1000 per node, something nobody could recommend investing in. Even though the bandwidth problems seemed to be headed toward elimination, you could still buy a lot of coaxial or twisted-pair cabling for far less than the cost of one or two FDDI node connections.

With the ten-fold increase promised from FDDI came the question, "Maybe this could be done using a fast(er) Ethernet?" After all, SCSI-2 seemed to be an answer in drive technology. People were already working on Ultra-SCSI and the prospect of video compression was going to reduce the need for additional bandwidth—probably in the

258 Video and Media Servers: Technology and Applications

same time frame that we would be installing that new network infrastructure anyway.

In the first half of the 1990s, some facilities that hadn't taken the leap to fiber began exploring a relatively inexpensive and probably interim step, to 100BaseT Ethernet.

CONGESTION AND CONFUSION

Once, or if, fast Ethernet was installed, problems with the network still didn't go away. Collision detection and network loading still are not uncommon in facilities where many hosts reside on the same network. The strange thing was how all this hype about moving video around an Ethernet network made any sense at all. We found that with overhead, network congestion, and lack of continuity in network interface cards (NICs), we still couldn't seem to get the average transmission throughput up above 6 to 8 Mb/s anyway. We found it was the applications part of the model that became the hardship and prevented many of the benefits in higher-speed LANs from ever being realized.

It seems that the digital *video* engineering folks still seem to be a few steps up on the distribution and loading problems associated with networks. Moving 270 Mb/s digital video around on a dedicated coaxial switched line is still the more efficient way to address real time video— but as we were about to see a few short years later, that was about to change.

SPEEDING UP THE ETHERNET

The newer 100BaseT Ethernet installations will require more careful installation practices than what we were able to get away with in standard Ethernet using Category 3 UTP. EMI was an often overlooked headache to deal with. How was it going to be possible to transmit 100 Mb/s over voice-grade UTP media and still meet the EMI regulations as defined by the FCC and IEC 801 specifications?

The new Category 5 UTP media (CAT-5), with its electromagnetic interference (EMI) mitigation, offers to make networks operate more quietly and efficiently. The process of getting CAT-5 to meet specifications was not taken lightly. Several approaches were proposed, including:

Connections at the Network Level   259

- Spreading the signal over four UTP wire pairs instead of two (as in 10BaseT)
- Limiting of transmissions to one direction at a time
- Silicon-based signaling techniques
- Increasing the current 10 Mb/s Ethernet to 20 Mb/s (by eliminating the collision detect portion of the protocol)
- Implementing full duplex Ethernet
- Using six pairs instead of four to spread out the EMI potential even further

Once it was determined that it was the twist of the wires that tamed the EMI tiger, the movement toward CAT-5 for all new installations went forward at high speed. However, the medium was not the only problem that needed an answer. Now the questions would revolve around new network hubs and routers, with careful system planning in order to function efficiently.

We will not delve any further into the inner workings of hubs, routers and the like, as that is a subject for another set of books entirely. We bring these subjects up because we're seeing manufacturers designing systems that require some level of network interconnectivity, even if it is only point-to-point. We'd all like to think that one network will operate the entire facility, but practical experience says differently.

The last thing we want to deal with is being forced to reboot the server just to get the next set of commercials on the air. For this reason alone, isolation may be the best medicine—at least for the time being.

EFFICIENCY IN NEW MEDIA INFRASTRUCTURES

Many facilities are facing a need to grow into the new media domain. The multimedia server and multimedia systems were expected to dominate the infrastructures of renovated facilities. Bridging those systems into true digital video will be a confusing and possibly avoidable prospect.

Many facilities are now incorporating multiple networks, interconnected by legacy LANs that consisted of older SNA or even asynchronous networks via dial-up connections. Several have significant

260   Video and Media Servers: Technology and Applications

numbers of Macintosh computers doing art and graphics, with Intel platform machines doing everything else. Those that have Novell networks are wondering whether to keep them or to simply start over with Windows NT.

As the facility expands and computers appear everywhere, efficiency in cost and in bandwidth are becoming important considerations. In order to define bandwidth efficiency, we need to understand what happens in sending data around in packets on any network.

HIDDEN PROBLEMS

Even though the 10BaseT makeup isolates the host from the hub, once on the backbone side of the network, data is running bidirectionally and can continually collide with each other. There are things that go on in networks that cause delays in transmission rates, regardless of the system bandwidth.

Any delay means that data is inhibited from smoothly getting from point A to point B. This results in slower transfer speeds. The average transfer speed of a connection can be grossly estimated by dividing the peak rate (10 Mb/s) by the number of hosts connected and/or sending data on the network simultaneously. With several hosts and large file sizes, it doesn't take very much traffic at all to slow even a 100BaseT network down to the level of a 10BaseT system. Couple these with the inefficiencies of applications and improperly balanced client/servers and you have a network headache that even the systems administrator can't do much about.

THE ATM QUESTION

Other high-bandwidth, high-transmission-rate solutions are still being developed and explored. One of the other interfaces being considered by some manufacturers was copper data distributed interface (CDDI) which also is in the 100 Mb/s range for transmission rates. Along with CDDI, the loudest buzzword of the mid-1990s was ATM, which stands for Asynchronous Transfer Mode (please refer to Chapter 25 for an overview of ATM technologies).

ATM is a methodology that splits up and sends data in packetized groups and then insures that it will get put back together

Connections at the Network Level 261

properly when received at the other end. ATM is very fast and was touted to produce transmission at rates far in excess of 100 Mb/s.

Keep in mind that fast Ethernet was not an intended replacement for ATM. On the contrary, it is believed that fast Ethernet is intended to enhance data traffic to the personal workstation or desktop. In sophisticated network installations, ATM is better adopted to support the backbone issues—making the marriage of ATM and fast Ethernet technologies ideal for many applications.

It was expected that ATM would eventually be brought direct to the desktop as multimedia and high data intensive applications continued to grow in all environments. As history will show, that was not going to be the case. Confusion and disagreement has left ATM standing in the wings except for some dedicated applications by some pioneering companies that saw ATM as the universal solution to everything.

This was the vision in early 1992–94, and it seems viable until other groups such as Gigabit Ethernet and Fibre Channel came along. There are still many good applications for ATM, mostly in the connectivity to public or private networks. It will be hard to tell with the advances in other types of networking and transport structures where this will all lead.

WHY GET INTO THE NETWORKS?

Therefore, we ask, "Why is the understanding of network topologies so important?" Basically, this is the backbone for the future of transporting multimedia applications and digital video technologies. As with any technology, industry must constantly cope with two significant factors as it grows. They are the preservation of existing investments and the support of lower-cost, more efficient methods or alternatives to solving tasks.

If your facility is just beginning its transition into digital media technology, you are urged to look further than ever before in analyzing all impacts of this advanced technology. Our next few chapters will look into packetized digital technologies, the future of the television technology industry for DTV. Recognizing that not all of the information to be presented is applicable directly to video servers, getting an overview of the systems, transports and data structures will help guide future decisions.

CHAPTER 24

INTRODUCING PACKETS

In the early 1990s, we began to realize that transporting digital data to locations beyond the actual broadcast facility would require new technologies outside of conventional analog or component digital video. As new technologies surfaced, we heard about solutions that were supposed to be the saviors for connecting facilities over long distances.

As some of those technologies developed, we also learned about how difficult it would actually be to extend the current real-time model beyond the broadcast or production facility we know today. Even though solutions developed for communications (such as ATM) looked good in theory, they would bring with them a new set of problems.

The next two chapters will introduce you to the concepts of packetized data. We will hear a lot about "packetized television" over the course of the transition to DTV, as it is how we'll get digital video, audio and data into and out of servers, and then through the narrow pipe it must fit through. The foundations of packetized data transmission come out of the telecommunications industry, where they have employed the transport methods for well over a decade. We present now an overview on some of the technologies that we will be dealing with in configuring and connecting the television and production facility of tomorrow.

We have heard as many pros as cons about each of the new networking and delivery technologies. In these still early stages, where digital television is almost in our reach, but we can't quite grab it, and we were left with two perplexing questions: "Which technology should I select" and, "When can I begin to use it?"

In the next generation of information delivery, we will be in for some real changes and certainly some difficult challenges. The integration of video servers has been explored thoroughly throughout the pages of this book. We have focused on technology, business models, and history. We've looked at servers mostly as extensions to video that we're familiar with and comfortable implementing.

Digital television will allow us to make use of new transports, methods for moving data around the facility and to the outside world. The boundaries of the video, combined with data and as pure data, can and will be extended beyond the walls of the facility through the use of new forms of digital video transport.

Where once we thought concepts were not extensible to video, we may see a revival of efforts to accelerate established transports for video and data delivery. The technologies that will follow, with certain alterations, may be applied to local, regional, and national/international system connectivity.

The need for video interconnection in the first half of the '90s never really surfaced outside of a few isolated high-volume communities. We have heard about Hollywood successfully enabling methods to move video as data between facilities, but not without a high cost impact. Yet, for the most part, the burning desire to circumvent satellites and Federal Express has not happened. Videotape continues to remain the predominant medium to store and forward program content.

The video professional of today is faced with addressing how to handle a vast number of choices in image processing, digital storage and transport. Since 1990, monumental changes in what "digital" means to the world of video media have occurred.

Most of us are aware of what JPEG compression has done for non-linear editing. Many of us are now addressing what MPEG can or cannot do in terms of transporting or storing of images. Some are facing the complexities of moving of images around both physical plants and wide area networks, which means we're dealing, once again, with the telephone company's far reaching powers of control over both rates and methods of data interconnections.

CHANGING DEMANDS

When portions of this original article were published in March 1995, the physical plant was undergoing some dramatic changes. The graphics unit demands workstations that are more powerful. Compositing tools have moved toward the desktop. The editorial process, whether at the news level, program level, or the entire broadcast plant, continues to grow. Nearly everyone is finding that the integration of multimedia, in one form or another, is becoming common.

Today, the agency and the postproduction facility's major request is to move data faster and more effectively between host computers or other regional facilities in a cost effective and secure way. Everyone is asking, "What is multimedia?" and, "What does it mean to me?"

Multimedia is also becoming an important part of the communications industry. For an application to be considered multimedia means that some combination of audio, video, data or images must be integrated in some form. For the most part, if you do not consider interactivity in your application, multimedia does not necessitate time sensitivity. However, as soon as you add interactivity, you must consider time sensitivity.

The media server is structurally a specialized computer and storage system configured to deliver media of a time sensitive nature. Multimedia, once it began integration with audio and video, lagged behind in achieving full-motion, high-quality video. However, network architectures have since begun to seriously address methods to move this time sensitive information in a real time way.

With new products aimed at delivering content over the Internet and within the local area network, we are continuing to see aggressive planning and implementation of pseudo-isochronous systems aimed at real-time delivery of video and audio content.

TIME SENSITIVITY

Video and audio delivery have always addressed time sensitivity. However, the computer platform, by its own nature, is not particularly time sensitive and thus presents a problem for interactive video media delivery. Internally the computer remains somewhat of a bottleneck for linear video delivery—but recently the implementation of MMX processor subsystems has helped to relieve at least the desktop complications.

Nevertheless, this bottleneck extends into the networking domain for any number of reasons.

Enter the telecommunications industry. Network design engineers are continuing to search for faster, more efficient methods of transporting data and media from one point to another. Ever since the invention of the telegraph, we've continued to advance telephone technologies with the goal of communicating between sources at maximum bandwidth, data at faster than real time rates, and over physical less cabling and interconnects than the day before.

In order to effectively distribute high-level signals, rich with voice, data or images, from any source to any destination, the communications industry is developing more and more advanced systems for routing, networking and processing. Simultaneous with the distribution (or transport) portion, the imaging industries, including both video- and computer-based domains, are addressing better, more efficient uses of image processing and compression in order to reduce the data requirements. Together, both facets of these industries will aid each other in reaching their goals faster and more effectively.

In 1995, development was progressing rapidly in a relatively new form of data transport. This transport had to that date evolved quickly at its own pace in order to keep up with the quest for interactive distribution of high level media related data.

At that time, and still in present tense, one of the most talked about methods of moving time sensitive data in telecommunications was called "Asynchronous Transfer Mode" or ATM. This data link protocol is used to transport data faster and more efficiently from one point to another, whether it is local or wide area–based.

A GOOD MATCH

Multimedia and ATM are intrinsically a good match—and here's why. What really matters in any media-based network solution is time sensitivity. The best way to get this message across is by a favorite example of early telecommunications.

If we return to the early days of telegraph, these early network communicators had to determine a technique to guarantee that the message sent from point A to point B was actually received. That simple method has carried forward into a terminology (and a technology) that is called "message switching." These techniques employ packetizing the streams of data so that it can be reassembled at the other end and will actually mean something.

In the telegraph days, machine-supported recording and storage did not exist. The primary governing factor became the operator's ability to send (encode) and receive (decode) the dots and dashes that made up the encrypted and otherwise compressed Morse code[33] transmission. To ensure proper receipt meant that each message packet had to contain a source "address" and a destination "address," plus the message (or "payload") itself. The addresses were similar in structure to the headers utilized in modern networking and data coding.

The principals employed in this delivery method aren't much different from the postal service (except they're faster) and have remained unchanged since the days of the Wild West. So long as the operator (a human) keyed in the correct addresses and the line between points A and B was unaffected by outside influences (fire, wind, vandals), the message probably got to the correct destination.

As technology advanced, it became necessary to establish a path (or "circuit") at the onset of any data transmission. This was fine as long as you were only sending data from one site within a city (such as from the Western Union office) to a similar and unique site in another city. Privacy and security were unheard of because everyone who was on the "wire" had to pay attention to recognize the address portion (the header) of the message to determine who was actually supposed to receive the message.

BRANCH AND SWITCH

Complicate this with a "branch" at some point in the wiring chain, such as a turn at some city to reach another, and this required the operator to make a "switch" so that the routing was correctly established. At first, in

---

[33] Morse code was one of the earliest examples of compression. The code assigned the least number of dots or dashed to the most frequently used letters in the English language alphabet. This "compressed" the amount of data (dots/dashes) the operator needed to send and sped up delivery of information over the telegraph wires.

268 Video and Media Servers: Technology and Applications

its crudest implementation, this switch was accomplished with a mechanical knife switch, made by a human within a telegraph office as the human decoded the address portion of the message prior to the actual message being sent.

As history and technology advanced, switching at branches advanced from the mechanical/human interface to electromechanical relay contacts to solid state electronics. Today digital multiplexing and now even optical switching employed in huge fabrics that resemble the weave of a large grid of electronically routed signals. This method became the root of what is referred to today as the "transfer mode."

As the telegraph moved aside for the telephone, it became apparent that a dedicated circuit needed to first be established and then maintained throughout the duration of the call. This was at first a manual patch panel and later used techniques we'll describe as *circuit switching*. Most of these operations did not change for several decades.

As the population and usage of the telephone grew exponentially, ESS or digital switching eventually followed. To preserve wire and bandwidth, it became advantageous to chop up the time periods to better utilize the physical cabling within the telephone network itself.

A new concept called a "point-to-point connection" began to evolve. The nodes where data either enters or leaves a processing point in the network are referred to as points, making point-to-point connections synonymous with node to node processing.

TIME SLOTS

The practice of utilizing time-slots became the next step in digital information transportation. Transmission between nodes is at 64 kb/s. This is determined by multiplying the inverse of the period times the 8-bit data to get a throughput of one 8-bit word every 125 µsec (microseconds).

$$\frac{1}{125~\mu sec} \times 8\text{-bits} = 64~\text{kbps}$$

TIME SLOT: EACH 125 µsec SEGMENT

Figure 24-1: Transmission time slot.

Introducing Packets 269

The process of "time division multiplexing" (TDM)—a principle familiar to those who truly understand how Betacam's chroma multiplexing works—allows several time slots to be interleaved, or placed together, into a unit called a "frame:. The frame is made up of several time slots. The frame is presented at a prescribed frequency so that a reference clock can be properly synchronized at both ends and used to *demultiplex* (disassemble) the frame's time slots and ultimately reassemble the packetized data in the proper order for decoding.

In order for this transfer mode to be effective, the traffic (data streams) must occur at regular intervals. This same concept is not unlike what video must do with respect to 3.58 subcarrier (analog NTSC), $4f_{SC}$ (14.3 MHz in composite digital) or in 4:2:2 sampling (27 MHz in component digital). The concept for relaying data in this form, and its appropriate hardware, is referred to as "frame relay."

Things are quite different for digital video. Even with compression, the transmission rate for video must be orders of magnitude higher than for telephony or voice. Therefore, this basic concept is great for digitized data, but it fails miserably for high quality-audio, video or higher data rates. We'll see why shortly.

Voice can be digitized using a process called pulse code modulation (PCM). Traffic of this nature occurs on a regular basis and thus is referred to as "isochronous", a word meaning equal in length of time. In circuit switching technology, the connection necessary to transmit PCM voice data is established and maintained throughout whether there is traffic there or not.

Silence, such as the space between words in the spoken sentence, is lost space and can be quite valuable. Leaving nothing there (no data) during these spaces is a waste of valuable bandwidth. When a specification states "bandwidth efficiency," it is thought of as the ratio of space that could be used compared to the space that actually was used to transmit the data. Therefore, PCM and circuit switching of this nature are considered bandwidth-inefficient.

Remember this basic concept, because it is the secret to understanding the principles behind maximizing the uses of the available transmission bandwidth. This principle is most important in moving high volumes of data such as digitized voice or video.

Data is somewhat "bursty" in nature. This means that it comes in short groups with time periods where there is no transmission of useful

270 Video and Media Servers: Technology and Applications

data. In this case, data need not be "synchronous" (having the same time periods between occurrences) in order to be intelligently reassembled on the other end.

These reduced requirements promote the ability to packetize data into small bundles and find time slots for these packets to embed themselves into the transmission path itself, without respect to timing or actual order. This is the key to this kind of transmission. Time and order are not necessarily important to collecting the data on the other end.

Packet switching is part of the transfer mode process that ensures that resources are occupied only during the bursty periods instead of continuously as in circuit switching mode.

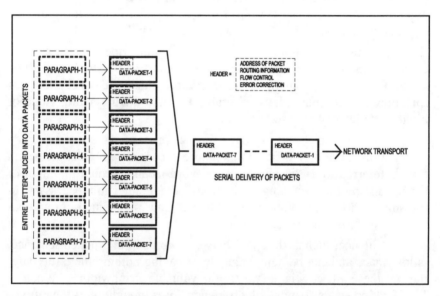

Figure 24-2: Letter example depicts packet assembly process.

To understand packet switching in a non-network perspective, we'll use an example of how information might be transmitted via snail mail—a.k.a. the post office. Say you had a lengthy document that had to go from point A to point B but you were restricted to only one paragraph per envelope. One method to get it there would be to cut the document up into pieces (Figure 24-2) and place it into individual envelopes, call them *packets*. At the start of each cut-up paragraph, you would place information that describes how to reassemble the document once you got all the envelopes at point B. This information is called the "header." This process is not unlike what a word processor can place at the top of each page or insert transparently in the document to allow pagination.

In the data domain, a header includes the information necessary to determine the routing and flow control for the packet and usually some form of error correction (EC).

Just like the post office which picks up mail at random time intervals and delivers it in much the same way, in a network transport, these packets arrive at random time intervals, or in bursts. Some bursts can come in large or small bundles; it varies depending upon how much "mail" came to the carrier at what time.

Here's the problem: Data can tolerate delays in delivery because for the most part, nothing is done with it until all the data arrives and it is reassembled with the aid of the CPU. Voice and video cannot tolerate time delays and as such, packet switching becomes unsuitable for real time traffic applications.[34]

THE NEXT STEP

Enter another form of transfer mode. A technology that is designed to carry voice and data is called broadband integrated services of digital networks (B-ISDN). Even certain video services are possible, but not with the quality levels we'd expect for broadcast use.

These network services offer different speed capabilities depending upon the service you specify or the bandwidth of the transmission system it is passing over (Table 24-1).

Frequency	Information Type	Relative Speed
<3 kHz	Voice, telemetry and control	Low-speed data
>3 kHz to 15 kHz	Hi-Fi and video telephony	Medium-speed data
>15 kHz	Low-resolution video	High-speed Data

Table 24-1: Information classifications.

---

[34] Since this article was written, in March 1995, a good deal of work has been done to circumvent this process for transmission on the Internet and via other fabrics. In most cases, a flexible buffer has taken up the random bursts and allowed the computer to reprocess the streams of data into intelligible mostly continuous structures.

The complications are that voice and video require both fixed rates and fixed delays. Packet switching poses problems because of variable delays and flow controls that regulate data rates at anything other than a constant. The only solution is to have the service provider install a dedicated lease line, a solution with long-term cost impacts that should be carefully weighed.

Trying to fully understand all the details in asynchronous transfer mode (ATM) is a little like trying to understand all the levels and profiles within each flavor of MPEG. In 1995, debate continued, but the prospect was that eventually ATM would simultaneously support several different kinds of traffic and will be tailored to specific objectives.

The development of ATM, like that of many other network technologies, is relatively recent. As early as 1991, the ATM Forum, an industry organization, was formed to orchestrate the deployment of ATM technology on an international level. By mid-1993, there were four classes of traffic being defined and supported. ATM is still considered a long-term solution to the transport of information that will make asynchronous delivery of time sensitive data a reality.

In our next chapter, we'll look inside ATM to uncover the parallels in other packetized delivery technologies and our own developing digital video infrastructure.

CHAPTER 25

INSIDE ATM

In our last chapter, we introduced you to the concept of packetizing information for transmission over networks or data systems. We were concerned about the time sensitivity of delivering that information, and had begun our introduction to asynchronous transfer mode with some broad definitions and fundamentals.

This chapter will study ATM at the detail level. We will look at classes of services, layers, headers, and internal workings of the cells themselves.

Circuit switching transmission assumes some level of synchronicity. As we stated in our introduction, in order to make efficient use of available bandwidth, data is chopped up and routed in an out of sync or "asynchronous" way. ATM, as a data link protocol, uses asynchronous methods based on the concept of "cell switching." Cell switching combines the better parts of circuit switching and another feature called "fast packet switching."

Packet switched data networks have variable delays and use flow control mechanisms to regulate their data rates. By allowing a set of classifications to be conveyed to the network switching system, control parameters that state the importance or sensitivity of timing can determine proper delivery options. Such options may let some time sensitive data pass through at a higher priority than other types of data for which delivery timing is not particularly sensitive.

ATM is a technology that strives to provide industry with a single network concept that will support different kinds of traffic and their respective idiosyncrasies. To accomplish this, ATM supports four classes of traffic (see Table 25-1) that permit the different levels of importance for traffic, as described in the preceding paragraph.

Class	Data Type	Applications
A	Constant bit rate, synchronous	uncompressed voice and video
B	Variable bit rate, synchronous	compressed voice and video
C	Variable bit rate, asynchronous	X.25, frame relay
D	Connectionless packet data	LAN traffic, SMDS

Table 25-1: ATM traffic classifications.

Digital transmission has been proven to show high reliability provided the transmission medium is of high quality. Fast packet switching makes use of these properties and reduces the need for error correction at each node in the network.

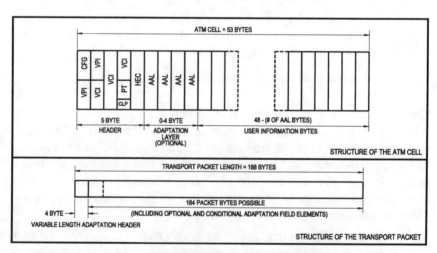

Figure 25-1: ATM cell and transport packet.

ATM uses small, finite-length packets called "cells." In our post office example (Chapter 24), we know that no two paragraphs are exactly the same length. This means that those packets have variable lengths, contained different weights of information (words) and might wind up in different-sized envelopes depending upon the size of the paragraph.

In ATM, no packet will be larger than 53 bytes—with 5 bytes being designated for the header and 48 bytes for the user information (payload). It has been mathematically determined that the probability for long delays to occur is greatly reduced because all the packets are both small and of finite length. The details of the 5-byte header that precedes the AAL are described in Table 25-4 and shown graphically in Figure 25-2.

The cell structure and the transport packet (Figure 25-1) are the primary elements of ATM. You will see later how and why ATM fits into the model of compressed digital delivery when we demonstrate how MPEG-2 maps into the ATM structure.

CONNECTION-ORIENTED STRUCTURE

ATM also requires that before information can be transferred from source to end user, a connection must be established. This defines ATM as "connection-oriented"—but it extends this to being either a logical or a virtual connection. Further, the network reserves these resources for the duration of the connection, and should the resources be insufficient, the connection is terminated or possibly rerouted.

Generally, the protocol for ATM follows an ISO seven-layer open systems interconnection (OSI) model. Since the ISO model is a complex structure, the full discussion of this model will not be included.

Layer	Layer Name	Description
1	Physical Layer (PHY)	Defines physical interfaces and framing protocols
2	ATM Layer	Specifies cell structure and flow of cells over logical connections
3	AAL Layer (ATM Adaptation Layer)	Segmentation and reassembly of variable data, timing control, detection and handling of lost or out-of-order ATM cells

Table 25-2: ATM base layers.

Within the ATM, there are three layers (Table 25-2). The physical layer (PHY or Layer 1) is segmented into two sublayers—necessary in order to decouple the transmission properties from the physical properties. In essence, to accommodate the various physical media expected in host to network interfaces.

The two sublayers of the PHY consist of the transmission convergence (TC) and the physical medium dependent (PMD) sublayers. PMD is used for bit timing and interface to physical media, and TC is responsible for a number of other functions (see Table 25-3).

Transmission Convergence Sublayer (PHY Layer)	Description
Transmission System Adaptation	Receives cells, packages into proper format
Cell Delineation	Extracts cells from bit stream
Cell Scrambling/Descrambling	Encoding for noise control
Cell Rate Decoupling	Inserts (or suppresses) idle cells to (or from) the payload—provides for continuous flow
Header Error Check (HEC) Generation/Verification	Uses CRC to determine error level

Table 25-3: TC sublayer of the ATM physical layer (PHY).

The TC calculates the HEC, which is a kind of automatic traffic cop. The TC compares bits received against the HEC value of the received cell. If the values match from cell to cell, then cell boundaries are correct—and if there is no match, then it is known that the cell delineation has not been found.

ATM LAYER AND HEADER FUNCTIONS

The ATM layer performs four functions. First, it must multiplex or demultiplex all the various cells from different connections. Then it must translate the virtual circuit identifiers (VCI) and/or the virtual path identifiers (VPI) at the points where switches or cross-connects occur. The third function is the extraction or insertion of the 5-byte header before or after the cell is sent to or from the adaptation layer. Fourth, the flow control mechanism at the User Network Interface (UNI) is implemented using the General Flow Control (GFC) bits of the header.

The ATM cell header (Table 25-4) is the key to getting the information through the network's highway. This 5-byte prefix is much like today's modern nine-digit postal zip code, except it involves dynamic calculations and basically occurs in real time. The ATM header is presented such that bytes are transmitted in increasing order and bits are transmitted with MSB (most significant bit) first, which is decreasing order.

GFC (General Flow Control):	Traffic controller across the UNI, prevents short-term overload conditions.
VPI (Virtual Path Identifier):	Depending upon whether it addresses the UNI (8 bits) or the NNI (12 bits), this is used to identify the virtual paths called up in routing the signal from source to end user.
UNI (User Network Interface):	The system that connects the host to the network
NNI (Network Node Interface):	The system that connects nodes in the network to additional systems within the network.
VCI (Virtual Circuits Identifier):	A 16-bit field used to identify virtual circuits.
PTI (Payload Type Identifier):	A 3-bit field used to identify whether the cell contains user or management information, as well as an indicator of network congestion.
CLP (Cell Loss Priority):	In resolving the state of network congestion, this identifies the importance of any given cell. This single bit is set by the AAL. When set at 0, the cell cannot be discarded, and if set at 1, the cell can.
HEC (Header Error Check):	Using CRC (cyclic redundancy code), this last byte in the header is computed over all the fields in the header as an error detection indicator.

Table 25-4: ATM cell header functions.

Each field consists of five bytes of eight bits in length each. A description of the functionality of the various portions of this header is provided in the following figure.

ATM ADAPTATION LAYER

With the header and the ATM layer described, the next requirement is to map the user, control, or management protocol data units (PDUs) into the 48-byte payload field of each ATM cell. Recalling the three classifications of services (shown in Table 25-5), the ATM Adaptation Layer (AAL) is charged with the task of transporting these different services over the ATM layer.

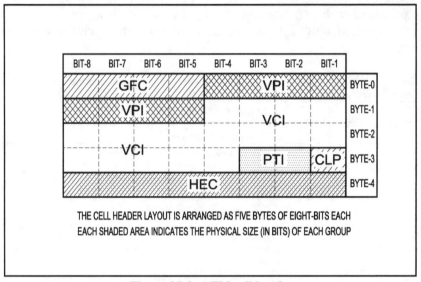

Figure 25-2: ATM cell header.

Criteria	Services
Time Sensitive	Video, voice, interactive real-time services
Bit Rate	Constant or variable services
Connection Mode	Connection or connectionless oriented services

Table 25-5: Information services classifications.

The AAL also has two sublayers, much like the TC and PMD sublayers in the PHY layer. Here, the AAL consists of a sublayer that segments and reassembles the cells, called the SAR (segmentation and reassembly) layer; and the other sublayer that performs multiplexing, cell loss detection, timing recovery and the like, called the CS (convergence sublayer).

There are five levels of AAL protocols (Table 25-6) that provide service to four classes of traffic. These levels are classified as AL-1 through AL-5 and have specific functions and designators that make them applicable to the specific types and functions of information being handled in the network.

Level	Data Type	Application
AL-1	Constant Bit Rate Data	Audio, Voice, Uncompressed Video
AL-2	Variable Bit Rate Data	Compressed Video
AL-3 or AL-4	Bursty Data	Non–time sensitive
AL-5	Short duration grouping (employs shorter header)	LAN uses, host to host

Table 25-6: ATM adaptation levels (AAL).

Additional AALs have been proposed to address the growing needs for data rates or applications that do not fit into AAL-1 through AAL-5.

VIRTUAL CIRCUIT SETUP

Unlike connectionless data service (such as IEEE 802—Ethernet), ATM performs routing only when a message to establish an actual virtual circuit is requested and properly formatted. The steps involved in ATM operation require that a connection must first issue a setup request, which is sent across the user network interface (UNI), then on to the network itself. This is similar in function to what happens when a fax modem is connecting to a remote machine and is readying to transmit a fax. In ATM, the request is processed all the way through the network to the end point's UNI, where it is either accepted or rejected, depending upon the available resources that were requested at the time of the setup call request issuance.

Upon acceptance, a "virtual circuit" is set up across the network —hence the function of the VPI and VCI. This effectively means that at both ends of the UNI and throughout all the switching nodes within the network, a virtual path and circuit are now in place, ready for the transmission of information.

Routing information, contained in the VPI and VCI fields of the five-byte ATM cell header, is kept in check throughout the connection process. The virtual channel connection therefore consists of two end or switching points connected together to form the complete channel. The virtual path is a combination of several virtual channel connections, switched together as one unit.

Once the complete virtual path is established, the ability to move information is no longer hindered by the several constraints placed upon connectionless interfaces. It also makes it possible to mix many different types and rates of information into the same medium, by guaranteeing that each can move as quickly and efficiently as possible from user to user.

NO SPEED LIMIT

Inherently, ATM has no upper speed limit. Following SONET[35] hierarchy, speed steps can increase from 155 Mb/s to 622 Mb/s, on up to some 4.8 Gb/s. Since the protocols remain the same, only the NIC (network interface card adapters), the switches or the media need to be changed.

As stated very early in this article, in order for a media server or multimedia application to perform at its peak, the network transit delay (or latency) must be minimized. Video e-mail and video teleconferencing both may have potential uses for ATM. Low-speed ATM networks are also being considered as a method of conversion from 10 Mb/s Ethernet to 155 Mb/s ATM.

Low-cost ATM is also available from some manufacturers. The physical interface for 25.6 Mb/s ATM, which uses NRZI (non-return to zero, inverted) encoding, is nearly identical to 16-Mb/s token ring, which uses a 32-MHz line frequency. The major difference here is that each data bit is represented by a single line transition. Further encoding is also employed where every four bits of data is translated into a five-bit data pattern prior to transmission. This encoding is called 4B/5B and is how the final 25.6 Mb/s rate is determined (4/5 of 32 MHz = 25.6).

MAPPING MPEG INTO ATM

The nature of digital video is one of massive bandwidth consumption. Bandwidth in itself is generally finite for either physical or economical reasons. The previous discussions have dealt with the understanding and makeup of the principles of ATM. To understand why this technology is important, when related to transmission of compressed digital video, we

---

[35] SONET: Synchronous optical network.

need to look into the means by which video will be compressed and delivered to users, servers, and transmission systems in the future.

It is not the intention of this book to spend time on the discussions and constructions of MPEG-2, except where necessary to understand how MPEG-2 will be applied to video storage and transmission. There are more tutorials and white papers on MPEG-2 available today than just a few years ago, and we will leave discovery of the details to the reader.

However, there are a few concepts that we should understand beyond the compression basics of I, B, and P frames and GOP.[36] The MPEG-2 structure is defined by three layers: systems, video, and audio.

The packet structure of MPEG-2 exists at the "systems" layer, which is where the multiplexing of multiple elementary streams (video, audio, and ancillary data) is accomplished. It is at this layer where these streams are combined into one "transport" stream that can be transmitted over a network or connection.

Each of the individually time sensitive streams must be given a guarantee that it can be presented to the decoder (viewer) at a fixed time. These streams are referred to as "packetized elementary streams," or PES packets.

The PES can be of variable length and are mapped onto "transport packets," which are fixed in length to 188 bytes (refer back to Figure 25-1). These transport packets are multiplexed together and create a transport stream (TS). Non-time-sensitive data and private data can be multiplexed at this level as well. The non–time critical data can be multiplexed directly into the transport packets and the private data can also be multiplexed at the protocol stack level.

A packet identifier (PID) is used to uniquely identify the different streams that make up the transport stream (TS). This makes separation of the streams (generally used for unique and different reasons) possible once the stream is finally received and demultiplexed by the decoder.

A timestamp of a program clock reference (PCR) can be carried by the transport packets' optional adaptation fields. The time stamps become the time base for synchronization of the individual frames or fields of a video stream. This PCR systematically provides the timebase

---

[36] I, B, P and GOP are initials and acronyms for frame structure behind MPEG.

282 Video and Media Servers: Technology and Applications

necessary for decoding the streams and allows for clock recovery support.

The PES header (found at the start of the PES packet) carries both a decode timestamp (DTS) and a presentation timestamp (PTS) that will be used to decode and present the video. These two timestamps refer back to the time base originally established by the PCRs.

Audio and other types of data streams, whether synchronized or not, can utilize their own DTS and PTS that can refer back to video's information and in turn synchronize all the streams to the video timebase. In this way, PES packets (elementary streams) can all be synchronized to each other even though they have been split up, packetized into different streams, and sent in varying order within the same transport stream.

Single- or multiple-program transport streams can make up the MPEG-2 transport stream. Inside sections of the transport packets is information that is necessary for a receiver to demultiplex the streams and parse out the stream of particular interest—letting the other stream(s) be ignored. The program specific information (PSI) is the result once the proper selection is made.

The program association table (PAT) is the first and highest level table and is always carried in packets of PID equal to zero (PID 0). PAT carries the program numbers for current programs in the transport stream multiplex. PAT further provides the mapping between program numbers and their associated elementary streams.

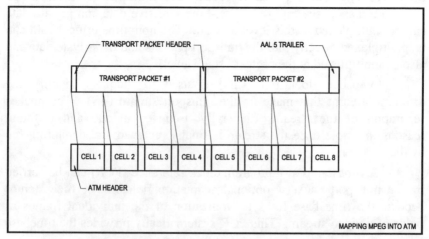

Figure 25-3: Mapping of MPEG into ATM.

Returning to our discussion of mapping MPEG-2 into ATM, the AAL-5 layer is used for this function. To understand the mathematics behind this process, we know that the payload of an ATM cell is 48 bytes (53 total bytes less the 5-byte header). Two 188-byte MPEG-2 transport packets contain exactly 376 bytes. In a sequence of eight ATM payloads (eight times 48 bytes) there are a total of 384 bytes. The difference between 384 bytes and 376 bytes is 8 bytes and is called the overhead or "trailer."

The point here is that MPEG-2 and the 19.39 Mb/s Grand Alliance (GA) system can interoperate with ATM through a basic process of nesting data packets within other segments of a data transport stream. This concept is being extended to several new facets of data transport for digital broadcasting. Not only can the principles of MPEG be extended, but soon we will see AES3, SDI, DVC, DV and other audio/video specific data structures making their way into new transports for distribution within and between facilities—thus becoming the next generation of data carriage for television.

WHERE DOES IT ALL FIT?

Finally, we stated that there is a reason for understanding what ATM is all about. Recall the statement about time sensitivity, referred to as "isochronous" information. Video servers are best thought of as specialized computers that deliver a constant stream of data in the form of audio and video information, essentially on a demand basis. Until a viable network system, such as ATM, is firmly in place, the ability to deliver all the possibilities and expectations of the vast amount of media information will be awkward, constrained or unpredictable at best.

ATM has been conceived as a solution to a variety of obstacles, many of which were discussed in this chapter. There are varying views on the potentials and life cycle of this ATM technology.

ATM, which is structurally a technology based upon telephony and networking, has been designed and influenced by an industry that has always had an alternative to isochronous data distribution. The principles employed in ATM seem, if nothing else, to be a domain that the traditional first-generation video engineers are just beginning to comprehend. As this and other new technologies begin to be implemented into broadcast plants, the systems will probably be configured by manufacturers or system integrators familiar only with the specific applications they are installing.

ATM, like other transports and the multitude of applications for compression technologies, will continue to develop—and with that growth, new choices will need to be made. ATM may be just one of the solutions to the network transport bottleneck. It may also be the method of bringing the dreams of the digital media pioneers to reality.

CHAPTER 26

RAID REVEALED

As information and storage technologies continue to span the entire spectrum from PCs to mainframes, RAID applications for the storage of digital media are being implemented by every major broadcast equipment company involved in producing media server products.

The concept of RAID has been widely used and abused. Early in the video server's entrée, RAID was used as a marketing tool rather than a technology per se. Confusion remains prevalent when trying to understand the numbering system applied to RAID levels. This has resulted in the myth that "the higher the RAID number, the better the performance."

This chapter begins our look into RAID. Throughout the next three chapters, we will detail the history, functionality, protection, reliability, availability, and fault tolerance capabilities of storing data on subsystems known as "redundant arrays of independent disks."

The reliability and the availability of data are the two most important requirements of a storage system. The more bulletproof you can make any storage system, the more reliable it becomes. The availability of data, on the other hand, implies that data can be stored and recovered in the most rapid, accurate fashion, and with a minimum of outside activity or interaction. Both of these requirements are essential to good system performance, whether that data is intended for the home PC, the mainframe computer, or the mission-critical media for a client's commercial that can air only once during the Super Bowl.

Since the beginning, video media, whether stored on magnetic videotape, optical or magnetic disk drive surfaces, has attempted to prevent corruption (unreliability) and make rapid access (availability) a feature set. As technologies in disk drive recording improve, we continually try to find more reliable and more efficient methods of storing the digital data.

The most recognized architecture for the storage of data, including media, is characterized by a well recognized, but generally misunderstood acronym. RAID, which stands for "Redundant Array of Independent Disks," has become one of the predominant and most recognizable of the disk storage technologies. While the technology is well entrenched in the computer data industry, RAID *terminology*, when used in the broadcast server context, seems to remain a source of confusion. The misconception about RAID numbering continues even after several years of explanation and implementation.

In order to focus on how this technology can be applied, a foundation of the terminology is required. We must understand what the properties of a media storage subsystem should be. We must understand why RAID is important and what the various architectures can do when applied to media servers in general.

THE IDEAL STORAGE SYSTEM

The builders of disk drive storage systems have a profound propensity to search for and achieve perfection. An obvious ideal storage system would be one that would possess all the properties of large scale storage expansion, easy upgrade and total assurance of data protection, and have complete, self-correcting monitoring—including fault tolerance and backup—while being economical, scalable, and extensible.

With that said, we all recognize that in order to achieve this nirvana, a level of practicality surfaces that necessitates trade-offs and compromises across the system.

To narrow the field slightly, the ideal disk storage system should mimic any individual disk drive that it replaces. The device, regardless of its size and complexity, should appear to the primary host system (the CPU) as an identical to any other disk in the system.

Downtime of any subsystem should be nonexistent. When any individual component fails, there should be no loss of data. In essence, the system should possess the capabilities to detect a fault, continue operations uninterrupted, signal that a failure has occurred (and pinpoint the failed component), allow for on-line repair, and finally implement a transparent, unattended rebuild of the data structure once the failed component has been replaced.

Replacement of failed components or the routine servicing of any system should be accomplished without interruption. The term "hot swappable" is applied to the remove and replace process whereby a drive can be taken out of service and a new one installed without shutting down or rebooting the system. The ideal storage subsystem should eliminate the need for unscheduled maintenance. This infers that maintenance can occur entirely at the user's convenience.

The popularity of RAID has increased for many reasons. One of the primary reasons is the continuing reduction in cost of and the growing availability of larger drives in the same physical form factor.

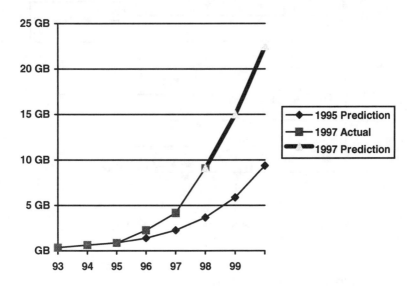

Figure 26-1: Growth in drive capacity.

In 1995, growth of drive capacities was predicted at 60 percent per year. That prediction was clearly in error, as Figure 26-1 shows, and

288 Video and Media Servers: Technology and Applications

now it is predicted that we will see drives exceeding 20 gigabytes before the end of the decade.

FASTER, BIGGER, CHEAPER

The plummeting prices for hard disk storage have made arrays more viable, thus encouraging the development of new storage capabilities, especially related to RAID. This trend is continuing as new drive technologies, new materials, and better software adaptations allow for faster, higher-density, and higher-capacity drives. Recently Moore's Law, which had previously applied to only silicon-based processor technology, was said to be closely approaching a new material product line—the hard disk drive. Disk drive prices and respective performance, are on the same kind of trend line as we've seen in computer power over the past 10 to 20 years (Figure 26-2).

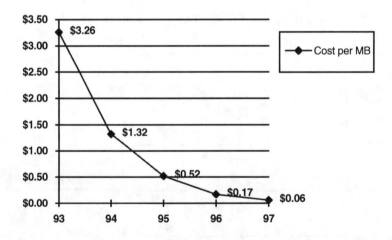

Figure 26-2: Disk drive prices per megabyte.

For definition, a "magnetic disk array" is a group of individual disk drives assembled to appear as one storage element or as a single drive. To meet the requirements for the ideal storage subsystem, you must insure that any drive array, made up of N drives, will mimic the disk system it replaces. In other words, the entire array must appear as one single large disk.

Next, a single drive failure in the array of N drives must not appear as a failure to the host processor and should continue to respond without compromise. The array must be capable of repairing itself without seriously degrading ongoing operations.

Finally, the user must be able to field replace any failed drive when it needs to—again without degrading or compromise of operations.

HISTORICALLY DEFINING RAID

In a 1988 paper, "A Case for Redundant Arrays of Inexpensive Disks," presented by David Patterson, Randy Katz, and Garth Gibson, the concepts for five methods of data protection on disk arrays were outlined. The methods were aimed to guard against data destruction due to failure of drives, and described techniques for mapping (or striping) data across arrays and their associate performance benefits.

The commercial development of the concept of RAID began its in 1989 when the first "nonmirrored" version with commercial viability was announced by Compaq. By the middle of the 1990s, the concept had become a virtual household word.

When first introduced RAID, stood for "redundant arrays of inexpensive disks," but the word "inexpensive" has since been modified to "independent," presumably because the word "inexpensive" is relative and the cost of all storage has literally become inexpensive over time. Although RAID is still considered an emerging technology, the principles employed are relatively simple and, by computer life spans, are not necessarily new.

The numeric numbering for RAID was an outgrowth of the first publication and was adopted over time by people to indicate certain "types" of protection and mapping schemes. In the researchers' first publication, only five "levels" were described, but a sixth form of protection technique was later added.

The primary principle advantage behind the RAID concept is that data is spread across a number of individual disks in such a manner as to provide an element of redundancy. Data protection is provided by the way that redundancy data is stored separately from the primary data. The redundant data may be a complete copy of the data or additional information that can be used to reconstruct the primary data, should some element of the hardware fail.

RAID makes it possible for multiple data records, when accessed simultaneously, to be located and recovered faster because elements of the records are spread over a wider area. Depending upon the RAID architecture, the rate at which data is recovered or deposited onto the disks can be substantially increased over single disk systems.

RAID technology can be applied to stand-alone independent subsystems for computer data, image storage, media servers and systems designed for video-on-demand. With the cost effective availability of small form-factor disk drives, promoters of RAID state significant improvements in reliability and maintainability. RAID subsystems also provide for higher performance than individual magnetic disk drives.

The opposition states that RAID can be expensive and unnecessary and that there are alternative methods to similar performance at a better cost point. Those arguments both have valid points bill will not be the focus here.

RAID ARCHITECTURES

The term "architecture" is meant to describe the physical structure of the elements in a RAID storage subsystem. The term "subsystem" is used because, for the most part, RAID is intended to be an element of the overall system that is comprised of a host (i.e., a computer and system drive), an input/output engine (i.e., a video compressor/decompressor) and a storage array of some significance.

Throughout this chapter and this book, we will use the terms RAID (followed by a integer number) and RAID Level (also followed by an integer number) interchangeably and with the same meaning. This is because some groups will say, for example, RAID 3 and others will say Level 3, and others will say RAID Level 3. It all has the same meaning and should not be confused.

However, one point should be made (and it will be reiterated throughout): the higher or larger the RAID "number" does not mean the better the performance. This will be more than evident by the end of these chapters.

SETTING RAID LEVELS STRAIGHT

To distinguish between different architectures of RAID, a number scheme was developed that was intended to identify the various combinations of drives, striping, and parity. In the early development of RAID, some expressed concern that the use of the term "RAID X," sometimes referred to as "Level X" (where the "X" is generally a number from 0 to 5),[37] as an attempt to capitalize on the incorrect notion that the higher the RAID level, the higher the reliability or performance.

Two organizations made up of industry and education have rallied to provide research and a set of guidelines and standards, that help to sort out the variances, definitions, and testing validity of RAID storage subsystems. RAID levels that are recognized by the original Berkeley papers and the RAID Advisory Board (RAB) have been categorized as RAID Level 1 to Level 6.

Other levels recognized by the Berkeley papers, but not by the RAB include Level 0, disk striping where data is mapped in stripes across the entire array. Combinations of existing RAID levels, sometimes called hybrid RAID Levels, include Level 53, a combination of Level 0 and Level 3 that combines disk striping and the features of Level 3 parity. Level 10, another combination, mixes disk striping and mirroring resulting in excellent I/O performance and data reliability.

Some have openly asked if this leads to confusion among the end users or if it is contrary to generally accepted industry practice. The formal answer was still under discussion in April 1994, but *The RAIDbook*[38] and the RAID Advisory Board (RAB) suggest that, "RAID levels be chosen with numbers that impart meaning, not confusion". So far, at least for much of the broadcast industry, the confusion surrounding RAID nomenclature continues.

Our next chapter will thoroughly cover the meanings of various RAID levels.

---

[37] Additional individual RAID numbers (such as RAID 6) and combinations of RAID numbers (such as RAID 53) are now also used (refer to Chapter 27 for more information).

[38] *The RAIDbook*, with at least five editions, is an industry publication that continues to evolve over time to reflect the "state of the practice in storage systems." The latest edition is available through the RAB (RAID Advisory Board).

CHAPTER 27

RAID BY THE NUMBERS

In 1993, the first edition of *The RAIDbook* was published by the RAID Advisory Board. This book formed part one of the RAID Advisory Board's objectives—the standardization of terminology for RAID-related technologies.

In 1997, the RAID Advisory Board (RAB) had so far recognized nine RAID implementation levels. Of the nine levels, five conformed to the original Berkeley RAID terms that were developed from the 1988 researcher's efforts. Beyond those five, there are four other RAID terms that are used and acknowledged by the RAB.

RAID 0, RAID 6, RAID 10, and RAID 53 were developed through committee work anchored by manufacturers, suppliers, and consumers. These additional extensions will be explained at the end of this chapter.

A drive array cannot be created by simply connecting a series of SCSI drives to a single SCSI controller. An array must consist of electronics that format, code and distribute data in some structured form across all of the drives. A RAID system consists of a specialized set of electronics and instructions that operate in conjunction with the various drives to perform protective and fault tolerance functions necessary to meet the level of RAID designated.

Most servers, whether on a Windows NT platform or for a digital video server, employ some form of RAID. This chapter will be concerned with hardware RAID control, although software RAID is available on some video servers in the form of patented and proprietary subsystem controllers (e.g., Leitch/ASC Video Corporation's VR-300 which uses a proprietary FibreDrive system to interconnect Fibre Channel disk drives).

Table 27-1 should be referred to as you read the explanations and details of RAID levels and terminology.

RAID 0: DATA STRIPING	
	Data is striped across multiple disks. No protection.
PROS	High data reads.
CONS	Lower reliability, no redundancy, no error correction.
RAID 1: DISK MIRRORING	
	Separate independent disks. All data is duplicated. No drive sharing.
PROS	Highest data reads and reliability.
CONS	Drive costs are doubled due to total redundancy.
RAID 2: HAMMING CODE REDUNDANCY	
	Data striped across multiple disks. Errors are detected and corrected in RAM or on a separate disk.
PROS	Transfer rates high.
CONS	Not commercially available. Low input/output request rates.
RAID 3: STRIPED ARRAY PLUS PARITY	
	Drives operate in parallel synchronization. Data is striped byte by byte across multiple disks. Separate parity-only drive which stores all redundant data.
PROS	I/O performance good—especially for large block transfers.
CONS	Slower read/write performance. Single I/O request execution. If parity drive fails protection is lost—but system still operates.
RAID 4: INDEPENDENT STRIPED ARRAY-PLUS PARITY	
	Data striped across all disks, separate parity-only drive, disks can work independently of each other.
PROS	Good large data I/O performance, good read performance.
CONS	Not widely available. Poor write performance. If parity disk fails, data protection is lost.
RAID 5: INDEPENDENT STRIPED ARRAY, DISTRIBUTED PARITY	
	Data and parity is striped across all disks.
PROS	Data reliability equals mirroring. High read performance.
CONS	Poor write-performance due to striping.
RAID 6: INDEPENDENT STRIPED ARRAY-DOUBLE PARITY	
	Data and parity is striped across all disks. Second parity drive is added.
PROS	RAID 5 features with redundant parity drive. High I/O and performance and reliability.
CONS	Poor write worse than RAID 5.

Table 27-1: Common RAID levels and descriptions.

RAID LEVEL 0

RAID 0 is also referred to as "striping," where all data is striped (distributed) across an array of disks without providing redundant information or parity. The striping of an array, although enhancing performance, especially in high transfer rate environments, has one serious downside. Since there is no redundant or parity information recorded and there are no redundant disks, the failure of any one drive results in the total loss of all data on the drive array.

The term "RAID 0" is seldom used even though the principle is commonly practiced. In earlier times, RAID 0 referred to the absence of any array technology. The RAB states that the term implies data striping—a means to evenly distribute data so that during the read request, the blocks can be rapidly recovered at random with a minimum of latency. Striping also is an efficient means to gain large amounts of storage by just adding drives.

Video servers, image storage arrays, and video disk recorders that do not want to incur the higher cost of RAID controllers or added parity drives, and do not desire the other fault tolerant properties inherent in RAID, have been using this concept for many years.

A familiar example of one form of striping was the early VDR (video disk recorder). The drives that made up the early 25-second (NTSC) and 30-second (PAL) recorders used one or more pairs of large form-factor disks to store chroma and luminance samples as data on a field by field basis. It was not uncommon to find, for example, one of the chroma-channel drives in the pair of drives that might develop a problem, appearing as a small dot (pixel loss) or flash frame. One could identify this problem by a frame (or field) that showed a loss of either chroma or luminance—but with paired drives, you would seldom find both samples missing simultaneously.

Fortunately, the formatting of the these early video disk arrays placed the respective chroma and luminance samples in the same location each time the recording process was initiated, which meant the user only lost that particular frame where the media was unrecoverable. The user could sometimes map out the bad frame or, in some drives, remap a portion of the individual drive itself, so that this bad frame would not interfere with operations.

While technically this principle is striping or RAID 0, it is more properly force-formatting the drive with data spread out over the drive in a repeatable pattern. Since it was always known where the video field was placed, and that the drives dispersed the data evenly around the drive pairs, instant and random playout of the video frames was always possible.

296 Video and Media Servers: Technology and Applications

Other uses of striped arrays were found in earlier graphics arrays when the used required large amounts of individual frames, usually written as single rendered frames that would later be played out in real time as video animation. In the early days of this process, the animation clips were usually relatively short sequences and the need for redundancy or protection of the data was far outweighed by the cost of the additional drives. In most cases, a backup tape drive was used as the redundancy form-factor.

RAID LEVEL 1

RAID 1 is the simplest, most reliable, and easiest of the RAID architectures to implement and to understand. All data is continually duplicated and managed as a redundant copy, residing on at least one separate drive (see Figure 27-1).

Level 1 is referred to as disk mirroring, shadowing, and duplexing. In all cases, when data is stored on two or more separate drives, total redundancy is achieved. In Level 1, both drives will operate in parallel with the host CPU reading from and writing to both (or all) drives simultaneously.

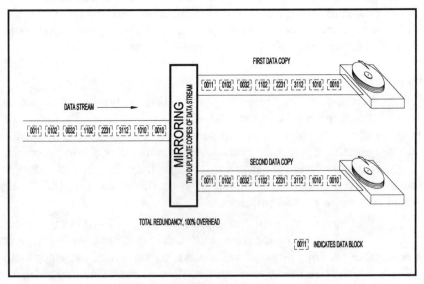

Figure 27-1: RAID 1—mirroring.

This approach also provides for a continual backup of all the data—thereby improving reliability while providing a high level of availability, especially when reading data. If one disk component fails or is replaced, all of the information can be restored without any down time because it has already been backed up.

RAID 1 writes as fast as a single disk. The systems are typically single I/O and nonscalable, which means that performance is not increased by adding disk elements to the array.

Some believe that for top performance and superior fault tolerance, RAID 1 solutions are the best choice. Yet the prime drawback to RAID 1 is that complete redundancy is also the most expensive cost per unit of storage to implement, since a duplicate drive is required for the each application.

RAID 1 is not prominent in servers designed for video purposes only. Here, the term "mirroring," in the world of video, is typically accomplished by providing two complete and independent video server systems that operate with an automation system or sequencer and always run in parallel.

Another form of non–RAID mirroring would be running a cassette-based videotape system (e.g., an LMS) in parallel and in sync with the server. This too is expensive and somewhat impractical given the number of other options available with today's modern server technologies.

RAID LEVEL 2

There are no commercially available implementations of RAID 2. This method uses error checking, such as Hamming codes, to produce higher data transfer rates. The technique uses RAM to detect and correct errors.

RAID 2 interleaves bits or blocks of data, and the drives usually operate in parallel. Typically the drive spindles are synchronized. The system will use redundant disks to correct single bit errors and detect double bit errors. Error correction algorithms determine the number of disks required in the array.

The concept behind RAID 2 assumed that mass storage devices, such as disk arrays, are prone to constant disk errors. This might have been true of larger parallel transfer disk drives, but smaller form-factor drives have proven much more reliable.

RAID 2 benefits are performance. Because their design is parallel in nature, this level, and RAID 3, provide the highest performance of all RAID levels. However, the drawbacks include low input-output request rates and low acceptance in commercial applications.

RAID LEVEL 3

RAID 3 is an array of N-drives with data bit- or byte-striped across all but one of the drives. A single dedicated drive contains redundancy data which is used to mathematically reconstruct the total data stream in the event any one member drive fails. The redundancy data is placed on a drive called the parity drive (see Figure 27-2).

RAID 3 employs parallel data access and has the advantage of extremely high data rates, which will increase performance over single disk structures. Each disk in the entire array is used for each portion of the read/write operation.

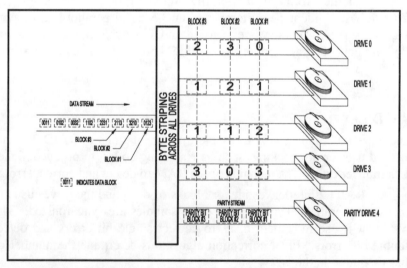

Figure 27-2: RAID 3 —striping with separate parity drive.

In a RAID 3 array, all of the disks are synchronized so that both read and write performance can be hampered. Each logical record from the data stream is broken up and interleaved between all of the data drives except the parity drive. Each time that new data is written to the data drives, a new parity is calculated and then rewritten to the parity drive.

RAID 3 is a preferred choice when large blocks of sequential transfers are required. Applications in imaging, CAD/CAM, and digital video or media servers, will generally select RAID 3 because of its ability to handle large data block transfers. In most streaming video applications contiguous blocks of data are written to the array so that during playback, only a minimal amount of searching is required. This increases efficiency and throughput.

One disadvantage that results, which is common to all striped arrays, is poorer level of write performance when compared with single or duplex/mirrored drives. This drawback can be controlled by proper buffering and sectoring during the write process.

By definition, the entire RAID 3 array can execute only one I/O request at a time—referred to as single-threaded I/O. This may or may not be important, depending upon the application. Some controllers and smart arrays have minimized this impact by providing intelligent algorithms and larger disk caches to buffer data temporarily while being written to the drive.

Protection is achieved by the use of discrete parity drives, so if a drive goes down, protection is temporarily lost until that drive is replaced and the parity information is reconstructed.

RAID LEVEL 4

This technique will stripe data across all the disks except the parity drive. RAID 4 lets the individual member disks work independently of one another.

Benefits include good input and output performance for large data transfers and good read performance. As with most striping, write times are extended because the data is dispersed across several drives in segments.

However, with RAID 4, where one disk is used strictly for parity, if a block on a disk goes bad, the parity disk can rebuild the data

300 *Video and Media Servers: Technology and Applications*

on that drive. If the file system is integrated with the RAID subsystem, then it knows where the blocks of data are placed and there can be a management control scheme implemented such that parity is not written to the hot disk.

In addition to the expected reduction in write time, another drawback is that extra steps become necessary to update check data and user data. If the parity disk fails, all data protection is lost until it is replaced and the parity drive is rebuilt.

RAID LEVEL 5

RAID 5 is used when independent data access is required. It is characterized by high transaction throughput and supports multiple concurrent access to data through independent actuator access capability. High read to write ratios are most suitable for RAID 5. Typically, the more disks in the array, the greater the availability of independent access.

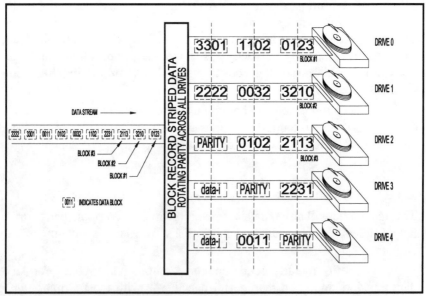

Figure 27-3: RAID 5—block record striping with rotating parity.

RAID 5 uses block or record striping of the data stream. The serial digital data word is spread across all of the disks in the array in block form. Independent access is available because it is now possible to extract the entire specific data block from any one drive without necessarily accessing any other. Latency and seek times are effectively reduced, resulting in performance increases.

As each block of data is written to the array, a rotating parity block is calculated and inserted in the serial data stream (see Figure 27-3). The parity block is interleaved throughout all the disks and can be recovered from any of the drives at any time. Because parity information is rotated over all of the drives in the array, the I/O bottleneck of accessing the single parity disk (see RAID 3) when concurrent accesses are requested is significantly reduced.

When a single drive in a RAID 5 array fails, the read and write operations will continue. Recall that data is block striped over all of the drives, so if the data to be read resides on an operational drive, there is no problem. If data is to be written to the array, the controller simply inhibits writing to the failed drive.

If data resides on a failed drive, the read process uses the parity information interleaved on the remaining drives to reconstruct the missing data. The algorithms that determine where the data resides are quite sophisticated and are statistically tailored to allow nearly transparent operations during failure modes. The host never sees the failed drive and the entire array remains transparent.

RAID LEVEL 6

RAID 6 adds a second parity drive to the structure of the RAID 5 array. The benefit is that two drives may fail without a loss of data. In the event one drive fails, the entire array can continue with a second level of protection while a replacement drive is found and installed.

With two parity drives, every write operation requires that two parity blocks be written. This makes write performance very low but keeps read performance on a par with RAID 5.

RAID LEVEL 10

RAID 10 is confusing to some, strictly from the terminology that it is made from. Some call this level RAID 0 and 1, as it represents a layering of the two elementary levels of RAID. This level employs data striping (from RAID 0) and equal data splitting between multiple drives.

The blocks are read from the drives using array management software that can further improve speed by tasking multiple operations simultaneously from mirrored drives (from RAID 1). Due to the mirroring function of this level, the costs for the drives is much higher and the physical storage doubles for the same amount of actual data storage.

RAID LEVEL 53

RAID 53 mixes RAID 0 (data striping) with RAID 3 (separate parity drive). All data is now striped between two RAID 3 drive arrays. The overall reliability of RAID 53 is very high because of the extreme fault tolerance, which exceeds that on any individual hard drive.

RAID 53's array capacity is the same as that of the drive array capacity comprised in the RAID 3 structure. Net throughput is improved by the foundation of the RAID 3 array, which has inherently good I/O performance, especially for large block transfers.

WHAT'S TO FOLLOW?

In the concluding chapter on RAID, we will investigate striping and performance issues related to the various RAID levels. We will offer an explanation of parity from a visual and a numerical representation, and discuss reliability factors the user should be aware of.

CHAPTER 28

PARITY IN THE PERFORMANCE OF RAID

The previous two chapters have explored the fundamentals of redundant arrays of independent disks from various angles. The details of RAID storage systems are becoming as fundamental as the makeup of videotape transports. Engineers of the DTV era should be conscious of the flavors and features of RAID, as they will come into contact with them in a variety of applications.

This concluding chapter on RIAD will deal with a few more of the fundamentals and applications we hear so much about in video server storage systems. Following this chapter, we'll begin to look into SCSI systems and the extension of SCSI to Fibre Channel and Serial Storage Architecture (SSA). Each of these relatively new systems is finding its way into our environment in the same ways that Betacam, BetacamSP, and MII have. The impact these new storage systems will have on the future of the facility cannot be under emphasized. Stay tuned!

RAID storage can be found in non-linear edit systems, Windows NT servers, Macintosh secondary storage systems and certainly in most of the video servers being marketed in the latter half of the 1990s. Video servers depend on drives that work relentlessly to read and write video information in the form of data day in and day out. Our future in DTV will depend heavily on the video server's never-ending drive to record, process, and retrieve data over networks and conventional digital video transports.

We will look next at how striping makes up the disbursement of data in RAID storage systems.

STRIPING WITH BYTES AND BITS

"Striping" is the process of dividing user data into blocks, generally in a consecutive string, and distributing the data over multiple devices. This is also referred to as "mapping" with the actual size of the block referred to as the "striping unit."

A host (computer) system has the capability to request read and write sizes of more than one block. To service this request, a need for parallelism is realized.

"Parallelism" is the process that allows more than one device to contribute to the data transfer and thus increase data transfer speed. If there is a mechanism to handle this parallel distribution, then larger requests for longer periods of time are possible.

Throughputs several times the speed of a single drive are possible, but a balance must be struck. This equalization can be achieved by reducing the striping unit—which in turn achieves more parallelism.

When the striping unit is decreased so that parallelism is increased, the simultaneous servicing of multiple smaller requests, called concurrency, is reduced.

Remember, as parallelism is increased, throughput increases but concurrency is decreased, which could slow throughput if the majority of servicing is for small data block requests. This becomes one of the balancing issues in RAID performance.

RELIABILITY, AVAILABILITY AND PERFORMANCE

There are performance values realized when replacing a single large disk drive with an array of smaller disk drives. RAID storage subsystems provide improved data reliability and availability.

The structure of an array automatically provides for multiple actuators that work in parallel. Instead of individual actuators constantly seeking tracks on a single drive, the parallel nature of two or more drives, each with their own actuator, results in a higher throughput and in turn offers the capability of multiple simultaneous access.

The Parity in Performance of RAID 305

RAID also provides for the even distribution of data across a wider range of disk drive surfaces. Part of the RAID system includes a controller, made up of hardware and software routines that manage the data throughput and the microcode instructions on the individual drives in the array. The RAID controller then, by its own nature, lends itself to the process of dynamic load balancing.

In a non-RAID system, when a CPU attempts to control the access to both single and multiple drives (not arranged in an array) the host resources (or CPU activities) get bogged down trying to address the dynamics of each disk drive on an individual request basis. On the other hand, the RAID controller will manage this process—resulting in less CPU activity or the need for a systems administrator to continually manage the disk balancing on a manual basis. This principle is one of the baseline fundamentals of RAID—making the multiple drives appear as one drive.

As already explained, each RAID level architecture has its own specific performance characteristics. For this reason an integrator or manufacturer might select one RAID level over another.

Reliability and redundancy is also an emotional issue. Some enterprises will insist that complete and total redundancy be the prime directive and make performance a secondary issue. In this case, mirroring of the entire system becomes important, even to the point that a second set of mirrored arrays might exist in an entirely different physical location.

RAID storage subsystems provide improved data reliability and availability.

PARITY PROTECTION

The process of error correction is simplified if you consider that, in binary, you can have only two digits, 0 and 1. It is straightforward to know that if the data is not correct, the only possibility is that it can be the inverse (or opposite) value.

Parity is a rudimentary principle in error correction. In the simplified example of Table 28-1, we show all the possible combinations in a two-bit data word. The total data will be three bits in length, with the first two places being data, the last place being a parity or redundant bit.

MSB	LSB	Parity Calculation	Data + Parity Word
0	0	XOR=0	000
0	1	XOR=1	011
1	0	XOR=1	101
1	1	XOR=0	110

Table 28-1: Exclusive OR binary functions, related to parity.

It is this parity bit that is either sent to the independent parity drive in RAID 3 or striped across the array as parity in RAID 5. We will next show how data can be reconstructed should one of the drives in the array fail.

Recall that the parity word constructed in our simple example actually is an indication of whether there is an even number of 1s (indicated by a 0-parity bit) or an odd number (indicated by a 1-parity bit). If you remove, for example, the LSB (least significant bit), you can easily tell by mathematics whether the missing bit is a 1 or a 0 by interrogating the parity bit. Table 28-2 shows the four combinations that result in the two distinct parity word indications.

MSB	LSB	Parity Calculation	Data + Parity Word	Meaning of Parity Word
0	?	XOR=0	000	even number of 1s
0	?	XOR=1	011	odd number of 1s
1	?	XOR=1	101	even number of 1s
1	?	XOR=0	110	odd number of 1s

Table 28-2: Parity interpretation.

Row 1 cannot be a 1 because it would contradict the parity bit, therefore the LSB=0. Row 2's LSB must be a 1 because that satisfies the odd number of parity bits, and so forth.

While this is the simplest of examples in error correction using parity, it is easy to see how this concept is applied to a RAID array where the entire drive is missing, and the data can be reconstructed, in the background, using the information on the parity drive (see Figure 28-1).

If the parity drive fails, then protection is lost, but the missing data can be regenerated once the drive is replaced by looking at all the data on the remaining drives and rewriting the error correction information to the new parity drive.

The downside is if two drives fail simultaneously. Here there is not enough information to reconstruct either a parity drive or a data drive, and the system fails.

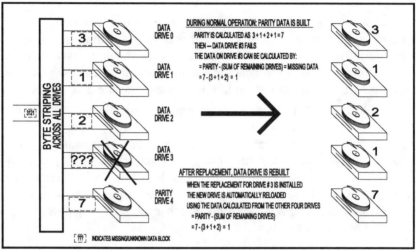

Figure 28-1: RAID 3 drive failure and data reconstruction from parity.

ERROR CORRECTION

Error correction coding (ECC) is generated from a mathematical equation or algorithm every time data is written into the drive array's front end write cache. Far more sophisticated real time algorithms, involving predictions and other statistical methods, are generally applied in the makeup of a RAID controller—principally to speed up throughput and add extra elements of protection to the data integrity.

In another method for ECC, a Hamming code (discovered by Richard Hamming) is used in the rarely commercially implemented RAID 2 implementation. This is the same basic method that is used in the familiar ITU-R BT.601 sampling structure for protection of the TRS (timing reference signal).

Complete data reconstruction can usually be performed as a background task with little to no degradation in performance. Error recovery algorithms, usually employed during the recording process, can be applied that should produce an identical image to that which was made when all the disk systems were fully operational.

Other designs and even later RAID levels, such as RAID 6, have produced sophisticated recording algorithms that permit data reconstruction even when two data disks fail. Technology like this is similar to the error correction and concealment techniques of digital VTRs.

When, as in RAID 5, parity information is distributed across all the disks in the array, the traffic jam of accessing (reading) a single disk drive during rapid, concurrent input and output operations is reduced. The job of assembling the data for meaningful output is now placed on the output frame buffers instead of the mechanics of the disk drives.

Recovery during failed operations is basically the same except that parity data is now spread across many drives, so in essence reconstruction performance may also be increased. Recording onto the remaining drives, in a fault situation, is straightforward—it's just spread among the remaining operational drives until full recovery is completed. In cases like this, however, writing to the drives is generally even slower than lower number RAID systems because computational and block recording requires more time.

THE READ/WRITE CACHE

The principle reason for a cache is to buffer data to compensate for latency in head seeks or rotation of the drive itself. All drive systems employ some type of cache. We will briefly describe some of the principles of caches to give only a conceptual idea of how they could be used in RAID or even single disk operations.

The cache is a sort of anticipatory buffer that makes predictions that would improve application responsiveness.

A read cache may be used to hold data it *thinks* will be read. In the example of a prefetch cache, data at the immediately preceding addresses has just been read, as in the contiguous playout of a video stream (compressed or otherwise), so it is logical to have the next block in the sequence ready in case it is called for. The probability of needing that next block of data in the sequence is high—so the data is prefetched and cached to a small buffer, eliminating the time it would normally take between when the fetch command is issued and the data is in the buffer and available to be read.

In a most recently used cache, data is buffered because it was recently read and might have another host error that would require a second read to verify the integrity of the block it originally requested.

A write cache is used to hold data, without necessarily immediately writing it to a drive, so that the write request can be acknowledged back to the host, signifying that the write operation was successful.

If an application had to wait for the actual period that it takes to physically deposit the data onto the drive surface itself, for every time a write request was made, the application would slow significantly. What really happens is that the I/O subsystem writes the data to the drive some short time later, when the drive is available or idle.

This process, called a write-behind cache, improves application speeds and allows read requests to be processed ahead of write requests, thereby improving overall system efficiency. This is not without its own problems, however.

The cache is made up of volatile solid state memory (RAM). If a power failure or fractional glitch happens between the time the write request is acknowledged to the host and the time the data is actually written to the drive, there is a risk of actual data loss and the host processor would never know about it. The complexities of the write-behind cache grow when considering parity RAID storage.

Each write request issued from the host will result in multiple writes to disks in the RAID array, including the parity drive. The probability of erroneous data being stored on multiple disks increases just by the liability associated with having more than one drive storing meaningful data.

Different manufacturers take different approaches to protecting against this possibility. Some will simply not employ write-behind cache, even if available. Others will suggest or actually include a method of protection to make the cache nonvolatile such as a UPS or onboard battery system associated with the cache memory modules proper.

When the write-behind cache becomes nonvolatile, it is referred to as a write-back cache, because it allows data to be written *back* to the media once the host has been notified the write was successful, but when or if a problem was detected (and corrected) prior to or during the actual physical write to media.

In RAID subsystems that use parity, write-back caches can increase input-output operations significantly by virtually eliminating I/O loading factors that typically slow data transfer between the host and array. This is especially beneficial to RAID 5 which suffers from poor write performance due to block striping and interleaving of parity throughout the media.

RELIABILITY FEATURES IN RAID

Besides protecting the data from loss due to failed drives, RAID architectures are typically combined with other subsystems to improve reliability.

The ability to "hot swap" defective devices and automatically rebuild those systems is probably the best known example of RAID architecture's self-maintenance process. Even though the reliability of discrete drives is high, other elements must be integrated into the RAID subsystem to make a completely reliable system.

Other features include redundant RAID controllers that parallel and transparently change over when the primary controller fails. These are typically entire circuit board subsystems that can be hot-swapped or ready on hot-standby. Hot-standby secondary RAID controllers are generally powered by separate power supplies as well.

Some video servers that use an external RAID chassis will make the secondary controller an (expensive) option; but the user should know if this is a hot-standby secondary controller or just a powered slot with a manual changeover. Beware that the change over time for a secondary hardware RAID controller (one that is not on hot-standby) can be significant. Changing to a different controller will usually require a shutdown of the drive chassis and restart of the entire server in the proper sequence. The entire process could take as long as a "first-birthday" (total reinitialization).

A redundant power supply, and if using a redundant RAID controller, even a third power supply, is not uncommon in a truly fault tolerant system.

Cooling fans and monitoring systems that allow for at least 50 percent overhead in cooling capabilities are also common in self-contained RAID storage systems. The fans should be replaceable without having to shut down the system, or at the very least, the system should allow for a minimum of several minutes of operation before self shutdown so replacement can be accomplished on-line.

MTBF

Often the most emphasized claim of a RAID system is its exceptional MTBF. Even though the claimed MTBF (mean time before failure) of a single disk drive can be expressed in terms of years, the weaker points in the system, such as power supplies, subsystem cards, software bugs and catastrophic failures are far more significant if there is no protection or redundancy.

MTBF decreases as multiple disks are added. Subsystem components tend to lower the MTBF more than the individual drives. Other environmental factors escalate the impact of MTBF. One should be careful in understanding what is included in the MTBF specification. In other words, the actual disk drives themselves will exhibit a much higher MTBF than the overall system—and sometimes it's the disk MTBF that is quoted.

RAID systems can be compatible with current single disk systems. Configuration, using standby and hot-swappable components, can provide high data availability and reliability.

RAID IN MEDIA SERVER APPLICATIONS

RAID is a reasonable solution for large arrays, provided the increased costs can be justified. In television systems, tape archival methods should be balanced with the degree and size of the (RAID) array itself.

Remember, too, RAID arrays do not provide for data integrity; they only reduce the possibility of data loss. Periodic backup, human data management and alternative archival methods still remain the only real substitutes for maintaining data integrity.

CHAPTER 29

THE HIGHS AND LOWS OF SCSI

This chapter begins with a look back on the perceptions of the SCSI interface as it emerged, matured and expanded its presence in the industry. As is common throughout this book, the historical perspectives have been updated to include present day concepts, and many portions of the original article have been expanded to delve deeper into the hows and whys of the technology.

In this chapter, we will focus primarily on SCSI-1 and SCSI-2. Following this, Chapters 30 through 32 will deal with SCSI-3 and the mapping of the SCSI command set to the new transport standards of Fibre Channel (FC) and Serial Storage Architecture (SSA).

In early 1995, when this article first appeared in *TV Technology*, there were two widely accepted platforms for the graphic designer and animation artist. The most popular personal computer for art graphics was the Apple Macintosh. For those more serious about animation or 3-D, and with lots of money to burn, the Silicon Graphics (SGI) workstation was a benchmark in capabilities.

These two platforms relied on external peripherals for storage, recording, inputting, and outputting. The Macintosh, born into the world with SCSI, used this interface for both internal disk drives and external add-on storage. Long before the PC was considered a viable alternative to the Mac for desktop publishing, Mac users had clearly mastered the very basics of the SCSI interface. This broad base of users understood SCSI—not from the technical whys or why-nots—but from the rudimentary principles associated with connecting external devices to the "bus."

Today, the PC has clearly demonstrated its capabilities, and the Mac has shrunk substantially in market penetration. At the same time, the use of SCSI devices has taken on a new visibility. The SCSI device of today has many new shapes and dimensions. Bridging the once limited capabilities in transfer speeds, we have seen exciting advances in speed, new applications using differential connections, and the extension of the root protocol to other mediums. The continuance of SCSI devices appears to be a reality for at least several more years, although the physical connection medium may contribute to significant changes along the way.

Several new contributions are helping extend the life of the SCSI bus, at least in principle, including the adaptation of the signals and controls over faster, more practical serial interfaces. While in the end it is expected that parallel SCSI interfaces will eventual succumb to higher data rate serial protocols, the basics of SCSI data protocols will probably remain for some time.

PERIPHERAL CONFIGURATIONS

Working in the desktop video, publishing or multimedia domains requires continual peripheral reconfiguration, typically with any number of various devices connected via an SCSI interface. Graphic artists, publishers, and agency designers familiar with Macintosh and UNIX machines, such as Silicon Graphics, have dealt with configuring external SCSI interfaces since they first began seriously using the desktop or workstation as a graphics and animation tool.

Until recently, the Intel-class personal computer only touched upon the options that SCSI-based peripherals and systems had available. The internal SCSI architecture for the PC was associated typically with connecting only a CD-ROM from a sound card to the physical interface. Once scanners, external removable storage devices, and the like became affordable; the impact of SCSI became more universally available and acceptable across all the desktop computer platforms.

As non-linear editing systems, media servers, mass storage arrays, DDRs and the like become more commonplace, a more thorough

The Highs and Lows of SCSI 315

understanding of the many flavors of SCSI is warranted. Regardless of the computer platform, understanding the highs and lows of connecting SCSI peripherals will be a necessity.

INTERFACES AND THE BUS

The SCSI bus and the IDE interface, both ANSI standards, are the two most common interfaces for computer peripherals. One must contrast the "interface" and the "bus" to fully appreciate the relative differences between device connections.

Computers typically utilize a number of internal buses. The movement of data between internal devices and controllers happens on these buses (see Figure 29-1). The groundwork for understanding the makeup of the bus lies in the layer models so often shown in networking and computer diagrams.

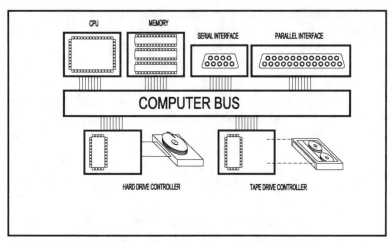

Figure 29-1: Internal computer bus and devices.

Computer buses are diagrammed with a physical layer, a bus protocol layer, and an optional device model, sometimes including a set of commands or extensions.

Three basic building blocks are associated with the bus. Many texts will diagram these blocks as separate, distinct paths, but they are essentially grouped together to form the foundation of the bus in its entirety. The three groups referred to contain addressing, data transfer, and control blocks.

Over the years of computer evolution, the bus has continued to expand. Early computers in the late-'60s and early-'70s had bus address widths of from a few bits to upwards of 20 and 22 bits. If the bus is 8 bits wide, it could address 2^8 or 256 individual locations. A modern 32-bit architecture can address 2^{32} or 4,294,967,296 locations (some four-gigabits plus of individually addressable locations).

It is important to understand how data can be moved efficiently over the computer bus. The identification location is the "address" and the process of getting to that address is called "addressing."

"Throughput" is another essential building block element in a computer bus. The process of transferring data from one location to another requires a certain amount of bandwidth. The amount of data a bus (or drive) can deliver or accept, on a sustained basis, is the throughput. The description of throughput is usually stated as either peak or sustained. In data transfers, over a period, we are most concerned with sustained throughput, expressed as an average amount of bits transferred in a unit time interval. The peak data rate is of little concern except when dealing with added interfaces or networking environments.

"Throughput" and "bandwidth" are synonymous, both generally reflecting a unit period of time measurement. The net throughput in a system is a product of two parameters: data width and clock speed, less some degree of overhead. "Data width" is the number of bits that make up one data word. The other parameter, "clock speed," indicates how many data words are transferred per unit period, typically expressed in bytes per second.

"Overhead" is a supplemental amount of data that is subtracted from the total throughput, which is the necessary protocol information required by the structure of the bus itself. The amount of overhead is different for all transport schemes from IDE through SCSI and on to ATM. Overhead is useless data to the application and should be kept to an absolute minimum for overall efficiency.

SCSI's Roots

SCSI (pronounced "scuzzy") is an acronym that stands for small computer systems interface. SCSI began between 1979 and 1981, when Shugart, working with NCR, developed SASI[39] as a scheme that would address devices logically rather than physically. The addressing would also be byte-based (parallel in structure) rather than serial.

To place disk drive development in perspective, remember that in 1975, the personal computer was only a hobbyist's dream and the incorporation of a floppy or a hard drive into the yet-to-be-invented IBM PC was still several years away. By 1980, the IBM PC was still only a white board concept. ANSI (American National Standards Institute), in 1980, had rejected standardization of the proposed Shugart SASI standard over the ANSI preferred IPI interface, which was thought to be more sophisticated. Arguments between developers included such controversial topics as using a differential versus a single-ended interface, 10- versus 6-byte commands, serial versus parallel, and the like.

Product development continued while the standardization process stalled—so typical in the realm of advancing technology. By 1982, the ANSI committee X3T9.2 was formed and the basis of SCSI, as it eventually became known, was developed. In 1984, the draft proposal was completed and submitted to ANSI. Subsequently, by this time, device controllers and host adapters were already on the market. Apple Computer introduced the Macintosh, based around SCSI for its storage and peripheral interconnects.

ANSI X3.131-1986 was finally published in its approved form as SCSI-1, but not without a rash of headaches that had surfaced as a result of nonstandard products already out in the marketplace. Between 1984 and 1986, as more SCSI drives came on the market, it was discovered that a separate driver had to be written for each type of drive. Eventually, as a means to clarify and quantify the many variations that were being identified, the evolutionary development of a Common Command Set (CCS) for hard disk drives was begun.

The CCS was essentially a means within a protocol that described how host devices would access and communicate with other devices (drives, scanners, printers, etc.) The early CCS proposals were

---

[39] SASI: Shugart Associates Systems Interface, the predecessor to SCSI.

not mandatory and each manufacturer figured it could interpret them as they needed.

Soon after SCSI-1 was standardized, work on the next-generation interface was begun. The SCSI-2 development incorporated the foundation work on CCS, as well as an attempt to address more than eight devices in a SCSI bus chain. While the greater-than-eight-device concept failed, synchronous transfers at 10 MHz and a 32-bit wide data bus did manage to become part of SCSI-2.

SCSI-2 formal proceedings began in 1989, but not until early 1994 did it become an official ANSI standard. While disappointing for some, many devices already were out that incorporated the principles that, eventually, would reside in the final ANSI standard.

In retrospect, the real purpose for what was actually a developing overall standard, separated by SCSI-1, -2, and even SCSI-3, was that standards needed to be published so that manufacturers and users could set a starting point. The sequence of SCSI numbers (-1, -2, and -3) then became chapters in the overall SCSI development story[40].

In November 1994, the third generation of SCSI protocol was being defined. Dubbed "SCSI-3," this protocol would extend SCSI applications in several important directions, such as serial communications and real time multimedia support. SCSI-3 specifications included downward compatibility with SCSI-1 and -2. SCSI-3 will have a new and diverse impact on device interconnection and communications. We will delve deeper into the many facets of SCSI-3 in following chapters.

To speak of standards in a very positive sense, the period between concept and product release can be quite short compared to the process of accepting an international standard. Often the standard begins as a manufacturer's presentation of a solution. Generally, a great deal of R&D has already been exerted before a proposal to industry or a standards body (ANSI or SMPTE) is even made. In some cases, a product may already be released before a recommendation for standardization is made.

As is the case in technological development, the standards process is an evolving one that must support extensions, adjustments, and

---

[40] Not unlike the development of MPEG-1, MPEG-2, and MPEG-4, where for marketing's sake some level of standard needed to get to industry before industry went off in its own direction.

improvements. SMPTE, AES, and even ANSI have all recognized that evolution requires the continual review, modification, recertification, and in some cases, abandonment of some standards. SCSI is, and has been, no different.

DEFINING SCSI

The SCSI interface allows a variety of devices to be connected via a single bus to a computer system. Incorporated in the interface are a series of established commands that communicate with the device(s) in order to convey various parameters and requests to and from host and device. The real plus about SCSI is its logical addressing, whereby the same command structure can request data or issue commands independent of the physical device attached to the interface.

Ideally, this makes the use of any number of devices—tape drives, CD players, scanners, printers and hard drives—quite straightforward. Although software drivers are typically added to the host drive set to convey device-specific functionality, the basic low-level command sets, including the physical media for interconnection, are common to most all SCSI devices.

The organization of data on the drive or peripheral is of little concern to the host computer as the interface presents the data to and from the host in a uniform and standard method. The peripheral can also manage its own housekeeping without bothering the computer host itself. Housekeeping functions such as control of media flaws, file allocation table structures, spindle and head/arm control, and others are performed as device-specific background tasks.

SCSI devices identify themselves using the SCSI ID, a unique number that addresses each unique device on the bus. The SCSI bus can address up to eight logical units (LUNs). Typically, the numerical ordering for the LUNs is from 0 to 7, and each LUN can play the role of either a target or an initiator.

The "initiator" will generally issue a command or request, and the "target" will execute that request. SCSI, by design, must consist of at least one initiator and one target device. Some SCSI devices can act as both initiator and target.

The initiator is considered by most to be the host adapter and is generally awarded the highest number, 7, as its SCSI ID. The first device, usually the system drive, is then given 0, and for the most part, these are the only two IDs that need to be in the system.

Generally, SCSI ID 0 has the lowest priority, and 7 has the highest. Devices with SCSI ID 1 through 6 are typically ordered such that the next number, 1, is the second hard disk drive, followed by any removable media, scanners, printers, and the like.

The computer may have more than one SCSI bus, usually referred to by letters (i.e,. bus A or bus B). In this case, the buses must be "balanced," that is each side must contain the same amount of SCSI devices on each bus. SCSI buses can effectively increase the number of devices on the bus. When 32-bit SCSI-3 is implemented, up to 32 devices can then be supported.

A single SCSI ID can also apply to an entire set of drives, such as in a RAID array. By definition, the RAID array appears as a single drive, although it may consist of dozens of drives in a single enclosure or group of enclosures.

SCSI is both a logical and physical scheme. At the high level, it contains a logical protocol used to communicate between computers and peripherals. At the low level, it defines the wiring scheme to connect a variety of physical peripherals.

The SCSI standard, although only slightly over a decade old, was most recognized at the personal computer level in the first Apple Macintosh computers. Although in 1984 when the Mac was born, their implementation was in a nonstandard form, SCSI was the primary means to connect (both internally and externally) the computer processor host to storage and peripheral devices.

We now recognize that SCSI technology extends into video-on-demand, mass storage, and media servers. The foundations of SCSI are employed in a variety of disk storage systems for computers, non-linear editors and other mass media and video operations.

SCSI's high-level communications protocol includes support for various peripherals, including scanners, CD-ROM, tape and disk drives, as well as media changers such as optical and tape libraries. It is being extended to serial communications for consumer and professional broadcast products ranging from cameras to VCRs, PCs to complete editing systems, and digital disk recorders to mammoth video servers.

SCSI Data Transfer

SCSI data transfers can be either asynchronous or synchronous. The SCSI parallel interface itself has no clock like that in a PC expansion bus. There are, however, strict timing relationships that are adhered to, which in practicality sets the limits on peak transfers through the system.

The asynchronous method is the most fundamental data transfer method. Asynchronous data rates typically range from 1.5 MHz through 4 MHz.

Synchronous SCSI communications are considered the preferred transfer method, especially in the high-performance drives. Peripheral command sets and information related to protocol are sent asynchronously. The greatest advantage to this operation is increased data throughput because individual devices are allowed to set up their own negotiations as to which method of communication is best for their specific application.

Synchronous SCSI-1 allowed data rates up to 5 MB/s. Varieties of other synchronous data rates have emerged over the years, including Fast, Wide, and Ultra SCSI (all of which are trade names with industry standard recognition). Table 29-1 identifies the state of data transfer rates in the 1995 time frame.

Device	Data Rate	SCSI Interface
CD player (2X)	500 KB/s	Asynchronous SCSI-1
Data tape drive	500 KB/s	Asynchronous SCSI-1
½" tape drive	1 MB/s	Asynchronous SCSI-1
Hard disk drive	5 MB/s	Synchronous SCSI-1
RAID array	20 MB/s	Fast SCSI-2
RAM disk	40+ MB/s	Fast + Wide SCSI-2

Table 29-1: SCSI transfer rates [1995].

Most modern peripheral devices, such as hard disks and CD players greater than 2×(double speed), no longer employ asynchronous SCSI-1. Furthermore, devices that use the synchronous option cannot reside on the same bus as those that do not use this option.

FAST AND WIDE

Improvements to SCSI-2 have continued since its introduction. In an effort to increase throughput, many manufacturers have devised methods to push the envelope. Device microcode, interface card, media and software, have all had to make substantial refinements in order to address the needs of real time streaming video and audio playback.

The low-level physical transport layer of SCSI communications consists of 8-bit, 16-bit or 32-bit parallel data transfer. Synchronous transfers were originally allowed at up to 5 MB/s as an option in the original SCSI-1 protocol. Fast SCSI, for SCSI-2, increases the data rate[41] up to 10 MHz and has essentially become a de facto standard for high-performance drives (see Table 29-2).

Bus width and version	Cable Type and Pin count	Bus Width	Peak Transfer Rate	Devices	Trade Name
8-bit/SCSI-1	A/50	8	5	8	Asynch
8-bit/SCSI-2	A/50	8	10	8	Fast
16-bit/SCSI-2	A/50 plus B/68	16	20	8	Fast + Wide
32-bit/SCSI-2	A/50 plus B/68	32	40	8	Fast + Wide
8-bit/SCSI-3	A/50	8	10	8	Fast
16-bit/SCSI-3	P/68	16	20	16	Fast + Wide
16-bit/SCSI-3	P/68 plus Q/68	32	40	32	Fast + Wide

Table 29-2: SCSI structures and implementations.

As discussed earlier, the synchronous nature of SCSI-2 has removed any interlocked hand-shaking requirements. The now more critical measure of bus throughput is based upon cycle time. The "cycle time" of the bus is comprised of a combination of the assertion period and negation period.

---

[41] "Data rate" is defined as the bus width (in bytes) times the rate in MHz.

The "assertion period" is the time it takes to ready the bus and deposit signals on it. The "negation time" is the time required to remove the signals. SCSI-1 allowed 90 nanoseconds for each of these actions, resulting in 180 nanoseconds for the complete cycle.

SCSI-2 reduced the total cycle time to 100 nanoseconds, increasing peak transfer rates to 10 MB/s, assuming no overhead. This is often referred to as Fast SCSI, symbolizing greater than 8-bits and/or faster than 5 MB/s.

Wide SCSI is a method incorporating a second cable ("B cable"), whereby 16-bit and 32-bit transfers can be accommodated, resulting in a doubling or quadrupling of fast SCSI data rates.

Nomenclature touting "Fast and Wide" (or "Fast + Wide") may seem rather redundant, as the term "Wide" already means an increased performance over "Fast." In essence, "Fast" signifies an 8-bit bus width with data rates in excess of 5 MB/s. The term "Wide" also signifies extensions of Fast as in Fast + Wide for 16- and 32-bit bus widths.

CABLING AND BUS WIDTHS

The nature of SCSI, being a parallel interface, necessitates data transfers of more than one bit at a time. As the wider buses appeared, the data rates also increased. Adding extra signals, beyond the 8-bit width, required additional wires to carry the signals.

The acceptance of Fast + Wide SCSI was originally somewhat low because of the requirement for the second cable beyond the traditional 50-pin ribbon or shielded twisted pair type of media. In SCSI-2, a new series of cables was required (called P and Q cables) to handle the additional bus width.

Considerations also must be given to cabling requiring a specialized connector like the high-density SCSI-2 to Centronics-50 or high-density 68 pin SCSI-3 P cables and adapters. In most cases, using these types of adapters is not recommended, as they are unrecognized by SCSI standards. The selection of any adapters or cables must be chosen to meet the needs and requirements for the fast data rates.

For serial transport, like that of future generations of SCSI-3, provisions for low loss cabling and connections must be made. Hot pluggable devices (employing varying pin lengths in connectors), copper and fiber-optic media (for Fibre Channel) and the ever increasing quest

324 Video and Media Servers: Technology and Applications

for even higher data rates are becoming part of this new, distributed and dynamic environment.

The traditional 50-conductor SCSI A cable permitted only 8-bit wide connections. The cable length and the connector were often misunderstood in early implementations, adding to confusion and poor system performance. In asynchronous SCSI-1, the combination of cable length propagation delay and overhead accounted for a lengthy cycle time. These resulted in cycle times being extended and transfer rates being reduced.

As explained earlier, synchronous SCSI does not depend upon interlocked hand-shaking. There is still a requirement for acknowledgments, but they can be delayed. Packets of data can be sent sequentially without having to wait for a return acknowledgment before the next packet can be sent. This will be important to understanding why SCSI-3 can be implemented serially.

Propagation delay experienced as a direct result of cable length between devices was a serious hindrance for asynchronous SCSI-1 (see Table 29-3). Even though SCSI-1 overhead remained fixed at around 160 nanoseconds, the propagation time increased the cycle time to the point where transfer rates slowed to under 1.5 MB/s when the cable exceeded 25m (82 feet).

Cable Length	Propagation Delay	Cycle Time	Transfer Rate
0.3m	6.3ns	166ns	9.0 MB/s
1m	12ns	181ns	5.5MB/s
3m	63ns	223ns	4.5MB/s
10m	210ns	370ns	2.7MB/s
25m	525ns	685ns	1.5MB/s

Table 29-3: Asynchronous SCSI transfer rates.

SCSI CONNECTIONS

Devices typically connected to an SCSI interface adapter are considered either internal or external. The internal device can be the system hard drive, expansion drives, CD player, backup tapes, or any other device that might be inside the host computer's chassis.

If the peripheral device has only a single connector, it is most likely a "dead-end" point in the daisy chain cable (or chain). The typical

The Highs and Lows of SCSI 325

SCSI cabling, generally made of 50-conductor ribbon cable, had a series of connectors along its length. A good cable has been designed for proper spacing of the cable, taking into account maximum bus length and appropriate taps for optimal signal timing. Table 29-4 identifies the more common SCSI cables, connectors, and their uses.

Type	Bus-Cable Configuration	Descriptions
50-place	**Internal Narrow Bus** — Two rows of 24 pins, 0.1" spacing, rectangular array. Signal assignment identical to A cable.	**Early SCSI-1**
50-place	**A cable** — Similar to Centronics printer cable. Ribbon or shielded cable. Standardized by SCSI.	**Single-ended differential SCSI**
50-pin	**D shell** — Same arrangement as D-type shells, two rows of 25 pins. Cable was 4" long, difficult to use.	**Not widely accepted**
25-pin	**D25-SCSI** — 25-pin D-shell, one row of 12 pins, second row of 13 pins. Eliminates many ground returns.	**Macintosh computers**
68-pins	**SCSI single ended B-cable** — Two rows of 34 pins, 0.050-inch x 0.100-inch centers. Provides second path for additional signals.	**Wide SCSI.**
68-pins	**SCSI single-ended P cable** — Primary cable for 16-bit SCSI, same configuration as B cable. Single-ended cable pinouts differ from differential.	**16-bit Fast+Wide**
68-pin	**SCSI single-ended, Q cables** — Secondary cable for 32-bit SCSI. Single-ended cable pinouts differ from differential.	**32-bit Fast+Wide, 2 cables**
68-pin	**SCSI differential P & Q cables** — P differential for first 16 bits, Q differential for remaining 32 bits. Differential versions of 16- and 32-bit SCSI.	**32-bit differential Fast+Wide, 2 cables**

Table 29-4: SCSI cables, connectors and uses.

The SCSI cable does not provide power to the peripheral devices, but it does carry the required data and control signals. The cables are generally keyed and indexed with a striped leading side conductor indicating pin-1 polarity.

TERMINATIONS

Proper termination of the SCSI chain is an important requirement to minimize electrical reflections along the chain. A "termination" is the final point in the chain. SCSI termination can be either passive or active.

According to the SCSI specifications, the first and last devices on the bus must be terminated. This requires care when multiple devices are connected to insure that double termination or incorrect termination does not take place.

"Passive termination" is what was originally used in SCSI-1. It consisted of two resistors connected in series between a voltage source and ground. The middle junction of the two resistors was where the SCSI bus terminated. Passive terminations were generally used when four or fewer SCSI devices were attached to the SCSI interface.

The "active termination" was implemented as a solution to noise reduction on the bus. Active terminations were required whenever long cables or greater than five devices were attached. Using a series 110-ohm resistor, and a regulated voltage source, insured a constant voltage appearance to the SCSI bus, regardless of the load.

The active termination not only controls the bus noise; it maintains signal integrity by clamping any overshoot voltages produced on the bus. This increases the reliability of the data transfers over several daisy chain loops in the SCSI chain.

Device manufacturers have been building in active terminations triggered by a logic gate instead of the older, outdated mechanical terminations. The SCSI-2 standard further recommends terminating both ends of the SCSI bus using active terminations.

Differential buses use a different termination method from standard, single-ended buses. Signals are terminated at both ends of the cable, using resistor networks that present a 122-ohm appearance to both the negative and the positive signal lines.

External terminators look like the same plugs as those on the cables, but without a cable sheath. Some external SCSI devices use integral terminations that require case disassembly to set or disengage.

CHAPTER 30

SCSI'S SECOND STEP

Not long after SCSI-1 was first introduced, its drawbacks became evident. Soon the need for a more advanced interface became obvious and the industry began what would be a long cycle of drive interface advances. The second step in our SCSI overview takes a brief retrospect of SCSI-1 versus SCSI-2 differences, then moves directly into the development and technology of SCSI-3—the next generation in connectivity and addressability.

Bandwidth, data transfer performance, and the number of large blocks of data being transported are all considerations for improving SCSI performance. While desktop, multimedia and video applications continue to grow, the need for implementation of enhanced bus performance becomes crucial. As SCSI transfer rate performance has increased, more applications are brought to market and the success of media servers for multimedia, non-linear editing, and video applications has soared.

Industry tests conducted in 1995 were already showing substantial improvements by using the newer and faster SCSI system protocols, especially as the block or file size transfer rates increased. The expectation of a future SCSI-3, under construction as early as 1993, changed the general perception that SCSI, as a command set protocol, may continue for quite some time.

THE NEXT STEP IN SCSI

The final acceptance of SCSI-2 came in 1994, when ANSI approved the X3.131-1994 standard. We recall that a major improvement from SCSI-1 to SCSI-2 was the inclusion of the Common Command Set (CCS). The final standard included some otherwise optional sections of SCSI-1, but now made them mandatory in SCSI-2.

Backward compatibility was included in SCSI-2. At the hardware level, SCSI-2 remained a parallel connection—but also added new hardware options. These included higher-density cables, multiple cables for wider bandwidth, and other termination improvements and electrical specification tightening.

Broader support for nondisk devices including tape drives, CD-ROM and scanners was now available. SCSI-2 also removed seldom-used options of SCSI-1, such as the single initiator option and the reservation queuing option.

SCSI-2 devices are now permitted to communicate with the host adapter using their own initiative. This means they can set their own operational modes depending upon their own requirements.

Another improvement was "tagged command queuing," which improves the host's performance by permitting several commands to be sent over the SCSI chain simultaneously. The improvement also allows a new command to be sent before the previous one(s) is completed.

As the development of a new SCSI-3 standards document, known as ANSI X3T10/1071D proceeded, other improvements in SCSI-2 and synchronous data transfer rates were made. Exactly where the overlap between SCSI-2 and SCSI-3 is remains a subject for discussion, though most of that was resolved once the base "layer approach" for SCSI-3 was finally adopted.

During the standards process, a new form of SCSI, called Ultra SCSI (referred to as Fast-20 and DoubleSpeed SCSI), came to be. Under the SCSI-2 hardware standard, higher data rates were achieved by marrying the timing requirements of Fast SCSI with the 32-bit bus width of wide SCSI. Ultra SCSI uses the familiar parallel interface as a full 32-bit connection with the 10 MHz timing parameters of Fast SCSI. At the same time, compatibility with current synchronous and asynchronous protocols is maintained.

SCSI has been interpreted, and consequently confused, to include broad generalizations in the protocol. Some still equate

Fast SCSI with Fast + Wide SCSI, and with Fast-20 SCSI. One should understand that all three of these schemes have differences in one another, fundamentally in terms of speed, bus width, bandwidth, connecting requirements, and ultimately in performance.

In the Fast-20 scheme (compared to Fast SCSI), there is an effective doubling of the data rate over current Fast SCSI bandwidths. If a narrow 8-bit SCSI bus is operating in Fast mode, data burst rates can reach 10 MB/s. Boosting this to Fast-20 means data can burst at up to 20 Mxfr/s (Megatransfers per second)[42].

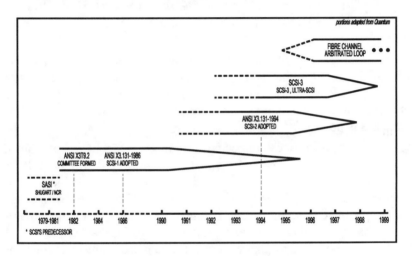

Figure 30-1: Time line of SCSI development.

Recalling once again that 16-bit SCSI is part of the definition of Wide SCSI, then Fast + Wide can burst at 20 MB/s and a Fast-20 bus can therefore reach up to 40 MB/s transfer data rates.

In 1995–96, it was also believed possible to adopt a Fast-40 system, with potential peak burst rates as high as 80 MB/s. Some of the potential development for parallel Fast-40 may have been consciously sidelined as serial SCSI-3 began its way through the standards process.

---

[42] "Megatransfers" (Mxfr), referring to millions of transfers (per second), is irrespective of bus width or word length, and is newer terminology applicable to disk drive and data bus performance or representation.

Performance Increases

From a performance standpoint, we have seen SCSI's bus speed double about every five years. Besides the data rate increases, the interface has become more robust by making significant improvements in many areas, physically and electrically.

Upgrading existing systems is an important function considering the tremendous volume of independent drives and arrays already in the market. The evolution of SCSI from SCSI-1 to Fast SCSI to Ultra SCSI has allowed legacy systems to coexist on new systems—a prime feature in extensibility.

The Ultra SCSI and Ultra2 parallel interfaces are extensions to the SCSI family that add significant performance gains while maintaining upgradability of an existing SCSI user base. The physical interface has been developed such that an easy migration to even higher data rates is possible.

From a devices-connected standpoint, Fast SCSI permits 16 loads and a total cable length of 6 meters. Ultra SCSI will support four loads on 3 meters of cable—or 8 loads on 1.5 meters of cable. From a connector standpoint, interior drive systems, such as servers with plug-in hot-swappable drives, use the SCA (single connector assembly), making it possible to eliminate interior cables. The SCA backplane and the SCA-2 (for hot plugging) are added for a more controller connected environment and result in greater throughput and higher performance on a system basis.

Very High Density Cabling Interconnect (VHDCI) allows for four Wide SCSI cables to be attached to a single interface card. This small form-factor permits a conventional PC chassis to be used for high-capacity drive arrays. This becomes valuable when connecting to Wide Ultra SCSI and Wide Ultra2 SCSI systems—which both permit up to 16 devices to be attached to the bus.

With VHDCI, cable size is also reduced, making it possible to add connections to Type II PC cards. This flexibility in configuration makes it easier and faster to assemble larger drive systems.

The automatic configuration of SCSI device IDs, made possible through the development of advanced SCSI protocols, has provided Plug and Play capabilities on a hands-off basis. Through a feature called

SCAM (SCSI Configured Automatically), the SCSI controller assigns IDs to SCSI devices on the bus for the user. SCAM devices are also fully compatible with other non-SCAM devices on the same bus. Additional flexibility is allowed when some devices permit manual jumper overrides of the ID settings.

Two levels of SCAM have been implemented. SCAM-1 instructs the system to automatically reset the SCSI interface whenever another device is added. SCAM-2 improves on SCAM-1 by enabling automatic ID resetting without resetting the SCSI interface.

ULTRA SCSI

Ultra SCSI became available to the market in full strength during the middle of 1996. The industry standard was complete. At the Fall COMDEX-96 exhibition, several vendors were already showing Ultra SCSI products with as many as 16 Ultra SCSI drives attached. At the same time, prototype Fibre Channel disk drives appeared, and the proponents of SSA (Serial Storage Architecture) were showing a complete range of products including tape, CD-ROMs, and disk drives. New groups had sprouted to market and support the efforts of both Fibre Channel and SSA.[43] The SCSI Trade Association (STA) had already included a charter to cover all the variations of SCSI, from SCSI-1 through Ultra SCSI.

Continued development of SCSI parallel interface (SPI) was ongoing, although a thread of the industry was weaving its way toward serial SCSI in a number of topologies. The push beyond Ultra SCSI remained evident as Ultra2 was proposed.

Ultra2 shifts from a single ended, unbalanced physical interface to a differential, balanced interface. The IEEE-485 interface uses high-voltage differential (HVD) drivers, making implementation on controller chips impractical. The Ultra2 transceivers, employing low-voltage differential (LVD) drivers, have low enough power consumption that they can be incorporated directly onto the controller silicon. By integrating the driver onto the same silicon, noise immunity, reduced EMI and extended cable lengths become possible. This practice also

---

[43] Fibre Channel Loop Community (FCLC) and SSA Industry Trade Association (SSITA).

reduces sensitivity to ground shifts often found when multiple devices each have their own power supply subsystems.

The levels of parallel SCSI performance were developed by various ANSI workgroups. Table 30-1 describes the SCSI Trade Association's designations of performance and "trade names", with their respective descriptive names and ANSI workgroup specifications next to bus width and bus speeds.

TERMINOLOGY (Trade Name)	Bus Width (bits)	Bus Speed (MHz)	Descriptive Name	Specification
SCSI-1	8	5	SCSI-1	X3-131-1986
Fast SCSI	8	10	SCSI-2	X3T9.2 375R
			SCSI-3	X3.277
			Parallel Interface (SPI)	
Fast + Wide SCSI	16	20		
Ultra SCSI	8	20	SCSI-3 Fast-20	X3.277-1996
			Parallel Interface	
Wide Ultra SCSI	16	40		
Ultra2 SCSI	8	40	SCSI Parallel Interface-2	X3T10/1142D
			(SPI-2)	Rev 11
Wide Ultra2 SCSI	16	80		

Table 30-1: SCSI Trade Association (STA) SCSI performance.

Despite attempts to persuade industry otherwise, parallel SCSI still maintains many advantages, including multitasking, low overhead, high bandwidth, wide range connectivity, and small form-factor interconnects. The command queuing mechanism in SCSI adds a powerful performance advantage over ATA, especially as these mechanics become applied to the ever increasing number of Windows NT platforms.

With all the enhancements presented as a result of expanding the interface, the user is cautioned to remember that care must still be taken when you begin to approach the maximum of 16 devices attached to the bus. The proper selection of cabling includes understanding both the capacitance of the cable and the spacing of the stubs. Mismatched capacitance adds to skew, which is a result of increasing propagation delays. In the faster bus speeds, the setup and hold times for data are decreased, and skew errors may cause data errors to accumulate, reducing throughput.

SPI-2 specifications also recommend an unloaded differential impedance between 110 and 135 ohms. A minimum stub-spacing requirement is stated as part of the standards document.

Another look at the parallel SCSI environment (Table 30-2) shows the entire gamut of possibilities for connecting multiple devices to the bus. The table describes single ended, high-voltage and low-voltage differential bus lengths and the maximum number of devices that can be connected to the bus. Note that single ended configurations are not defined for any of the Ultra2 bus speeds.

Trade Name	Maximum N° of Devices	Maximum Bus Length (m)		
		Single Ended (SE)	High-Voltage Differential (HVD)	Low-Voltage Differential (LVD)
SCSI-1	8	6m	25m	-
Fast SCSI	8	6m	25m	-
Fast + Wide SCSI	16	6m	25m	-
Ultra SCSI	8	1.5m	25m	-
	4	3m		-
Wide Ultra SCSI	16	-	25m	-
	8	1.5m	-	-
	4	3m	-	-
Ultra2 SCSI	2	unsupported	25m	25m
	8	unsupported	12m	12m
Wide Ultra2 SCSI	2	unsupported	25m	25m
	16	unsupported	12m	12m

Table 30-2: SCSI performance by configuration.

The continued details of SCSI get far more complicated from here on out and there are plenty of white papers, books, and trade association documents that get down to the inner workings of SCSI. We will now move on toward the latest advances in SCSI-3, including the serial interface.

SCSI-3 ADVANCED ARCHITECTURES

Even before the final X3.131-1994 standard was approved, work had begun on SCSI-3 (1993).

You may recall that SCSI-3 provides a means to use the Common Command Set (CCS) on several hardware and connection schemes. SCSI-3 has also brought to light a serial connections method, opening the door for many new physical transport and protocol layer implementations.

Those that are familiar with the network communications models, layering models or architectures already know that the same types of models are gaining acceptance in the area of storage and communications protocols. A clean method of describing hierarchies, such as that of the OSI[44] model in the data communications world, is now being described in SCSI-3.

OSI COMMUNICATIONS MODEL	
LAYER 7	APPLICATION
LAYER 6	PRESENTATION
LAYER 5	SESSION
LAYER 4	TRANSPORT
LAYER 3	NETWORK
LAYER 2	DATA LINK
LAYER 1	PHYSICAL

Table 30-3: Open Systems Interconnection (OSI) model.

Layering is now described in the SCSI-3 Architectural Model (SAM), such that all proposed SCSI mappings are based upon the same common understandings of SCSI itself. The family of standards that comprise all of SCSI-3 were broken up from a single document into different layers and command sets. By looking at SCSI-3 from a layering model, it is possible to separate physical media, transports, protocols and primary commands from the principle commands familiar to us in the OSI Applications Layer (See Table 30-3).

---

[44] OSI: Open Systems Interconnection, is the OSI layered communications model that traverses the Physical Layer (Layer 1) through the Application Layer (Layer 7), generally equated with Ethernet (IEEE 802.x).

Another reason for taking the layering approach is that development work can proceed in smaller "byte-size" steps, rather than waiting for all the issues to be resolved before any portion of the standard can be approved.

Some examples of current SAM mapping schemes, those step-added protocols that are part of the SCSI-3 lower-layer model processes, are illustrated in Table 30-4. The SCSI-3 Architecture layer model will be described in the following chapter on serial extensions and systems.

Protocol Mapping Schemes	Physical Transports
SCSI-3 Interlocked Protocol (SIP)	Parallel SCSI
Fibre Channel Protocol (FCP)	Fibre Channel (FC)-serial
Serial Storage Protocol (SSP)	Serial Storage Architecture (SSA)-serial
Serial Bus Protocol (SBP)	IEEE 1394 (FireWire/SSI)-serial

Table 30-4: SCSI-3 Architectural Model (SAM) mapping schemes.

In 1996, SCSI-3 proposals included several new concept extensions to the expanded flavors of SCSI-2: support for graphical commands, support for SCSI-2 commands, and a general packet protocol for serial interfaces. Hardware extensibility included a low-voltage differential parallel interface, CD-ROM command sets and algorithms.

Differences between SCSI-3 and Fast + Wide SCSI-2 include improvements in both the physical (media, connector base and electrical) makeup and the signaling structure. Wide SCSI-2 required two cables (P and Q) to do 16-bit wide transfers. SCSI-3 defined a single cable with a single REQ/ACK[45] for 16-bit wide operations.

We spoke earlier about the "family" of standards. One of the families includes the SCSI-3 SPI (serial parallel interface) which defines the mechanical, timing, phases, and electrical parameters of the familiar parallel cable. Some of the electrical and cable parameters are tightened/improved over SCSI-2.

---

[45] REQ/ACK stands for REQuest (where the transmitter device, or host, makes a request for data); and ACKnowledge (where the opposite, receiving device, returns a data word or byte that confirms it received the data).

TIGHTER TOLERANCES

In order to get systems such as Fast-20 to operate effectively and reliably, many crucial signal parameters have been tightened or modified. Much of these revised technical specifications are set into the hardware components themselves.

For example, the power supply tolerance, left unspecified in SCSI-2, is set at ±5 percent in Fast-20 SCSI-3. Single ended characteristics are also tightened, with the termination impedance strictly specified. The 220/330-ohm network of SCSI-2 is prohibited in Fast-20. Characteristic impedance is set at 90 ohms [±6 ohms on the REQ and ACK signals and ±10 ohms on all other signals] instead of 84 ohms [±12 ohms] for all signals in the looser SCSI-2 standard.

Ground offsets for Fast-20 should also be less than 50 mV, making cable connection interfaces and isolation much tighter. All of these parameters make the SCSI interface less susceptible to radio frequency emission, cross-talk, and general signal degradation. For the service technician and system integrator, much greater care must be exercised in selecting interconnect components and wiring practices than ever before.

Other SCSI bus timing parameters have been severely tightened, as one might expect would be required whenever doubling the data transfer rate of any system. Internally connected drive arrays usually follow the recommendations on cable stub length and spacing because the cables are generally supplied by the peripheral manufacturer itself. However, most external system problems that occur when migrating to Fast-20 are a result of cables with wrong lengths, poor manufacturing or use of passive instead of active termination. Data transfer errors or other electrical problems result in timing slew errors, improper hold times and other similar problems. All these add up to poor performance or no functionality at all.

Still implementing Fast-20 essentially requires only a hardware change. Consideration for the long-term goal of SCSI remains essentially in tact from when SCSI-1 was initially adopted, with supplements and performance enhancements being the primary modifications.

SINGLE AND DIFFERENTIAL

Single ended SCSI devices have a radically different set of rules than differential systems. In a single ended application, four devices (i.e. host-plus-three) are permitted so long as the overall cable is left fewer than 3 meters in length. When using eight single ended devices, this cable length reduces to 1.5 meters. Even spacing of devices along the path is required, and a stub length of 0.1 meter is the limit.

Differential SCSI—typically used for remote storage, in noisy environments, and where improved system reliability is essential—allows for eight devices over 25 meters (using HVD). In this application, the stub length increases to 0.2 meter as well.

In moving from differential Fast or Fast + Wide SCSI to differential Fast-20—with the one exception of synchronous data rate transfer adjustments—practically nothing is changed in the differential mode. The result is that the migration from higher-end, differential fast-SCSI, to Fast-20 is almost transparent.

Systems integrators configuring drive arrays and component manufacturers, even though aware of the requirements for various SCSI configurations, may take short cuts in assembly or construction. The end user, neglecting to consider the additional external interfaces, may find unexplained system problems. For example, in considering Fast-20 bus construction, nearly all single-ended Fast-20 designs are typically within the host drive array chassis. Differential mode Fast-20 should be used for external bus and peripheral interconnections. Furthermore, mixing wide and narrow Fast-20 systems is not recommended.

Getting the correct number of drives for the expanded SCSI bus requires an understanding of why the SCSI bus must be balanced and how it is achieved. This task is generally left for the systems integrator to specify and configure. Even still, the end user should know the basics so that they can be certain the correct entry-level storage devices will allow for growth and extensibility in the future.

CLASSES AND STRUCTURES

Advances in SCSI-3 have necessitated new commands that in turn have developed, like most technology, a new language of their own.

We will next briefly explore some of the base principles of disk drives and their supporting codes and physical makeup. SCSI supports various devices, from tape drives to printers and scanners. Some devices, such as scanners and printers, are essentially one-way devices—they either send or receive data. Data is exchanged between SCSI devices in block format. Some devices, such as printers, will receive data in varying lengths. To identify the types of devices and how they must be addressed, device classes have been assigned (see Table 30-5).

It is not our intention to cover the many acronyms and associations with SCSI or other disk drive protocols, but we show some of the more familiar ones in Table 30-6.

Code	Device Class
00h	Disk Drives
01h	Tape Drives
02h	Printers
03h	Processor devices
04h	WORM drives
05h	CD-ROM drives
06h	Scanners
07h	Optical discs
08h	Media changers
09h	Communication devices
0Ah-1Eh	Reserved
1Fh	Unknown device

Table 30-5: SCSI device classes.

For each device class, SCSI defines a model, a command set and a parameter page used for configuration. The model for an SCSI drive consists of a physical construction and an organizational structure of data on the medium. Disk drives are considered direct access devices because they enable direct access to any logical block. Diskettes, RAM disks, and MO-drives are also direct access devices.

SCSI-3 Protocol/Command		Description
SCSI-3 Interlock Protocol	SIP	No changes from SCSI-2 except for additional messages. Defines the messages and how the phases are invoked.
SCSI-3 Architectural Model	SAM	Common set of functions, services and definitions for how a physical transport properly gets commands, data and status exchanged between two devices– complete with error handling and queuing.
SCSI-3 Primary Commands	SPC	All commands executed by any and all SCSI devices (e.g., REQUEST SENSE, INQUIRY).
SCSI-3 Block Commands	SBC	Disk commands.
SCSI-3 Stream Commands	SBC	Tape commands.
SCSI-3 Controller Commands	SOC	RAID box commands.
SCSI-3 Multimedia Commands	MMC	Essentially for CD-ROM.
SCSI-3 Fibre Channel Protocol	FCP	SCSI commands over gigabit Fibre Channel.
SCSI-3 Serial Bus Protocol	SBP	SCSI commands over IEEE 1394 High Speed Serial Bus ("FireWire").

Table 30-6: SCSI commands, protocols, and models.

The set of physical terms that drives describes the makeup of a drive includes sectors, cylinder, interleave, and logical blocks. The drive is generally divided into a sequence of logical blocks that store data. Logical blocks are sequentially numbered, beginning with logical block number 0 (LBN 0).

The size of a logical block will vary from 1 to 64 kilobytes, with the most common being 512 bytes (as in DOS). In theory, each block could be of a different size, but that is not a common mode for most SCSI disk drive configurations. A continuous sequence of blocks is called an extend.

PHYSICAL DRIVE MAKEUP

Each physical disk surface (medium) is divided into a set of concentric tracks (see Figure 30-2). The disk drive's read/write heads are positioned vertically over each drive surface and can address any of the tracks. Tracks are divided into radials called "sectors." A sector is made up of a number of different fields, which make up a sector format. A "cylinder" is a set of vertical tracks that can be addressed from a single position. The specific location of a single sector can be identified by its cylinder, head, and track number.

340 Video and Media Servers: Technology and Applications

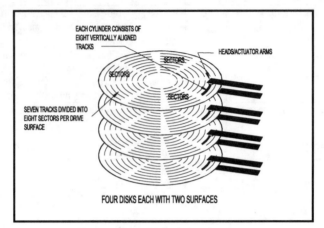

Figure 30-2: Disk drive structure.

The sector format is constructed in similar fashion to packets found in data transmission protocols. A header, a sync locator, a data payload, and some form of error correction code (ECC) make up each sector. Gaps are dispersed throughout the sector to allow for variations in motor speeds and to aid in data recovery from the heads. The header, ECC, and gaps consume between 40 and 100 bytes, depending upon the sector format (Figure 30-3).

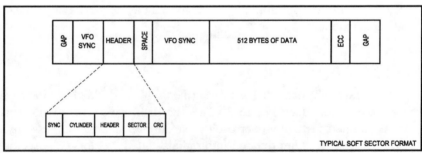

Figure 30-3: Generic sector format structure.

The Chapter 31 and 32 will discuss the extension of SCSI into other protocols and structures, beginning with Fibre Channel and continuing with Serial Storage Architecture (SSA).

CHAPTER 31

FIBRE CHANNEL DISCLOSED

Fibre Channel, at the conclusion of 1997, was viewed by many as the most probable method of intra-facility interconnection for the high-speed transfer of video and data around the broadcast plant. Nearly ever major video server manufacturer has now incorporated Fibre Channel into their product in some form.

The portion of Fibre Channel currently used in video servers is only a small subset of the entire range of capabilities, and futures, of Fibre Channel.

The work on Fibre Channel began in 1988 as an extension of work on the Intelligent Peripheral Interface (IPI) Enhanced Physical standard. It wasn't long before the original scope took on many new dimensions.

Some of the objectives the working groups focused on included extensions to the layer model, where the logical protocol being transported would be separated from the physical interface. This, in itself, was an intended replacement for parallel SCSI.

The concurrent transport of multiple protocols over a common interface was another goal that made practical sense considering the new interface would permit high-speed transfer of large amounts of packetized information.

On a larger scale, Fibre Channel advocates further see Fibre Channel as a replacement for HiPPI and ESCON. On the smaller scale, at the end of 1996, there were already proprietary fiber optic interfaces on desktops and workstations.

CHANNELS AND NETWORKS

Part of the understanding of peripheral protocols for device communications is in knowing the basics of networking functions, topologies, and architectures. Fiber Channel attempts to combine two opposing methods of communications, channels and networks, to create the best of both worlds.

Channels operate in a closed master-slave environment. Channels are used for transferring data[46] with error-free delivery and a lessened concern over transfer delays. Networks operate in an open, peer-to-peer and unstructured environment. Networks transfer both data and media[47] related information that is time dependent—where error-free delivery is secondary.

The spelling of Fibre Channel was chosen ("Fibre" as opposed to "fiber") because the physical serial media[48] can be either copper wire, coaxial cable or fiber optic cable. The word "Channel" in this case, does not reference channel protocol as described above.

Fibre Channel uses its own special device, called a GLM (global link module), as an avenue to easily convert between fiber optic cable and copper wire as a connection medium. The GLM is also the solution for upward migration to 100 MB/s (quarter-speed to full speed).

"Nodes" are Fibre Channel devices, each of which has at least one port. Workstations, disk arrays, disk drives, scanners, etc., can all be nodes. Ports provide access to other ports on other nodes. A "port" generally consists of an adapter and a certain level of software. When there is more than one port on a host node (i.e. a primary CPU or server), the software may be shared across all the ports—making the distinction between port and node somewhat fuzzy.

---

[46] Data is defined as files of information many kilobytes long.
[47] Media is defined here as voice and video, generally but not necessarily in a streaming or continuous delivery fashion.
[48] Media in this sense, refers to the actual cabling or material used for the transport of data from one node to another.

"Links" are pairs of signal carrying media (two fibers, a pair of twisted copper wires, or two coaxial cables) that attach to the node ports. One carries data into the port, the other carries data out. The topology of the Fiber Channel system consists of the physical media, transceivers, and the connectors (e.g., a GLM) that connect two or more node ports.

Fibre Channel can use several kinds of ports. Table 31-1 describes the current definitions of these ports for the three topologies (point-to-point, Fabric, or arbitrated loop).

The ANSI open standard X3T11 conveys the guidelines for an open architecture for storage interfaces that can be used in a network interface for protocols such as TCP/IP.

Port	An access point in a device (node) where a link attaches.
F_Port	Port in a fabric where an N_Port or NL_Port is attached.
FL_Port	Port in a fabric where an N_Port or NL_Port is attached (loop connection).
L_Port	Arbitrated Loop Port (may be an NL_, FL_ or GL_port).
N_Port	A port attached to a node (used for point-to-point or fabric topology).
NL_Port	A port attached to a node (used for all three topologies).
E_Port	An expansion port that enhances the value of a device attached to a Fabric.

Table 31-1: Fibre Channel port nomenclature.

Fibre Channel is an evolving standard. In the past few years, we've seen it become scaled to some "standardized" speeds and topologies. FC is capable, by present design goals, of reaching speeds up to ~1 Gb/s—or 100 MB/sec in each direction simultaneously. In a full duplex mode, FC can support 400 MB/s.

A FLEXIBLE TOPOLOGY AND PROTOCOL

From a topology standpoint, FC includes connection methods for point-to-point, Fabric (a network type similar to switched systems) and Arbitrated Loop (FC-AL). When compared to SCSI's distance limitations of 25 meters, FC can support distances of up to 10 kilometers, which is four times greater than SSA.

Fibre Channel uses a flexible address scheme, much like ATM does. Data transmitted does not need to know the route, in advance, to the destination. The address of the destination is carried in the Fibre Channel packet and the route is determined by the network. This makes

Fibre Channel Fabric implementation possible, as the Fabric concept makes the routing choices independently.

Fiber Channel protocol characteristics do not use a command set as IPI-3 and SCSI-3 do. Instead, Fibre Channel becomes only a carrier of those command sets. Fibre Channel does this in a way that informs receivers of the distinct information they need to discern. This is a bit like the mail carrier, who can carry letters for an entire neighbor, but the receiver (the house's street address) only gets what is addressed to it. The carrier, within limits, may deliver newspapers, packages, and letters. Fibre Channel may carry data of a diverse nature but always concurrently with other data.

Multiple command sets may be used simultaneously because the I/O operations are separated from the physical I/O. Fibre Channel is based upon the definition of a common physical transfer mechanism for the various commands sets (i.e. HiPPI, IPI-3, SCSI, I/P and IEEE 802).

Presently for applications related to video server technology in the professional broadcast domain, it has not been possible to carry multiple command sets on the same physical medium or link. While the Fibre Channel Community continues to define and refine the protocols and other issues, manufacturers have determined it best to leave their systems as dedicated (and proprietary). It is expected, however, as the FC-AV[49] definitions and standards progress, that a commonality or at least connectivity option will eventually permit more than just one manufacturer's data to traverse the same links. It may be that recently balloted standards for SDTI[50] may have a translator or configurer that will permit digitized video information to ride simultaneously on the same physical media, but that determination has not happened as of yet.

---

[49] FC-AV is Fibre Channel Audio/Video and is presently in the development stage. It is expected that FC-AV will be operating in the Class 4 (fractional bandwidth) communications strategies that were designated for voice and video.
[50] SDTI is the Serial Data Transport Interface – a proposed SMPTE standard, was, as of October 1997, moving through technology committees and on to formal standards processing within SMPTE and then on to ANSI.

FIBRE CHANNEL STRUCTURE

If Fibre Channel were considered just a network topology, then one could use the same terminology that is applied to networks. In network lingo, information is to be transmitted is called a "packet."

A "frame" is a block of bytes packaged together that includes control information, similar to addressing. A packet may be larger than one frame, since it is constructed without previous knowledge of the method through which it will be transmitted. Therefore, a packet may bridge one or more frames, and is generally sent to the transport layer as a sequence.

Figure 31-1 shows the makeup of the Fibre Channel frame, which consists of multiple transmission words. The length of the frame may be up to 2,148 bytes—with the payload up to 2,112 bytes.

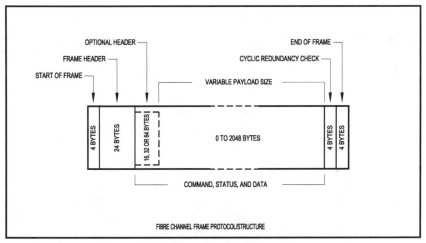

Figure 31-1: Fibre Channel frame format.

A frame header is used at the transport level (FC-2) to carry control information. Inside this header is data that identifies the sender, the destination, the protocol, and the type of information. This header, which is four bytes wide and can consist of six words, contains information required by the receiving port for and additional routing information used by the Fabric.

Fibre Channel adds another level of complexity to the aforementioned "network" model of packets and frames. Fibre Channel frames are grouped into sequences; and sequences are grouped into exchanges.

A packet, in Fibre Channel, may be quite large. In the case of digitized video-media data, the file may be as large as several hundred megabytes.

Fibre Channel must handle this large file at the operating system level by the port driver software resident in the OS. This software handles the entire video-media packet as a single exchange. This driver will divide the exchange into several sequences, and then further divide each sequence into several frames.

The ports on the Fibre Channel interface handle the addition of error-checking and reassembly on a frame by frame, sequence by sequence basis. Next, we'll begin our look at the three topologies of Fibre Channel.

FIBRE CHANNEL TOPOLOGIES: INTRODUCTION

Fibre Channel is not bound by dependency upon any one topology. This differs from what many are used to in network topologies such as Token Ring (IEEE 802.5), Ethernet (IEEE 802.3), and FDDI. Nearly all network transports are prohibited from sharing the same physical mediums because their access protocols were constructed strictly for one use only.

Fibre Channel was intentionally designed to support multiple topologies. Inherently a closed system, Fibre Channel relies on ports logging in with each other. If the topology is a fabric, it must trade information on characteristics and attributes so a harmony exists from the start of the session.

The important principle to realize is that management issues related to the topology have no bearing on Fibre Channel. Whether the fabric is a switch, a hub, a LAN/WAN, or a combination of the above, Fibre Channel expects the management to be handled externally from the process of configuring and moving frames and sequences around the connected network.

The selection of a topology for Fibre Channel is based upon system performance, budget, extensibility (growth) and hardware packaging. Fortunately, the choice of one N_Port design remains compatible with all fabric implementations.

FIBRE CHANNEL–ARBITRATED LOOP TOPOLOGY

This method was the state of the art for production video servers as of the spring of 1998. It should be noted that while more than one manufacturer is producing a Fibre Channel–based system, there are marketing attempts to sway the vision of Fibre Channel in different directions based upon the extremely closed loop environments present today.

As we explained earlier, the closed loop environment is an important principle in the systemization of Fibre Channel. "Closed loop," however, in this scenario has two different meanings. More appropriately, when addressing media or video servers being manufactured today, the proper term should be "closed *system*." This means that server manufacturer A could not connect their equipment to the same topology as server manufacturer B and expect any degree of compatibility.

Eventually we should expect this to change. However, given the speed at which products are developed and brought to market, expectations are that any course of integration will be much like that of motion-JPEG. After years of struggle, little interoperability between one JPEG scheme or the other has materialized.

The Fibre Channel–Arbitrated Loop[51] (FC-AL) concept is more involved but less complicated than a switched fabric. FC-AL is a low-cost solution for the attachment of multiple communications ports in a loop, without having to purchase a hub or a switch. Although most video server manufacturers specify the purchase of a Fibre Channel hub[52], the basic underlying principle of FC-AL is connections without external signal management.

In FC-AL, each port has the minimum needed functionality to connect to another port, making FC-AL a shared-bandwidth, distributed topology. The specification states that up to 127 L_Ports may be in the participation mode on any given loop. Node port connections can

---

[51] FC-AL is the more common nomenclature for print purposes and has the same meaning as the fully annotated wording "Fibre Channel–Arbitrated Loop."

[52] Relying on a hub adds another single point of failure to the system, one that has not been addressed by most server manufacturers due to the lack of hot-standby Fibre Channel hubs.

include printers, scanners, disk arrays and, of course, the host computer or server.

Each port is assigned a hierarchical address with which to negotiate an appearance on the loop[53]. The lower the port's address, the higher its status. The negotiation process is called "arbitration" and is where Fibre Channel–Arbitrated Loop derives its name.

Only one pair of ports may communicate at any one time. Therefore, once one L_Port is opened, only one other port may open, completing a unique bidirectional point-to-point circuit. This pair of L_Ports now controls the loop and both ports must release their control before another point-to-point circuit may be established.

Remember that an L_Port ("L" refers to an "arbitrated loop" port) may be any or all of the three described in Table 31-1. This may add slightly to the confusion when you begin attach two or more different types of ports to the same loop, which is a perfectly valid condition.

One drawback to FC-AL is that bandwidth is shared. With only two ports active at any one time, all the remaining ports act only as repeaters. If the operating system is set up such that other activities (such as real time I/O in a video server) take precedence over FC-AL activities (which is also the case in most video server applications), then the Fibre Channel transfer becomes the lowest priority. When the system is a large-scale environment (e.g., many drive arrays tied to the Fibre Channel backbone), throughput from a disk array node-D to a disk array node-C could slow to a crawl.

The example given in Figure 31-2 is an Arbitrated Loop topology with five ports, one being a Fibre Channel hub that connects to another portion of the Fibre Channel backbone. There are five nodes shown, indicating a single node (hub, workstation, or server) that may have more than a single port.

While not yet an adopted practice in media or video server applications, it is possible to have point-to-point, arbitrated loop, and fabric topologies all coexisting on the same loop. This would be the case when the FC-AL sub-group is required to communicate with a WAN via a fabric switch. Since the loop is self-configuring, it may operate with or without a Fabric present, guaranteeing equal access for all NL_ports.

---

[53] "Loop," for brevity purposes means, "arbitrated loop" in FC-AL.

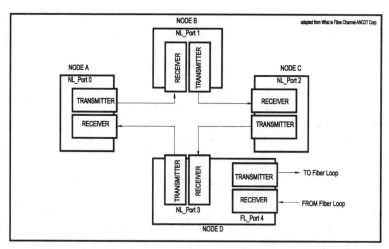

Figure 31-2: Arbitrated Loop with four nodes and five ports.

FIBRE CHANNEL FABRIC TOPOLOGY

The second of the three possible topologies is Fabric. A good example of a fabric topology comes from the foundation of fabric technology, the Telephone Company. The terminology of "fabric" is representative of a woven cloth, where many threads (fibers) enter the weave and many will be leaving. Every point in the weave where a thread intersects is analogous to a switch that can redirect the signal to another thread (path). Sometimes the signal traverses a near straight line and sometimes it makes many turns (switches).

It is difficult to predetermine which path to take until the actual circuit is "ordered" and the digits from the call reach the actual fabric. This is the same in Fibre Channel fabric topology. The Fabric is constrained only by the number of N_Ports that can be identified in the destination address field, found in the head of a frame. That limit is 2^{24} (~16,777,000), which is the number of ports that can be simultaneously logged on to a Fabric with the 24-bit address identifier.

In the telephone company model, it is of no value to the telephone instrument (and in some cases even the local exchange) how a long distance call may be routed through the wide area networks of the various LD carriers. All the traffic knows is that when you supersede the prefix of a telephone number with a "1+area code," you have given an instruction to the local exchange to route its digital packets or frames onto a switched fabric with a predetermined destination. The local carrier supervises the hand-off of the call to the Fabric and then releases any further action (except to disconnect).

In Fibre Channel, the Fabric makes dynamic interconnections between nodes through ports. The port in a Fabric, denoted by an FL_Port or an F_Port, is connected to a port in a node by a "link." This makes it possible for any node to communicate with any other node through the Fabric; yet the actual path is never conveyed to any of the nodes or ports.

Fibre Channel does not particularly care what the Fabric really does. Internally the Fabric may have multiple routes and may even use SONET or ATM networks as a means to move traffic through the system. By using conventional and in-place transports, such as ATM, wide area network connections may be possible—employing Fibre Channel as the underlying local backbone. This service is akin to a peer to peer network, and is a design feature of Fibre Channel that is not the same as the SSA or SCSI structures.

POINT-TO-POINT TOPOLOGY

In this, the simplest of the three topologies, only two ports are used. Addressing is straightforward and the probability for availability of the circuit is near 100 percent. Figure 31-3 depicts an example of point-to-point topology, the simplest connection between two nodes with a direct link between them.

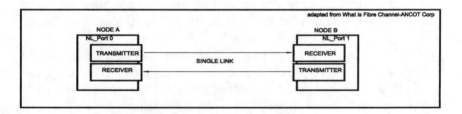

Figure 31-3: Point-to-point topology.

When considering a strict point-to-point interface, compatibility between the two ports must be maintained. This indicates that the media, protocols and speed must all be similar.

FUNCTION LEVELS OF FIBRE CHANNEL

The layering model for Fibre Channel is diagrammed in Figure 31-4. Although the model is incomplete, from a standards point it is not unlike the OSI model for networking familiar to most network specialists.

The interface connection enters at the Physical portion of the model, shown at the bottom, and moves upward through the various levels until routed to appropriate other channels or networks. The principle in these levels is that the signal path becomes bidirectional. Each successive hand-off from one level to the next has introduced all the necessary information to make flow possible, regardless of the connection on either side of the level boundaries.

Notice that the bracketed area on Figure 31-4, called Fibre Channel Physical (FC-PH), has been standardized. The first review of FC-PH began in January 1993 and the second review was begun in October 1993. FC-0 accepts the physical media input over specific speeds from 131 Mb/s through a gigabit per second. Provisions for growth beyond 1.062 Gb/s are left open.

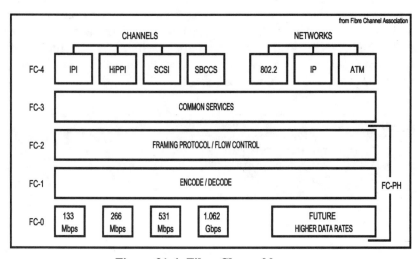

Figure 31-4: Fibre Channel layers.

From this point, the system makes the proper translation into the FC-1 level where the transmission protocol is defined. At FC-1 level, 8B/10B encoding, error detection and order of word transmission is set up. From here, in FC-2, the signaling protocol and framing, sequences and exchanges are established.

FC-2 concerns itself with defining functions within a single N_port. At FC-2, functions such as login and the rules for frame header content and layout are determined. Once the properly configured data leaves the FC-PH standard, it enters the common services level, which are considered the upper layers (levels) of the Fibre Channel architecture.

FIBRE CHANNEL UPPER LAYERS

Fibre Channel's level FC-3 and level FC-4 are the two top levels of the Fibre Channel architectural model.

Level FC-3 is the level intended to provide common services and those necessary for advanced features. Striping is one of these proposed features. The intent in striping is to multiply bandwidth using multiple N_ports in parallel to transmit a single information unit across multiple links.

Hunt groups, another of the common services in FC-3, are a set of associated N_Ports attached to a single node. When assigned a special identifier, any frame containing that identifier can be routed to any available N_Port within the set. This effectively permits more than single N_Port to respond to the same alias address. By decreasing the chance of reaching a busy N_port (one that has already been connected to another port), efficiency will be improved.

Protocol	Protocol Definition	Type
SCSI	Small Computer Systems Interface	Channel
IPI	Intelligent Peripheral Interface	Channel
HiPPI	High Performance Parallel Interface Framing Protocol	Channel
SBCCS	Single Byte Command Code Set Mapping	Channel
FC-LE	Link Encapsulation	Channel
IP	Internet Protocol	Network
IEEE 802.3	Logic Link Control (LLC)	Network
AAL5-ATM	Asynchronous Transfer Mode-Adaptation Layer 5	Network

Table 31-2: Protocols, mapping, and interfaces.

Multicasting is the delivery of a single transmission to multiple destination ports. This includes broadcasting to all N_ports on a Fabric or to a sub-set of the N_ports on a Fabric.

The highest level in the Fibre Channel structure is FC-4. This, like the highest layer in the OSI model, defines the application interfaces that can execute over Fibre Channel. Here mapping rules for the upper layer protocols are defined so that they may be properly used on the levels below. Because Fibre Channel is capable of transporting both network and channel information, concurrently, this makes it possible to address a variety of existing protocols.

You may also refer back to Figure 31-4 for an overview of all the Fibre Channel layers.

CLASS OF SERVICE

Fibre Channel uses the term "class of service" to describe communications strategies for making connections or conveying information for a wide range of needs (see Table 31-3).

Class 1 is based on hard, circuit-switched connections. The complete path is established before data transfer begins. When a host and a device are linked, that path is not available to other hosts. Connection time is short and this method is used for rapid, high-rate–high-volume data transmission.

Class 2 is a frame-switched service without a dedicated connection. Delivery is guaranteed and a receipt (or acknowledgment) is returned. If data cannot be delivered, a "busy signal" is returned and the transmission must be retried.

An optional mode, called Intermix, can be selected. Intermix reserves the full bandwidth as a dedicated connection—but allows excess bandwidth, idle Class 1 time, to be allocated for connectionless traffic. This is an efficiency plus in Fibre Channel and maximizes the use of links for more than one class of service.

Class 3, or datagram service, is similar to Class 2, but is connectionless. By eliminating the requirement for confirmation of received data, delivery can be made to several devices attached to the Fabric. This one-to-many form of transmission does provide notification to the sender that data has been received. If one or more of the user's links are busy, the system cannot know to retransmit the data.

Delivery time for these services is a key factor in deciding which class of service to employ. When the transmission distance is a factor and time is key, it may be prudent to determine which service is best for

354 Video and Media Servers: Technology and Applications

the application required. This is where the final class of service, Class 4, fits in.

Class 4 uses virtual connections rather than dedicated connections. It is still a connection-oriented type of service, but Class 4 distributes the bandwidth of a port among several destinations. Class 4 is also referred to as fractional bandwidth, because is parses out bandwidth based upon need.

Class of Service	Connection Type or Name	Description
Class 1	CONNECTION ORIENTED	Dedicated connection.
Class 2	CONNECTIONLESS	Frame switched – without dedicated connection.
Class 3	DATAGRAM	Connectionless – without confirmation.
Class 4	FRACTIONAL BANDWIDTH	Virtual-Connection oriented – managed bandwidth.

Table 31-3: Fibre Channel classes of service.

More classes of service are being proposed and developed as applications for them arise. In a closed system, such as a Fibre Channel RAID storage system, the manufacturer has already selected and satisfied the engineering application for the class of service that is best. The network engineer may have another set of tasks on their hands—as the need for real time, voice and video, isochronous delivery is realized.

FIBRE CHANNEL FUTURES

Manufacturers are continuing to develop various portions of Fibre Channel for their own and others' uses. Debate continues, as the convergence of SSA and Fibre Channel–Arbitrated Loop has spawned a variety of new high-performance serial interfaces for the Upper Level Protocol (ULP).

Since late 1996, there is work ongoing in the community on Fibre Channel-Audio Video (FC-AV). As of late-1997, most of the work was completed and it was expected that in early 1998 a viable proposal, suitable for standardization, will be made. As of this writing, the concept of using Class 4, with fractional bandwidth allocation, seemed to make the most logical sense for multicasting and broadcasting of time sensitive, isochronous data.

During 1997, at least one video server manufacturer was producing a Fibre Channel structured RAID system. Trademarked FibreDrive[54], this system pioneered the use of Fibre Channel disk drives in a software RAID configuration (called RAIDsoft) for video servers. Using software to construct the RAID protection parity and backup schemes eliminates the costly hardware RAID controller that can be viewed as a single point of failure.

Attempts in ANSI technical committees to harmonize similar and/or parallel developments in Fibre Channel and SSA (such as gigabit SSA and Fibre Channel-Enhanced Loop [FC-EL]) are being met with emotions running from hot to cold. The process continues and the marketplace either waits for the standard or tests the infant products.

When the concept of Fibre Channel was introduced on the coattails of SCSI-3, the door remained open for solid technical advancement and a progressive migration to serial architectures for the transmission and storage of information. Getting it all to work together is proving to be another monumental task.

FIBRE CHANNEL COMPATIBILITY ISSUES

Compatibility remains a very large issue in Fibre Channel. Due in part to the fact that Fibre Channel is actually a number of individual specifications, providing a level of interoperable products (e.g., disk drives and host adapters) may prove to be more difficult than desired.

SCSI is still the predominant means of interface. Manufacturers may actually be reluctant to push development of anything other than their best sellers for risk of loosing that market. This has certainly been the case with SSA and is proving so also in early Fibre Channel product development.

In an effort to verify cross interoperability and compatibility, it was suggested a means of certification would be in order. Yet, as of

---

[54] Leitch/ASC Video Corporation produces the VR-300 video server architecture which uses FibreDrive and RAIDsoft for their disk array management and protection scheme.

March 1997, the Fibre Channel Loop Community (FCLC) still had no plans for a "certification sticker" that might indicate compatibility for product. The FCLC trade association did at least offer to provide a laboratory for the testing of products.

CHAPTER 32

SERIAL STORAGE ARCHITECTURE

The past several years have placed an increasing burden on I/O subsystems, the direct result of an insatiable appetite for increased storage requirements for more data. This creates a growing number of applications, demanding larger blocks of data, based largely upon client requests addressed over networked servers. Dealing with this demand has spawned new protocols, new hardware, and renewed interest in serial storage architectures. This chapter will deal with the functions and technical details of Serial Storage Architecture—a serial connectivity implementation for the transfer of large blocks of data, based upon established SCSI protocols.

Server data is expected to be available all the time. And this requirement applies to both enterprise and global servers. The same is true for the video server, which has mission-critical implementation with 100 percent up time.

System reconfigurations are not possible any longer when the clients are always awake and active. Still, the users expect scalability, improved performance, virtually no down time, and ease of use. The result is a new generation of serial storage I/O technology that can be applied to network computing as well as digital video and media storage for servers and non-linear editors.

New Demands

To meet these newly defined and certainly expected requirements meant some extensions to the key technology of SCSI-2. SCSI-3 offers several improvements over SCSI-2 in many areas.

SCSI-3 increases the number of devices or nodes on the SCSI bus. There is new support for Ultra SCSI, Serial Storage Architecture (SSA), and Fibre Channel. Devices and interfaces now have automatic configuration, without jumpers, using SCAM—and a new connectorization scheme including SCA (Single Contact Assembly) and SCA-2, which offers hot-plugging capabilities.

Very High Density Cable Interface (VHDCI) is offered for maximizing the number of bus connections on a given interface or PC card. This includes support for Type II PCMCIA cards giving rise to mobile computing applications with faster throughput drives.

SCA has also been implemented into the backplane, permitting multiple drive bays to be backed by a PC-type board connection scheme. This can be found in some of the recently released servers from the major manufacturers as well as the stand alone drive arrays positioned for RAID or other applications. At Fall COMDEX-97, brand new systemization chassis are now being touted that will offer a "no-need-to-open" PC chassis with plug-in openings that will accept a variety of industry approved peripherals, storage, and I/O devices. The promise of 15-second software configuration is an attempt to truly take the words Plug-and-Play[55] to a proper dimension.

The Ultra Approach

As we described in the previous chapter, parallel Ultra SCSI employs the SCSI-3 protocol, effectively doubling the maximum bandwidth to 40 MB/s. This is accomplished by new signaling and clocking techniques, permitting Ultra SCSI to coexist along with SCSI-2 devices.

It should be understood that when mixing unlike SCSI devices using the same controller, the bandwidth becomes limited to the slowest device on the bus because of the poling and acknowledgment functions that occur. Ultra SCSI adapters and drivers for the major manufacturers

---

[55] The original Intel-Microsoft Plug-and-Play specification for ISA was released in May 1993. The first commercial PCs arrived in late 1994 and the first operating system to incorporate it was Windows 95, released in August 1995.

were expected due in late-1997. Windows NT drivers are also expected in the same time frame.

Ultra SCSI uses faster clock cycles and is therefore more stringent on many parameters when compared to existing parallel SCSI devices. Connectors, cables, terminations, and the number of devices on the bus must all be carefully considered when migrating to Ultra SCSI.

It is possible to retain the slower legacy SCSI devices by moving them to separate SCSI controllers, thus preserving the integrity of the new Ultra SCSI bus. Newer technologies, such as LVD[56] are expected to relieve some of the problems with bulkier cables and single ended systems. It is also foreseen that Ultra SCSI may approach another doubling of bandwidth, to 80 MB/s, in the future.

With the considerations still growing for parallel SCSI and Ultra SCSI, there were new serial storage interfaces already on the market in mid-1997. Some of these devices have already been planted into digital media storage systems in both non-linear editing systems and broadcast-quality video servers.

To migrate from conventional and advanced parallel SCSI interfaces, new serial transport technologies are being developed. Unlike the parallel interfaces that send data over a series of individual, parallel wires, serial interfaces send the data in packets over a single set of wires, usually shielded twisted pairs (STP). This method simplifies transmission and reduces overhead, thus improving speed.

SERIAL ARCHITECTURES FOR MEDIA SERVERS

Before we begin the discussion of the second relatively new serial architecture for the transport and exchange of media- and non–media-based data, we need to spend a few moments discussing the broad overview perspectives of serial connections.

Until just about the last two years, around 1995, SCSI was predominantly the only method of connecting disk arrays, CD-ROMs, scanners, and storage devices to what we'll call "processors" or "engines." Dating back to the late 1980s, users of media related workstations—intended for the editing and manipulating of video or still images—thoroughly understood there to be basically only one overshadowing method to connect a hard drive, and that was SCSI.

---

[56] LVD: Low Voltage Differential drivers (see Chapter 30)

Considering the advances in processor speeds, drive throughputs, graphical user interfaces, and other display accelerators, this approach seemed logical and acceptable for the masses. Only a few could afford superengines such as Silicon Graphics (SGI) workstations and IBM video servers, which used both SCSI and HiPPI[57] as a connection interface. Although drive array integrators and manufacturers continued to apply their own techniques to products designed to improve storage throughput; they were also inhibited by the many factors that reduced overall performance. SCSI, although it was doubling its bandwidth every five years, just wasn't keeping up with the growth in processors and the insatiable demand for improved media applications in the industry.

Once the introductory concepts of SCSI-3 were presented, it became apparent that a new dimension to connectivity was on the horizon. After only a few short years of product development, applications in serial storage architecture for media servers became evident. The roller coaster ride is beginning to settle and industry is starting to reap the benefits of both SSA and Fibre Channel. Manufacturers of video servers and storage arrays, which include OEMs' newest forms of disk storage devices, are beginning to employ the new technology to their own products.

While advocates of both SSA and Fibre Channel continue to press their beliefs and disbeliefs upon industry in general, some manufacturers have already taken a stand and made a choice. The purpose of the previous chapter and the remainder of this one is to give the reader an opportunity to open the door to new a concept in storage technology.

Since we all expect to be deeply entrenched in digital television (DTV) before the start of the next millennium, it is important to understand the foundations of this technology. Whether you stand on one side of the fence or the other, somewhere over the course of the next few years the technician, engineer or manager will be making choices on drive and storage technology—including the interface and topologies that support it.

---

[57] HiPPI: High Performance Parallel Interfaces, commonly used in high performance super-computer platforms with multiple processors and drive arrays.

That choice will have a long-term impact on extensibility in the physical plant in the same manner that choosing between BetacamSP versus MII did in the mid-1980s. This time, however, the choices are more critical and less reversible than selecting a videotape transport format. The decision to incorporate one form of connectivity versus another for servers, drives, and arrays will also meld into the choice of the compression format, the topology and the commitment to one particular manufacturer for life.

With that reminder, let's look at the industry's alternative approach to Fibre Channel, called SSA. Serial Storage Architecture and Fibre Channel have already begun to address some of the inadequacies of parallel SCSI implementation. Fibre Channel offers some features that include the ability to use either copper cabling or fiber optic cabling for extended distances and high noise immunity. SSA uses a similar concept, but a different approach.

Fault tolerance is an important and critical element to ponder when designing any mission-critical application. For serial storage devices, one element is achieved by the use of dual porting or dual pathing. This method provides either bidirectional loops or two physically distinct SSA or Fibre Channel ports that allow data to flow over two paths, protecting the entire data network in the event of controller, interface, or media failures.

Although it may seem that by sending data in packets the throughput is slowed, the fact is that serial technology will outperform parallel technology significantly. We will all become intimately acquainted with packets and packetized delivery of television information, as it is expected to revolutionize the transmission end of the industry. The SMPTE has had working engineering committees developing standards (such as SDTI) for several years. We must recognize and understand that packetization will be the method of transmission for digital television before the end of the decade.

AN INTRODUCTION TO SSA

Serial Storage Architecture (SSA), is designed as a peripheral interconnect interface for all types of devices—CD-ROMs, tape drives, printers, workstations, servers and most any storage subsystem, from mainframes to personal computers.

IBM's Storage Subsystem Development group pioneered the research and development along with other vendor/partners. Real products have only been available a few short years. When SSA development started, the stated goal was: "...to achieve high performance for a majority of applications and to provide a road map to growth in device data rates."

SSA claims high reliability and comprehensive error detection and recovery, and it provides for good fault isolation as well. The system is implemented in low-cost CMOS that uses little power. SSA does not require the expensive and fragile cabling like that of SCSI interfaces. SSA uses a single chip as the only real interface point, and that can be built onto both the drive controller and the host interface card.

While speed remains a broadly defined and relative specification, what really matters is performance. The parameter to be most concerned with remains net total throughput, after overhead. Some of the issues in defining throughput performance are bandwidth, read/write ratio, record length, and the actual interface architecture itself. SSA is placing a new perspective on performance by addressing and solving several of the issues associated with peripheral interfacing.

SSA opens the door for a new storage interface standard (through ANSI X3T10.1) designed as a replacement for parallel SCSI, with a long-term goal that includes replacement for older and slower interfaces as well. One of the features of SSA is its ability to independently address different devices without being affected by other devices on the network link. To accomplish this, each SSA fundamental building block is made up of a single port with a capacity to carry on two 20 MB/sec conversations simultaneously. The ring-based links in a SSA network have a built-in feature called "spatial reuse," whereby each link can operate independently.

Links that are not involved with one particular data transaction are available for use in another. Spatial reuse permits device A to talk to B while at the same time permitting device C to talk to D. Furthermore, none of the device sets are hindering or detracting from the other device sets—even when they're all on the same network ring. This is not possible in bus architecture (e.g., SCSI-1 or SCSI-2) because each node connects to essentially the same pieces of wire end to end.

SSA's independent paths allow for increased total throughout. Each node or SSA device can support two loops each with two links in and out. Spatial reuse allows concurrent data transmission on both loops.

Serial Storage Architecture

This principle increases bandwidth without the requirement for arbitration between data paths.

Concurrent data transmission is an important element in getting systems to work independently and efficiently. The upgrading of SCSI-2 and SCSI-3 standards has enabled hosts to send one set of data without waiting for return acknowledgments before the next group is sent. One of the ways that SSA can continually transmit data is due to SSA's alternative paths (loops) that provide an additional route for different data packets.

A typical SSA subsystem consists of an adapter with a 20 MB/s read path and a matching 20 MB/s write Path (see Figure 32-1). The typical peripheral connection point usually has a pair of dual differential ports that connect into and out of each peripheral device. The net effect, with spatial reuse, is an 80 MB/s total throughput with no single point of failure.

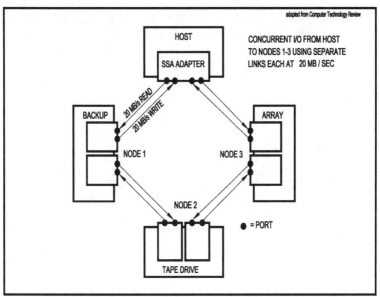

Figure 32-1: Connecting SSA nodes and ports.

Limited overhead and smaller frame bytes also increase system performance. The SSA network links are full-duplex, meaning that data travels in both directions on the bus, and frame multiplexed. The maximum bandwidth on any given link is 20 MB/s, which is the same speed as Fast + Wide parallel SCSI.

Connections between nodes are much simpler, reducing latency and the complexities in error correction. Error detection is placed on both the driver channels and the receiver channels. All these elements continue to aid in reducing the overall cost of implementation.

Physically, the SSA node to node connection is made using two pairs of differential cables, with a port capable of handling two signals in each direction. A simple shielded four-wire cable makes up the interface between two 20 MB/s point-to-point links.

SSA dramatically increases the number of devices placed on the network compared to parallel SCSI. The addressing scheme further supports different topologies. In SSA the "adapter" is considered the base device (see Figures 32-2 and 32-3) and there are essentially three other supported functions.

The "string" is a linear network consisting of up to 129 nodes. The "loop" consist of one host adapter and up to 128 dual-port nodes (127 devices). Finally, a "switched" topology offers up to three levels of switching at 126 ports per switch and no single point of failure.

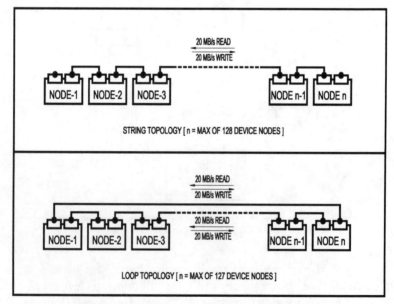

Figure 32-2: SSA string and loop topologies.

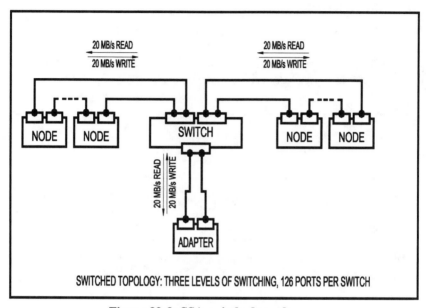

Figure 32-3: SSA switched topology.

Most adapters have built-in support for two loops. Other host adapters have provided connections for up to four loops, as in IBM RS/6000 systems. Cable size is smaller and the length between ports is much greater. An SSA cable diameter is less than 6 millimeters and the physical external connectors are microminiature leaf connectors with six pins. At a base of 15 meters between ports, the maximum raw error rate is 1 in 10^{12} bits transmitted. For comparison, SCSI has 31 signal wires (68 total pins on a P or Q cable), has a much shorter interconnection distance and requires a complicated connecting scheme.[58]

Disk drive synchronization complexities are further reduced in SSA. With a large disk array application, where the drive spindles are typically synchronized via an external cable, SSA has eliminated the requirement for extra cable by placing the spindle sync signal into the SSA protocol.

---

[58] Some implementations for Ultra2 SCSI, using LVD (low voltage differential) interconnects, permit up to 25 meters, but are limited to two devices on the bus.

SSA Frame Structure

A frame is organized into 128-byte groups, which yields a very low protocol overhead, ensuring excellent high-load capability with high bandwidth. The minimum overhead is about 6 percent of the entire frame, which averages around 8 bytes.

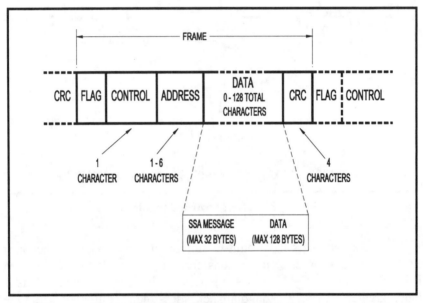

Figure 32-4: SSA frame characteristics.

SSA FRAME COMPONENTS			
FRAME	1		Indicates the frame type (divided into a control, privileged or application reference).
ADDRESS	1-6		Used to direct a frame to the correct node.
DATA	0-128		Must be in multiples of 16 bytes for each sub-group.
CRC	4		Cyclic redundancy check—similar in function to a parity verification or check sum.

Table 32-1: SSA frame components.

A single flag acts to separate each frame and provide a start point for the next control character. The flags remain neutral and are shared as a common dividing point between the end of one frame and the start of the next (see Figure 32-4 and Table 32-1).

MIGRATION AND IMPLEMENTATION

From a cost standpoint, SSA claims to be much less expensive to implement. Adapters, hubs, interface cards and other product lines for networking are established.

The migration from SCSI to SSA was of high concern in the architectural design. The SSA protocol for storage is still that of SCSI-2. There are little to no changes in the application code, and only some minor changes in both the device driver code and the file codes.

The SSA serial interfaces have significant advantages in many other areas as well. One of these is distance, as SSA can be implemented using media ranging from simple twisted pairs through fiber optics. As mentioned, when running on copper, the distance between each device can be up to 15 meters with some cases designed to allow up to 20 meters, with no performance degradation.

Using optical extenders, a short-wave laser multimode interface extends stable performance to 680 meters at full bandwidth and up to 2.6 kilometers with acceptable performance degradation.

SSA can coexist with other interfaces as well. Functions such as database management or media server RAID subsystems can be implemented for FDDI, FCS, ATM and host bus, making SSA a logical route from parallel SCSI systems to serial interface for several RAID device applications.

INTERFACE ADAPTERS

SSA is now being implemented on the PCI bus using adapter cards. Product specifications indicate a maximum data transfer rate of 132 MB/s (PCI burst rate), with a maximum sustained data throughput of 50 MB/s (50/50 mix of read/write cycles). Greater than 3000 I/O operations per second are possible when 4K block transfers with sustained 70 percent read and 30 percent write balance is averaged.

Two SSA ports support a total available bandwidth of 80 MB/s. The interface provides for SSA primary initiator capabilities and multiple

SSA primary hosts. Error rates for the interface are one error in one trillion bits.

There is also an IBM SSA RAID adapter for PCI that combines software selected RAID levels 0, 1, 3 and 5 for higher performance and high data availability. Uninterrupted data streams, in RAID 0 and RAID 3, are possible for multimedia and digital video applications, with implementation already available by prominent desktop video editing systems[59]. Demonstrations have shown sustained data rates of over 70 MB/s using RAID 0 and 30 MB/s, with no data loss, using RAID 5.

INDUSTRY PROS AND CONS FOR SSA

SSA is well suited for clustering and multihost configurations, especially where connectivity and large aggregate storage bandwidth are required. However, SSA is not well suited for single stream, bandwidth intensive applications. SSA has not been widely accepted by industry when compared with the accelerated movement in Fibre Channel activity.

SSA also requires knowledge of the path to the peripheral device and should that change (e.g., should a storage subsystem be removed from the loop for maintenance), the address must be recalculated. SSA sets up the path at the start of a session (boot up), and with that path, it directs the specific data along that path and only that path. With this scheme, optimization of data flow becomes limited should the path have to change for any reason.

There have been many comparative arguments about SCSI versus SSA. Some state that SSA advantages include hot-plugging (not founded; see the discussion on SCA-2). SSA requires no discrete terminations (true, SCSI must be properly terminated, but in many cases, SCSI devices now have built-in automatic terminations). There are no physical address switches in SSA (some still remember SCSI's manual address switch settings, which are also not necessary in Plug-and-Play[60] compatible systems). Moreover, as a last example, SSA permits 127 devices per loop or string. (SCSI allows for 15 devices per string, but

---

[59] At Fall COMDEX-96, Scitex Digital Video demonstrated their desktop editing system using SSA and shared media storage.
[60] Plug-and-Play is extended to include SCSI devices–using the SCAM configuration methods standardized in SCSI-2.

when adding multiple controllers or bus expansion devices, this number increases.)

SSA has not had the exposure in media communities that Fibre Channel had throughout much of 1997. Despite the low penetration, SSA is not just a design; it is a reality. Chips sets are being manufactured by at least six manufacturers. The ANSI Task Group of Technical Committees of the Accredited Standards Committee X3 has produced the documents (ANSI X3T10.1) for full standardization.

Future mass media storage continues to depend upon increased speed and lowest-cost implementation. In 1995, IBM suggested that SSA links could be close to 100 MB/s by 1996, allowing for a 400 MB/s interface theoretically once fabrication on thinner silicon becomes cost-effective. This has not moved significantly in the marketplace and with lower-than-expected industry acceptance, it may not come into being for some time.

NETWORKED MEDIA APPLICATIONS

SSA is a powerful storage architecture, but it is not the complete equation. To complete the solution, switched channel and network interfaces should be added. Figure 32-5 represents a conceptual networked system.

Presently, these subsystems are being developed for data communications networks that attach SSA cache and RAID controllers to channels and networks. The market has expressed good support for IBM's SSA products. Many digital video and prepress companies are already using SSA interfaces to connect storage subsystems to processors and host computers.

In some cases, third-party SSA interfaces from companies such as Pathlight are connecting well-established non-linear editors and workstation-based graphics systems to both drive arrays and other workstations, over a serverless networking system. There are even video buffer cards that provide all the interface to and from the workstation over SSA, making a complete solution possible.

SSA interface and networking products are available on the entire gamut of computer platforms, including Macintosh, UNIX platforms, Sun and Windows NT. This makes the sharing of data on a common storage library practical and efficient.

Networking the SSA system is not unlike using Fibre Channel. A typical hub would connect loops of disk drives and discrete SSA host adapters for nodes. Nodes can be systems, storage components, or other peripherals.

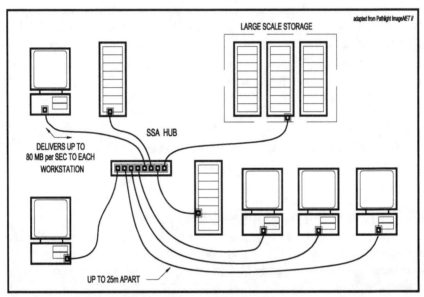

Figure 32-5: Basic SSA-based networked system.

Professional video products, including both the Media 100 and the Scitex Digital Video Sphere platforms, are selecting some implementation for storage systems using SSA. Scitex states, as do many other major players in this market, that the architectures of their platforms are designed to take full advantage of high-speed networking solutions, such as those in SSA. The common goal is to simultaneously handle media transfer and creative tasks without exiting the editing application or tying up the entire system. In turn, this delivers exceptional connectivity and creative freedom to users.

Fast Multimedia has chosen to support networking using SSA based architecture to combine MPEG-2, DV and uncompressed 10-bit digital video for digital end to end production. The advantages of working with native digital video in a networked environment can only be realized with extremely fast, reliable and shareable storage subsystems like SSA.

While the hope in the end is for an SSA physical drive, the move toward Fibre Channel, for both drives and networking, may force SSA drives to take a back seat. Only a few vendors currently are in serious development, even though the SSA development community remains strong. In the hard and fast case of business and product development, the market place will make the final judgment.

CHAPTER 33

SOLID STATE DISKS

Traditional disk storage has not kept pace with processing power performance. As computer power continues its upward spiral, Moore's Law holds merit—but it doesn't seem to apply to magnetic disk storage. In mid-1995, it was stated that magnetic disk performance was only increasing at about 5 percent per year while CPU power is doubling every year.

True, by 1996 and even more in later 1997, the intense I/O operations-based applications continued to escalate with every new product or software package. To avoid what might actually be a plateau in overall system throughput, the search continues for improved disk I/O performance that is comparable or beyond that of SCSI-based systems.

Of equal importance are backup, protection and data integrity for computer storage activities, as exemplified by the profundity and desirability of RAID control architectures. In this chapter, we will look at a rebounding solid state disk technology that has had an expensive and not particularly attractive record of accomplishment—at least until now. We will examine how this technology might find its way into the growing media server marketplace and why it remains a complicated and seldom implemented storage subsystem.

DISK CACHE BASICS

The mainframe market of the '80s pushed forward a demand for data transfer speed. As the boundaries of system processor memory gave way to more intense disk/data activities, the notion of a solid state disk emerged. This then-conceptual technology was quickly subdued as cost and volatility made it nearly impossible to integrate solid state disk subsystems into even high profile, large-scale mainframes.

Later in that decade, as the PC became a viable solution to localized computer activity, the primitive disk cache was born. Then, during the later part of the 1980s, as memory size and memory upgrades taxed the base cost-to-performance level of most large systems, the concept of a solid state disk (SSD) was again considered. Technology innovators studied SSD as an alternative to a costly central processor or front-end processor upgrade.

In the early 1990s, as multimedia, video-on-demand and other near-real-time, stream-intensive applications come on-line, data transfer rate increases are shedding new light on the concept of a memory-based storage subsystem. The goal of this process is to ride out the disk seek and latency problems sometimes inherent in all data store-and-forward architectures.

One probable solution lies in the comeback of the once not-quite-ready-for-prime-time SSD. In this concept, memory is used as a dedicated resource that appears as a logic drive in physical DRAM memory.

Today, solid state disks are being developed for applications such as on-line transaction processing for bank ATMs or credit card database servers. This concept works when the application(s) that are called frequently have the same base processes and are repeated nearly identically every time there is a request by a remote on-line host.

The solid state disk is far more expensive compared to the conventional magnetic disk. In the coming years, as parallel and multiple processing engines, such as the Texas Instruments TMS320C series DSP-based processors now being developed for studio and video processing applications, come into being, that may become a different story.

As the appetite for these newer and faster CPUs grows, it will certainly place even greater demands on data I/O and will force new advances in disk subsystems. This will surely drive costs down and integration up. At that time, the SSD may play an important part in the makeup of other laser and optical storage systems.

The Magnetic Drive's Fundamental Problem

There are fundamental problems associated with magnetic disk storage subsystems. Designers are aware of them and understand that they must be improved upon or overcome.

One of these problems is that the conventional disk cache, has just about run out of steam. The disk is an electromechanical device and even when coupled with relatively large memory-based caches, it doesn't come close in I/O performance to what storage in dedicated solid state memory can achieve.

The perfect magnetic drive, the ideal mechanical storage subsystem, would require zero mechanical latency and disk cache hits that were 100 percent for all I/O requests. The mechanical latency has a finite factor that can be improved with more spindles, more heads and faster drive speeds, but can never be made to approach zero.

The cache hit ratio performance obligates a different set of requirements. Write requests with a 100 percent cache hit ratio are achieved by directing all write requests to cache, then announcing to the host that the data has been verified and that it has been written to cache.

This task can even be achieved in the background and seem transparent to other CPU or storage activities. However, achieving read request cache hits of 100 percent is an entirely different and complicated matter. The basic flaw lies with the fact that the serial-type read cache, based upon the first in first out principle (FIFO), suffers from write-over once the cache is filled. It takes good data management software to maximize cache hit ratios because constant evaluation of all data requests is required in real time.

A typical disk cache system performs best when data is positioned with patterns that allow for good read recovery. That is, the data is relatively contiguous and local to the head positioning for the given data being read from the drive. Disk cache memory only holds blocks that were recently accessed or blocks that are local (adjacent to physically on the drive surface) to those recently accessed. This means that the types and forms of data that are accessed during a disk cache become key to the performance of the application.

Lookup tables or indices that must be consistently accessed as pointers to other larger segments of data are another candidate for the cache. Simply stated, it is a little like the CD-ROM game where the animated cursors and sound sprites are sometimes stored on the hard drive and even sometimes in memory. These sprites are generally used more frequently and they would be far less meaningful to the game action if they had to be extracted at 150 kb/s from a CD-ROM.

The depth of any application can sometimes dictate the intelligence of the cache. The management of the cache size and hit ratio—or percentage of accesses that generate a cache hit—depends of the specific locality of the data payload. Again, locality physically refers to the degree or amount of data that is around any given block of data that will be accessed.

HIGH SPEED ACCESS

Fast access time remains the goal and makes the solid state disk a practicality. In a solid state disk, some of the data normally resident on the disk is now kept in memory. The data I/O management process considers the mechanical limitations of access time (seek and rotational latency) and diverts those latencies into software controls that actually intercept SCSI commands and replace them with memory fetches.

According to disk drive manufacturer Quantum, in a solid state disk data access time improves up to 30 times versus a conventional cache integrated with a magnetic disk. When data I/O becomes free of mechanical liabilities, as with a solid state disk, it is likely to achieve an access time at or below one millisecond when compared to an average magnetic disk access time of nearly 15 milliseconds.

To cite an example: In conventional data retrieval, a significant amount of the I/O activity occurring to or from the disk drives is focused on about 1 to 3 percent of the actual on-line data transfer. This, of course, depends upon the operating system, because some OS's are more focused on I/O activity management than others.

Studies by industry and researchers show that small I/O activity is typical for most operating systems and applications, which is where the solid state disk excels.

CACHE HIT IMBALANCE

An imbalance in cache hits is another problem that is associated with I/O bottlenecks. This can occur most frequently because of a "hot file," which is an overactive data set that continually demands a high percentage of I/O requests.

If the size of these frequently accessed files overflows the cache, then it and other data that must be regularly accessed, such as root or index files, must be purged and/or reloaded regularly. This causes a reduction in I/O performance.

Reducing the access time for hot files is important. This is another example where solid state disks have benefits where access time becomes one-tenth that of the fastest magnetic disk drive.

PROS AND CONS

Objections to solid state disk implementation extend beyond the prohibitive costs. When inactive data is stored in memory the overall inefficiency of the data I/O system becomes an important factor and effectively reverses net performance improvements. This balancing act is difficult to control unless the applications are quite specific and well understood by the data management controllers.

Products are now being developed that constantly monitor data and thus maximize efficiency by storing the most active data in cache memory and the inactive data on the disk drives themselves. By dividing the memory appropriately, be it magnetic or silicon, the overall effect is a reduction in the total amount of memory required for temporary tables and transaction log files. Heavily accessed tables and indexes can then be stored on a silicon-based storage subsystem in solid state memory.

Cache pollution is another problem that results when data is only used once or very infrequently. Advanced algorithms, used in data cache or solid state disk accelerators, are being implemented for the control of protected and normal data space segments. Essentially this process divides a virtual cache of memory and parses it out depending upon the need. Here again, dynamic manipulation of memory minimizes the effects of an unbalanced cache and will improve I/O throughput. This positive feature thus reduces the impact of a more costly SSD implementation.

NOT ANOTHER RAM RECORDER

The concepts of a solid state disk we've examined are different from what we might remember of the video RAM recorders of recent years. At that time, RAM disks for video had very short clip storage. It is unlikely that when RAM disks for video were first being explored, that the concept of the video server would be such as it is today.

Video disks did not possess much storage either, and because satisfactory compression technology was unavailable, any video on hard drives had to be stored full-bandwidth. Both systems required a lot of RAM or magnetic disk space. Now, with a lot of experimentation in high-volume data transaction and on-line processing, plus the improvements in compressed video, there may be a potentially new and important place for solid state disks in the future development of the streaming media server.

The RAM disk may also play a part in the integration of metadata, where short-term caching could be accomplished in silicon rather than continuing to tie up disk resources.

It is expected there will be many instances where dynamic storage of compressed video data will tax storage subsystem I/Os, and this is one case where throwing in more gigabytes of disk storage won't be the only answer. Someday we may actually see solid state disks become a portion of the media streamer's specification as a means to tackle the inevitable changes in switching latency brought on by packetized television transmission.

CHAPTER 34

INSIDE THE DRIVE

The conclusion of 1995 marked both the 40th birthday of the hard drive and the silver anniversary of the microprocessor. These two technological advances, and their paralleled growth over the years, have made it possible for media storage to transition from live to tape to disk recording, and will probably allow it to move beyond. The factors and the technology that are contributing to the changes in disk drive and storage technologies will be the subject of this chapter, portions of which appeared in the *Media Server Technology* column in September 1995 and January 1997.

Not too long ago IBM offered a 10-MB drive as an option for its first-generation x86 IBM-PC. Just a few short years ago, during the era when 180-MB drive was considered impressive, growth in disk drive capacity was only around 30 percent a year. Like so many areas of technology, this growth curve has dramatically changed once again,. We recognize that in 1997, a gigabyte (GB) hard drive is considered the low end of the scale.

Recent postulations stated that drive capacity growth would continue at nearly the same pace at which the power of the microprocessor has increased. It appears that Moore's Law can actually be applied to storage devices as well as to silicon compute devices. Given the recent increases in storage size and reductions in cost for magnetic disk drives, some have stated that the new Moore's Law should be changed to 6-months for disk storage instead of the 18-months used in the 1970s computer power prognostication.

Between 1980 and 1996, the PC industry consumed 850 million disk drives made up of all capacities and all form-factors. Of those, 530 million drives were scrapped, meaning that of approximately U.S.$169 billion in total sales over the period, around U.S.$106 billion worth of drives are no longer in service.

Advances in several areas of magnetic recording technologies have allowed the principle goal for the drive technologist to be realized. Fundamentally, that goal has been to increase capacity, reduce cost and improve performance.

Industry keeps pushing drive capacity upward at a rate of around 60 percent per year. This increase forced some significant technological changes between 1994 and 1997. The ultimate result for both the user and the manufacturer is that the OEM can now produce better, more efficient products to meet the needs of the ever increasing thirst for the storage of data. These factors have been part of the reason that the storage of digital media has become more cost effective and in turn more accepted.

SHIFTING TO THE UNIPROCESSOR

It is expected that chip counts on drives will continue to drop as the onboard, dedicated disk microprocessor is phased out. This shift to a uniprocessor architecture is possible because designers are using digital signal processors (DSP) to perform functions that were once handled by microcontrollers and DSPs independently.

This transition is by no means trivial. Even though new silicon fabrication techniques are allowing for higher-density transistors and switching, the level of microcode that is being shifted away from discrete component processors and on to single or uniprocessors, such as the DSP, is substantial. Some predicted [in 1995] that "It might be six or seven years before we see a complete shift to the uniprocessor."

With the increased demand for digital media storage, emphasis grows on speed or throughput. Throughput of a system includes getting more data on and off the drives, as well as moving the data through the pipeline with higher accuracy and improved reliability. Together these factors all contribute to performance.

At the core of advanced media storage design comes some interesting developments in drive technology. New disk drive head and media properties are producing more advanced processing and new silicon requirements. New improvements in data packing are allowing system designers to get higher system performance and in turn give the video server designers greater flexibility in reaching their goals. The balancing process of systems design notches upward with every new stride in storage technology, leading to new products that come from emerging technology.

CHANGES UNDER THE HOOD

As stated at the end of 1995, "During the period from early-1996 through mid-1997, disk drives were expected to exhibit some major under-the-hood changes that would aid in advancing the capabilities of media storage and retrieval." Other expectations stated in 1995 that "By 1996 the mainstream desktop system will exceed 100 Mb/s." Manufacturers remain poised to provide high volume solutions to the demands for faster, higher data density disk drives. Many [in 1995] expected the 140 Mb/s barrier to be broken by mid-1996, and that's not just for a few systems—it may actually become the norm.

By the end of 1997, the 300 MHz Pentium processor was out of the ovens and into the test labs. To keep up with the demand for the increased storage demands will be a challenge. The computational capabilities and the amount of data that 300 MHz and above desktop machines will produce brings a new set of complex tasks for memory makers, software designers, and storage device manufacturers. We can only extrapolate to a steep, upward slope over time.

What does this do to the drive manufacturers? For now, a once dormant technology may be coming back with a new market-driven demand. Chip manufacturers are reacting to the strong request for faster read times from disk drives–brought on by faster processors, bus speeds, and the masses of desktop platform becoming available.

At issue in the mid-90s are new methods of internal disk drive architectures that allow better and faster throughout from the read standpoint. There are at least two primary technologies available for the disk-read process, referred to as the "read channels." The most common method is the "analog peak-detect" read channel. The other technology is partial-response, maximum likelihood (PRML). PRML has been around for quite some time, yet until recently remained somewhat stagnant and little used.

382   Video and Media Servers: Technology and Applications

PRML Emerges

At the Fall COMDEX-93 exhibition, only three manufacturers showed PRML drives. In the early summer of 1995, somewhere around ten disk drives were employing PRML from manufacturers, including Seagate, Quantum, Fujitsu, IBM and Western Digital.

As far back as May 1994, almost a generation ago in technology life cyles, designers had become so well versed in pushing the analog peak-detect read methods to their limits that they began to seek new methods. At this same time, the industry also became aware that to get the extra boost needed to cross into the new zone—to get that 30 percent boost in performance—meant a whole new world of problems in a gradual migration away from peak-detect to PRML.

PRML read channels have provided a methodology to go beyond the 90 Mb/s of peak-detect read channels, considered state of the art in 1995. PRML changes the electronic circuitry that resides between the heads and the disk controller, which makes up the read channel. PRML is a digital technique that can tell whether the disk's magnetic head is reading ones or zeros, with a predictive algorithm that increases speed.

In the analog peak-detect read, the analog read channels require a relatively consistent peak voltage in order to find clearly defined peaks in the signal. As the bit density grows higher, the peaks and valleys come closer together and voltage levels no longer appear constant. Detection errors grow and the plateau of throughput is reached at around 80 to 90 Mb/s.

PRML works by extracting a sample clock from the analog signal, and then it samples the data signal peaks at that channel rate. It basically says, "Only look for this peak signal at this specific time and ignore all others." Going further, the technique compares the digital sample against a predetermined threshold to ascertain if a flux reversal did or did not happen. Even if the response is only partial, considering that the platter is spinning at over 7000 rpm, the circuitry can determine what it is *most likely* to be. Thus, PRML—partial-response, maximum likelihood—gets its name. This "maximum likelihood" prediction comes because the system is also looking at the bits surrounding the response and comparing its level to signal levels that preceded it.

The challenge will be to make not only the drives but also the silicon that processes the read data meet the demands. PRML inherently costs more and requires more power than peak-detect. The silicon also increases from 50 to 75 percent in physical size. In early 1994, the mainstream drive players stated they weren't too concerned about PRML technology because they didn't think that the need would hit the average desktop until late in 1996. Still, the pay off is in higher speed and higher density when using PRML.

BANDWIDTH, CODING, AND FILTERING

Bandwidth comparisons have also revealed more speed increases when using PRML. Employing new digital filtering and shaping technologies to the signal waveforms leads to new data encoding schemes. This new data encoding scheme is actually more efficient than that used in peak-detect encoding.

PRML encodes eight user bits in nine bits of data, referred to as a 0,4 encoding scheme. Peak-detect uses a 1,7 encoding scheme where two data bits are mapped into three encoded bits. If a read channel were running, for example, at 150 MHz in peak-detect, you would get a 100 Mb/s rate. The same 150 MHz channel with PRML would give 133 Mb/s of throughput, which is a 33 percent increase.

Already one device is manufactured that has a throughput of 148Mb/s. These rates were reached using the silicon side of the equation. Drive heads must still be matched to media in order for this overall performance increase to be realized.

Faster throughput demands are now being achieved with higher drive rotational speeds and greater drive densities. In order to achieve the plus-130 Mb/s ranges, some hard drives are now using a magneto-resistive (MR) head makeup, which was invented by IBM. This MR head construction is necessary to reach beyond the 80 Mb/s data rates, which is where thin film read/write heads are claimed to have peaked out.

The proper head-to-media balance is still difficult to achieve. It is expected that designers will continue working on the silicon portion that will take throughput in read channels to the 220 to 240 Mb/s range.

VIRTUAL CONTACT

There is another class of heads that promises higher capacities without the shift to magneto-resistive heads. So-called "virtual contact" recording claims to offer similar performance because the new heads fly closer to the disk, thus increasing density. Virtual contact heads can achieve the desired number of bits per square inch of disk media, or areal density. Virtual contact heads can also use a derivative of the existing read channel technology as part of their system.

As a perspective on how close the contact is for a disk drive head, in 1995 drive heads on disks were flying around at just 2 to 3 microinches off the surface. The virtual contact head, also known as a "proximity recording head," flies around at 1.5 microinches above the disk media's surface. Interestingly, it has always been thought that when a head contacts a media's surface, you've crashed the drive, rendering it useless. In virtual contact head drives, there is often a light contact with the drive surface. Of course any contact at all will cause friction, yielding to debris collecting on the surface from either the media or the head. So the question then becomes, "How much debris is too much?" and thus, where or when is the failure point reached?

Promoters explain that by providing a newer media coating and an extremely light head load, and by using an air bearing between the head and disk, performance is achieved at a cost point and reliability factor similar to other methods.

Opponents and proponents of both virtual contact and MR drives admit that the pure sales volume of each respective drive will become the determining factor in acceptance. So experiments are now being conducted that might combine virtual contact and MR technologies together.

The end of 1996 marked the beginning of final development for the next generation of PRML technology. PRML has advanced to become a technique that allows for statistical prediction of where and what the next set of magnetic data might be. PRML now presents marked differences in the way data is retrieved from the disk drive media.

Industry originally thought that PRML would be used in only high-end systems, but by late 1996, that was not the case. Following another year of development, it seems that PRML digital techniques now make more sense in mid-range storage systems.

AREAL DENSITY

Areal density has long been a baseline measurement factor for drive manufacturers. When the physical footprint of the drive, and housing, is fixed, the quantity of platters and the areal density become the principle factors in setting the amount of data stored on a given form factor. As stated earlier, when PRML was developed, it brought about a 30 percent increase in areal density versus the older analog peak-detect techniques. With additional digital advances in PRML, it is expected that another 10 percent increase in areal density will result. These next-generation digital drive technologies were expected to begin showing up in hard drives by late 1997.

Other goals being sought by today's engineer include lowering the signal to noise and designing for greater diversities in head materials and recording media. Even though the next-generation PRML has resulted in an improved overall areal density, increases in drive speed have imposed other engineering challenges. As data bits are lifted from the drive's media the read channel electronics must process them faster. This becomes more significant as the both the rotational speed and the areal density increase. When both factors are addressed properly, the net result is improved performance all around.

CHANGING THE HEADS

Another major feat to be accomplished by manufacturers revolves around a physical change in the size and types of the heads and the mechanics of the slider arms that support them.

Magnetic head technology has all but accomplished the process of converting from inductive thin film to magneto-resistive (MR). IBM began incorporating MR heads in 1994 and since then, others have followed with their implementation in 1994 and 1995.

MR heads are much harder to integrate than inductive thin film heads. Still, much of the 2.5-inch drive market uses MR heads, most notably in the mobile market because of their lower power consumption and smaller disk to disk spacing.

SLIDERS

The slider is the mechanism that supports the heads in a disk drive. This slider arm must move extremely fast in order to get the head position from the inside to the outside edge of the drive media or platter. Angular speed and precise alignment are very important when drives are rotating at upwards of 7200 rpm and access times are in the low teens of milliseconds.

When discussing the slider's mechanical specifications, drive manufacturers make reference to a percentage of the physical properties (size and weight) of a slider supporting older ferrite—the lower the number, the better.

The sliders for modern drive heads are being replaced with "70 percent-sliders," mechanisms that are 70 percent of the size of those used in the 1970s drives. Advances in slider technologies have lead some disk manufacturers to use "30 percent-sliders" with the norm actually around the 50 percent size. The reduction in mass contributes to moving the heads into position faster and more accurately while consuming less power and risk of failure.

FLYING HEAD HEIGHT

The distance between when the head is not picking up data (i.e., seeking) and when it is in close enough proximity to extract bits is the "flying head height." Much older sliders used to move the head into position, stabilize, drop the head slightly, read the data, then lift and move on.

Today, flying head heights are the key to getting more driver platters closer together. This dimension will continue to be reduced as head design technology matures even more. Proximity recording, another technique where the heads actually sit on the media surface, is still on the horizon but with little anticipated progress during 1997.

HIGH PERFORMANCE DRIVES

Non-linear video editing (NLE) demands a high data rate for the capture, editing, and display of full-screen broadcast and near–broadcast-quality 30 frame per second operations. No single disk drive can provide the 20- to 27-MB data rate required for broadcast quality video. For this reason, compression, picture display size, and frame rate trade-offs are often used in order to reduce the effective data rate.

The newest of the high-performance disk drives incorporate many of the features explained in the preceding paragraphs and some more. Spindle speeds have traditionally climbed up the ladder from 5,400 rpm speeds to present-day implementations of 7,200 rpm. Beginning in late 1997, new drives were available that increase the rotational speed to 10,000 rpm, yielding a remarkable increase in data transfer rates.

With these higher spindle speeds, data access time is significantly improved. These new drives are capable of internal formatted transfer rates of 11.3 to 16.8 MB/s—an impressive improvement 40 percent improvement over 7,200 rpm drives and 85 percent over 5,400 rpm drives.

Along with the increased 10,000 rpm spindle speed, the average latency time is reduced to less than 3 milliseconds. This is a reduction of nearly 1.2 milliseconds in average latency from comparable performance previous-generation disk drives. Add in the capabilities gained from embedded servo technology, and the seek time is also reduced similarly to an average of 7.5 milliseconds.

The family of 3.5" form-factor Ultra SCSI parallel transfer interfaces has also moved into industry-standard Fibre Channel–Arbitrated Loop (FC-AL) serial interface. These drives are now fine-tuned for applications that benefit from high-performance systems that include:

- Transaction processing
- Scientific and graphic-processing
- Network file servers for the enterprise
- Professional audio and video applications
- Video servers

This new drive family, pioneered in late 1996 and introduced in January 1997 by Seagate, shares many of the common attributes including:

- Magnetic resonance (MR) Heads
- Positive response-maximum likelihood (PRML) channel
- Embedded servo

This drive family became available during the middle of 1997 and is provided in both a 4.55-GB and a 9.1-GB ultra-high-performance capacity.

The use of embedded servos removes the problems associated with thermal calibrations (TCAL). This is extremely important for video servers, video-on-demand, and digital NLE applications.

The following 8- and 16-bit parallel SCSI I/O configurations, using the Advanced SCSI Architecture (ASA II), give designers several choices for implementation:

- Single ended Ultra SCSI
- Wide single ended Ultra SCSI
- Wide differential Ultra SCSI
- Wide single ended SCA Ultra SCSI
- Wide differential SCA Ultra SCSI
- SCSI-3 Fibre Channel SCA

Reliability and data integrity are continually being improved. The power on MTBF specification for the Seagate Cheetah series drives is now one million hours. Inherent in this architecture is support for SCAM Level-1 and Level-2 compliance, which includes self-monitoring of the drive systems with onboard internal diagnostics.

New high-performance drives incorporate support for physical drive carriers for interchange and replacement. The industry recognized nomenclature is Single Contact Assembly (SCA) and has a later performance criteria specified as SCA-2, which offers hot-plugging capabilities.

At least some server manufactures are incorporating the reporting capabilities to give the user advanced warning of potential problems. As seek errors and other indications are tracked, a sudden increase or change in these errors gives a heads-up notification that can result in replacement of the drive before total failure or possible loss of data occurs.

FIBRE CHANNEL DRIVES

The newest extension of the SCSI-3 command set has been extended to the hot topic of Fibre Channel. To avoid the problems associated with parallel SCSI interfaces, including high cost and restricted connectivity, new drives are beginning to come into the marketplace. The new drives emerged in early 1997 and began to be incorporated into video servers about the time of the NAB-1997[61] conference.

The new 9.1-GB and 4.55-GB series 3.5" drives are now available in the first truly open industry-standard serial interface for Fibre Channel-Arbitrated Loop. The FC-AL interface addresses the performance issues necessary in making large system arrays, such as those used in video server applications, easier to implement. SCSI protocol is supported in FC-AL drives, providing significant savings in SCSI software development, which is being captured by at least one video server manufacturer.

Burst data transfer rates for FC-AL of 100 MB/s per port are achieved. Since FC-AL is a loop technology, a dual port is included that in turn provides for excellent fault-tolerance and high connectivity. Up to 125 drives can be connected on the loop, giving an addressable storage space of nearly 1.14 terabytes. RAID 5 command support is also built in to the FC-AL interface and a 1-MB multisegmented cache (for FC-AL) is included on the new Seagate Cheetah series FC disk drive products.

---

[61] "NAB" is the National Association of Broadcasters, whose annual national conference and exhibition is held every spring in Las Vegas, Nevada. This is the premiere conference for the broadcast industry–having the same impact in size and draw as COMDEX, the Computer Dealers Expo, also held in Las Vegas, during the fall.

CHAPTER 35

PREVENTING DISASTER

Most facilities employ some type of fault tolerance systems management. This may be in the form of two video cart machines, dual records of programming, a primary and a secondary satellite feed, etc. As we move toward the reliance on a reduced-videotape structured environment, the question always arises, "How do we operate when something as important as our video server goes down?" The answer is not really as simple as it might sound.

Providing 100 percent redundancy, with two servers and dual playback of the same material, does not always offer 100 percent insurance. In fact, it might even complicate things more if the proper ancillary equipment is not brought into the equation.

Preparing for all the possibilities of disaster is very difficult and costly. Disasters could be categorized as earthquakes, floods, power failures, failed drive arrays, server crashes, or even someone accidentally unplugging the network hub in another portion of the building total removed from the server itself.

Down time is another costly evil. Unscheduled down time results in either lost revenue or additional expenses for repairs. When the time or availability cannot be recovered in any form, then management asks, "Why weren't we prepared for this," or "What can we do to prevent it in the future?"

There are many elements involved in protecting the assets of the facility. The schemes go beyond just dealing with what is now being stored on the video server. You may have purchased additional drives so as to configure the storage array in some level of RAID. An off-line data cassette system may have been included in the system. You may be relying on making a backup/break tape that stands ready to jump in if the primary systems should fail.

All these preventative options were most likely based upon budget and operations at the time a server was being considered in the first place. This chapter will deal with some of the hardware and subsystems used in preventing disasters when a server-based video store and forward environment is being considered.

CLEAN ENVIRONMENTS FOR YOUR SERVERS

Disk drives, although they seem rugged, are not built to the same standards as we've grown to appreciate in professional video terminal equipment, transports or television cameras. Protecting your investment requires attention to a multitude of physical areas in the facility.

The facility engineer should consider several issues. Is the server unit insulated from heat, moisture, vibration, and dust? Is the equipment room thermally and random particulate controlled? How close is the ventilation duct, both the supply and the return? What is the power regulation? Will an uninterruptible power system (UPS) be required? How susceptible to static and lightning is the facility? Is there a trained technician on site who can deal with networks, systems administration, and video compression technology?

Operating system corruption, lost data due to read failures during a crash, and other expected problems can wreak havoc on the data structures. Protecting data in the event of an operating system crash is a function of how well the software was written. When the power is removed from the system and it hasn't completed its write cycle, this could create disaster, not just for the piece that was being recorded at the time of the failure, but possibly for the entire drive array, depending upon what else was being addressed at the moment power was lost.

Placing the server on its own UPS is a good idea if the balance of your facility is not already on one. Having that extra few milliseconds to complete a write cycle during a feed or transfer could mean the difference between losing just a portion of the data during the failure and potentially losing it all.

All the disk drive protection in the world, including two sets of parity disks or multiple RAID 3 drive arrays, won't help if there is only one power supply that feeds the entire system. If the system is on the same plug strip, or there is no UPS available, all the efforts taken for disk protection are lost if there is a power failure during a read or write from the storage system. Be sure you put all the elements associated with the server (RAID, hubs, and controllers) on the UPS so no single element is compromised.

Routine maintenance of the physical portions of the server (filters, power supplies) and keeping the network operating system updated and free of unnecessary software are other wise strategies for operational integrity.

FAILOVER STRATEGY

Today, RAID protection remains the most accepted means of data protection. However, RAID cannot protect against complete disaster, especially when some single point of failure has not been provided with an alternative. A RAID storage system should be more appropriately described as *resistant* to disaster, rather than protective of disaster. We should further understand that employing RAID can increase data *availability* and *integrity*, but not data *accuracy*.

"Failover" can be defined as the processes that occur during a potential or actual failure—and how they respond during the recovery. Failover, although not desired, is certainly planned for in RAID architectures. Fans and power supplies, controllers and processors, monitoring and diagnostics all account for elements in failover strategies.

The most difficult item to deal with in RAID is also the most expensive. RAID controller failover is far more complicated than power supply or cooling fan failover. In a RAID system, the host computer must communicate with the RAID controller. Most conventional operating systems are not designed to handle a disk controller failure in a graceful manner. When the SCSI device is suddenly gone, a great number of error messages are generated. This is not how a RAID controller should failover. Most video servers have their RAID array connected to the hose via an SCSI connection. This is how most RAID controllers are connected to the CPU (host computer), but the RAID controller should appear transparent to the host computer when it does fail.

Most RAID controllers today were originally designed to run as stand-alone. The concept behind RAID is such that the entire array appears to the server as one single drive. Software is usually written into the operating system to make the failover of the RAID controller appear graceful. This, in the data server world, is referred to as a host-based failover strategy.

TOLERANT DISK SYSTEMS

The RAID Advisory Board (RAB) affords classifications to systems based upon certain guidelines they consider relevant to the integrity, protection and sureties of drive systems. The simplest system classification that RAB recognizes is Failure Resistant Disk Systems (FRDS).

FRDS is designed to protect against loss of data due to any single component failure within the system. In addition, FRDS requires protection against the loss of *access* to data due to the failure of any single disk.

The second level of protection recognized by RAB is called Failure Tolerant Disk Systems (FTDS). This protects against both the actual data loss and the loss of data access due to failure of any one of the components. For example, to have FTDS classification requires a separate dual-controller RAID system, with redundant power supplies and cooling subsystems in each.

Both classifications have extended variations (identified by a "+" sign following the acronym), which include additional data availability and enhancing qualities not specified in the standard forms. A more in-depth description can be found by contacting the RAID Advisory Board.

The most demanding of the classifications is Disaster Tolerant Disk Systems (DTDS). This class requires both FTDS+ and FRDS+ criteria certification, and must provide undiminished, continuous access to data, even if several faults should occur simultaneously.

In portions of this book, we have described conceptual video server systems that come very close to meeting classifications similar in function and protection to what RAB offers through their certification processes. Unfortunately, in the broadcast industry we must rely more on our own intuition and engineering judgments, along with the reputation and demonstrated performance of the industry's manufacturers, to obtain this level of insurance in system design.

RAB offers suggestions for DTDS that center on mirroring functions that are performed either by host software or by cooperation among multiple disk controllers. Other methods that might be developed in the future are left open and RAB offers channels that do not preclude other techniques from being implemented.

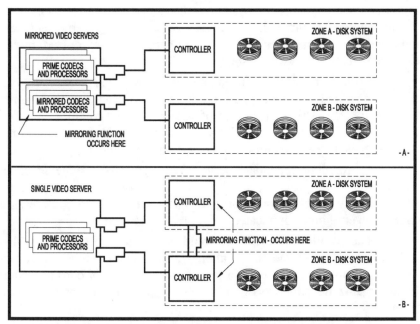

Figure 35-1: Disaster tolerant video server and disk architectures.

To meet the fundamental requirements for disaster tolerant systems, the broadcast scenario uses automation systems that include "air and protect" operating procedures integrated into stand-alone subsystems that come from existing and variant levels of server performance and fault tolerance. Mirrored servers and mirrored disk drive subsystems (see Figure 35-1) are alternatives that follow the RAB classifications, but can be extended to apply to the specifics of the broadcast industry.

MIRRORING

The act of protection is the prevention of disaster. One of the more common methods of protection in a video server system is called "mirroring." When speaking of mirroring a video server, we consider this differently from the RAID 1, which is also known as, but more appropriately called, "disk mirroring."

The principles are basically the same—a mirror is a complete duplicate copy of all the data or, in the case of a mission-critical video server application, the server system itself may be completely duplicated. Complete server mirroring requires two completely redundant and independent servers, each operating in parallel, synchronized by an external software/hardware system.

Each of these independent and discrete servers, however, can contain mirrored drives (e.g., RAID 1 disk mirroring), or some other level of RAID. Although the use of RAID 1 is not common in video media servers, the use of "RAID's concepts" for other than just disk mirroring, as in redundant or mirrored servers, definitely is.

Mirrored or redundant servers will typically operate in exact parallel, therefore providing a total redundancy of all the components. By parallel, we include power supplies, data storage and compression engines–everything is duplicated.

The parallel to mirroring would be to run two identical LMS-style video cart machines in parallel, with detection software that can make an automatic changeover in the event of a failure or corrupt media. Large organizations, networks and the like typically will operate in this fashion.

Mirroring of servers is expensive and some feel strongly that this is not the way to construct a system. Where budget is of concern, providing a totally separate system—whether running in parallel or not—is not generally something that will get the approval of a manager who's task it is to be efficient and careful with the dollars.

TAPE ISN'T DEAD–IT JUST CHANGED ITS ROLE

There are a few different schools of thought about backup systems. Many prefer to keep the video masters on the shelf as protection in the event something is erased. This is not very effective and is protection that exists on the surface only. It is highly unlikely that having the masters around will do much for keeping spot revenue continuing if the majority of the transports (or the labor force) have been reduced or repurposed.

Preventing Disaster 397

Others prefer to use their existing video cart machines to keep backup material on-line and readily accessible. This is expensive and probably less reliable because maintenance is reduced and operator familiarity is marginal.

The third up-and-coming method is the data tape robotics library system. A data tape robotics library system consists of a computer controlled picker arm, a carousel-type set of storage bins that house the data cassettes, and any number of data tape transports for recording and playing back of data cassettes. A computer instructs the picker arm to retrieve a cassette and load it into an available transport, and directs the transport to shuttle to the correct location on the tape to start a recording or playback.

Long before there were robotics-assisted systems, there were discrete data tape transports that were used for the purpose of archiving and backup data from computers or hard drives. The extension of this principle to video server backup systems has evolved over several years to the point that it is now an accepted peripheral of most video server systems.

To realize how much data will be generated or will need addressing in servers, we should take a brief diversion to understand just how big the task at hand is. If we consider that 600 hours of programming needs to be stored at a compression rate of 24 Mb/s (with 8-bit coding), you will need 6,480 gigabytes (6.4 terabytes) of storage. If there is RAID 3 parity striping, this amounts to 7,200 gigabytes of disk space.

$$\begin{aligned} 600 \text{ hours of storage} &= 2{,}160{,}000 \text{ seconds} \\ \times\ 24 \text{ Mb/s} &= 5.184 \times 10^{13} \text{ bits} \\ \div\ 8 &= 6{,}475 \text{ gigabytes} \end{aligned}$$

If all this media were stored on-line, on disk, this would be a very large disk array indeed. So the real solution will warrant a method of keeping the digits on a medium that is far less expensive and provides higher density. The combination of compression and data tape is the ideal solution for today.

Tiers of data storage are necessary if cost control and physical data space are to be managed. The most expensive of these storage tiers is on-line—which provides immediate access to limited amounts of content in the form of video, audio, and timecode. The life span for online storage is relatively short.

Near-line is the middle ground for storage of large amounts of data. Near-line's life span is usually a matter of hours (used for program time shifting) or days, when replay on second channel or next day/later in the week is required.

Off-line is the where data is stored that can allow for from several seconds to several minutes for retrieval. It is typically the least expensive and the highest density. Far-line is off-site or remote-storage not easily accessible for routine operations.

STREAMING AND SCANNING

Tape was the first magnetic mass medium for computer data storage. Tapes for the PC environment have grown from the original 30 MB capacity per tape to the modern systems that pack gigabytes into similar-size cartridges using compression. Most tape systems are considered "streaming," although that is a somewhat inaccurate definition. The streaming *mode* describes a specific method for recording that requires a continuous and uninterrupted flow of data.

Streaming tape is limited only by the speed at which the tape runs and the packing density of the bits on the tape. In a streaming operation, the tape does not have to stop between blocks (with blocks being segments of data with specific start-stop points per segment). This allows for faster transfer of data and in turn lowers the cost of the transport, and extends the life of the media.

Other forms of data tape mechanisms include start-stop, which handles data in blocks–starting and stopping between each recording while the transport recognizes the data and writes information about the block before moving on. Parallel recording places data onto parallel tracks across the length of the tape. And, finally, in serpentine recording, instead of utilizing the parallel tracks for writing bits in parallel, the data is serialized and written to the tape first on track one going forward, then on track two going backward, and so on for as many passes over the tape as there are tracks are used.

Helical scan recording, like the Type C and BetacamSP formats, have also become quite popular in the data industries. The two scanning systems most popular today include 8mm and digital audio tape (DAT).

Helical scan recording uses the entire tape area, providing for higher density. Some helical tapes use guard bands, areas between tracks where no data is written, and some write the edges over one another (as in DAT).

Several format options are available for storage of data on linear tape. One product is the conventional QIC (quarter-inch cartridge) tape, developed by 3M around 1972. At the time, there were no PCs and the expectations for success in QIC were small. The first cassette (3M's DC300A) was about the size of a paperback book and held 300 feet of tape. The 30 inches per second recording time used phase encoding and put one track at a time on the tape, with a serial stream of 1,600 bits per inch.

Modern data cassettes of this type and size now record upwards of 13 GB on a cartridge and use multitrack heads, higher recording densities, error correction and much more sophisticated servo driven transports. After 25 years, the principles have changed little and the formats continue to be widely used for the storage of compressed data.

Since its inception, QIC has evolved into various other forms. The minicassette forced its way into a 3.5" drive bay while at the same time grown from 40 MB to 15.5 GB under current QIC standards. The other format, DCC, for digital compact cassette, has not kept up with QIC and is currently undergoing a new marketing plan, renaming it DCC Data.

The hot new name in tape storage is DLT, digital linear tape. DLT claims to have some five times the equivalent life of DAT formats. A single cartridge holds some 20 GB and that can easily double with compression. Automation or "jukebox" implementations will generally use DLT for several reasons.

Tape speeds of 100 inches per second and up make these tapes ideal for mass storage. When coupled with jukebox automation systems, this can amount to terabytes of storage in a near-line/archive arrangement. Sustained transfer rates of 10 MB/s are obtainable, gained by nonmoving heads and high-density recording.

DLT tapes are rated for 500,000 passes as opposed to 2000 passes in helical scan systems. DLT has less tape tension than helical or 8mm systems, and because DLT uses several tracks (64, 128 and 206 tracks on ½" tape), shuttling backwards and forwards will reduce the tape's expected life to 10,000 uses.

One drawback to DLT is its cycle time—the time between issuing a command to find a file, when unloaded, to the time it begins playing that file. Due to the higher-density track spacing, the tape must align itself each time it is loaded into the transport. Two calibration tracks are recorded on the tape during formatting. These tracks are used by the heads for automatic alignment. This process is a stiff penalty to pay, as it can take upwards of 90 seconds to load and calibrate before shuttling off at high speed to locate a data track. Most DLT systems will specify a time to locate any track from any point with that maximum generally around 90 seconds.

Since the invention of helical scan recording by Ampex in 1961, the principles have been extended to all forms of video and computational data storage. One other form of tape backup that has evolved out of the development of what broadcasters would quickly recognize as digital videotape transports, is the helical scan, digital linear tape transport. Employing 19mm linear tape cassettes, these recorders are being used as instrumentation recorders, data storage and retrieval systems, and data/telecommunications recorders. Coupled with automated cartridge libraries, these tape systems can hold over 2 terabytes of storage in fewer than eight square feet of floor space.

Cartridges range in capacity up to 330 GB each and utilize a cartridge partitioning format that allows for hierarchical storage management execution. Sustained transfer rates of 15 MB/s, with burst rates of 20 MB/s are achievable. This specification allows video, recorded and compressed to 24 Mb/s, to be recorded or played out at five times faster than real time. The tape search speed is 1.6 GB/s at 300 lineal inches per second of tape speed.

The high costs of these transports can more than cover the labor spent waiting for tape to transfer at real time between server and library. A typical one-hour program can be off-loaded and stored in less than 15 minutes, opening valuable on-line storage space.

Other features important in high-bandwidth tape backup systems include read-after-write verification, which includes automatic rewrite. The user can set the block size for any file or file segment when writing the file. A block size can be one, which is called byte-stream mode, or any even number usually between 80 and almost 2,000,000 bytes. Buffers built into the transport smooth out the transfer of data to and from the tape, and build the logical block structure to and from the physical blocks on the tape.

Tektronix is using a product manufactured by Exabyte that houses four 8mm scan transports that can operate at a data rate of 3 MB/s. A total of 80 cartridge bins each hold one cartridge with 20 GB of data for a total of 1.6 terabytes, the equivalent of 110 hours of play at 24 Mb/s using motion-JPEG compression.

The stated access time for a file is less than ten seconds with a worst-case retrieval time of three minutes. Cartridges can be partitioned for individual block addressing.

As the transition to video servers continues, the departure from conventional videotape advances, but the future of magnetic tape for storage grows. The nature of data storage will prevent the conventional videotape concept from being applied to the storage of data files. Similarities do exist however. High-performance data tape still uses longitudinal tracks to identify data addresses, in a similar fashion to the way ANSI/SMPTE 12M-1995 timecode is recognized. This is where the principles depart.

Once video is transformed from linear video into packets, it can no longer be thought of as a continuous stream. Videotape, as we know it today, remains somewhat segmented and frame bound. Very simply, for example, in analog tape the divisor is the vertical and horizontal sync signals. In digital tape, it is the EAV and SAV codes. With servers and compressed digital, a different structure takes form for storing video as data. Inserting video into an existing data tape is not the same any longer, principally because insert edits are no longer possible. What occurs on a streaming data tape is an append (add-on) or overwrite, with both requiring that records be kept and updating be cataloged.

PARTITIONING

Partitioning is a process that reserves specific segments of the digital linear tape to hold data as though they were envelopes in a long string tied end to end. The computer industry speak for partitioning is "volume stacking," where each tape is divided into many sections. In an application where a data tape is used to hold multiple 30-second commercials, partitioning would predivide that tape such that the location of each 30-second spot would be known to either an on-tape database or a database held in another device.

Standard partition sizes must be constructed such that the reserved space is adequate to handle the average minimum commercial size, once compressed and transformed into digits. When data needs to be replaced, added, or erased, partitioning permits manipulation of only that single partition, without the need to erase or reformat the entire data cassette. This principle permits inserting and deleting of specific spots rather than resorting to an "append-only" profile in nonpartitioned data tape.

Up to now, we have discussed the storage of relatively short segments of video data onto partitioned digital streaming tape. System architects need to be considering the storage of larger and random length segments in the form of an on-line archive for the production and airing of programs. The video information provider (VIP) will need to flow material to clients in an asynchronous mode. The staging of the production process will inevitably require that several recordings of digital data onto tape will be necessary between acquisition and distribution. It is also quite probable that distribution will be in the form of digital data tape, rather than baseband video, once some universal interchange formats, such as MPEG-2 4:2:2P@ML, become popular.

Partitioning now takes on a different process, and the need for smaller segments disappears. Program-length recording, if stored to relatively small data cassettes, should probably be stored on individual tapes to allow for easier management of media over the long term.

Tape still remains plagued with a few problems, such as too many standards and a lack of consensus on those that are already in place. Software also always seems to stay a step behind hardware. Latency, the time from issuing a command to locate a tape, until it is ready, is another difficulty that is usually addressed from an operational perspective.

Regardless of the difficulties, tape is an insurance policy against disaster–even though the thought of implementing tape as a storage medium in today's age of digital servers is a rather dull one.

HIGH-SPEED BENEFITS

Tape systems must be integrated with the capabilities of the server. Having a five times transfer speed is of high benefit, but an even greater benefit is being able to play the data straight from tape without having to transfer it, in its entirety, to the server first before it is available for playout.

Preventing Disaster 403

As the model for the server matures, the disk recorder features will be expanded to act as a temporary buffer between tape drives and real time video playback. Already, Hewlett-Packard is stating that with their MPEG-2 MP@ML server, a data tape needs only a few seconds of transfer from tape to server before the media can be played out as video. HP is implementing their MediaStream product with a DLT archive as are all the major video servers including Tektronix's Profile, Leitch-ASC's VR, and Philips-BTS's MediaPool.

In most cases, the server manufacturer has partnered with or would associate with another manufacturer, such as StorageTek, to provide the ancillary off-line and archive peripherals. More often than not, another, third party vendor, would collaborate with both the server manufacturer and the off-line storage manufacturer to provide the automation and/or database management system that integrates the two product lines into one system solution.

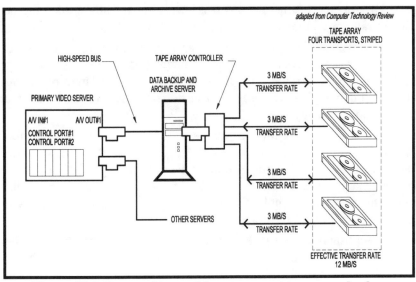

Figure 35-2: Data archive and four transport tape array backup.

Promoters of tape library systems are beginning to play their cards one by one. They are starting out with the myth that a data tape library is natural replacement for the videotape-based cart machine. We contend that until direct playout from the data cassette is possible (and that may truly not be far away), then we have difficulty stating that there is a one to one replacement for video-to-data tape library systems. Today, the data tape model still requires a cache—something the videotape cassette model doesn't necessarily require, but it does help.

TAPE ARRAY TECHNOLOGY

Like disk arrays, tape arrays are designed to use multiple transports in a way that offers redundancy, capacity, fault tolerance, and performance (see Figure 35-2). Tape arrays are considered when the time to transfer data exceeds the window for completion of the transaction, given the current transfer rates and disk or tape storage capacities.

One technique for tape array implementation is to use software that drives concurrent tape transports. This decreases backup times, but adds overhead to the system because typically the SCSI bus must share the processor bus, which is also operating the disk array. Sharing the SCSI bus for many simultaneous tasks will dramatically reduce the net transfer rate for all the devices occupying the bus.

However, if you look at a tape array in a similar form as we look at RAID for disk drives, we see some interesting corollaries. A major benefit to a tape array would be striping data across multiple drives simultaneously. This increases the transfer rate equal to the number of drives in the given array.

Typically, for data files, this amounts to between two and four transports. Using 8mm type tape drives, if you considered a sustained transfer rate of 3 MB/s, times four, it would yield 12-MB/s in native format. Capacity also increases at the same time. If the cassette holds 20 GB then four would hold 80 GB. Adding data compression increases this by the multiple of the compression ratio.

Assumptions must be made, however. We must first consider that each drive controller must have a separate SCSI bus for each tape drive. Not doing so would bring us back to the same problem, described earlier, of the SCSI bus bottleneck. One solution to this would be to have a separate server that did nothing except handle the tape drive processes. This would be not much more than a PC with a Fibre Channel interface card and a high-speed SCSI interface to a controller that is connected to each of the tape transports.

FAULT TOLERANCE RESTORE

You can further extend this concept to other fault tolerant features such as mirroring and "fault tolerance restore." The latter feature allows an automatic default to the mirrored drive in the event a failure of a primary drive occurs. If there is striped data across multiple transports, then each group of transports must be capable of switching if the primary set of drive transports fails.

The main difference between tape transport and disk drive failures is apparent. If the tape transport fails, the data on the tape is still usable (provided the tape is not physically damaged). In a disk drive, the drive *and* the data are lost. When applying RAID to tape drives, this adds confusion to the model. In tape systems, adding the extra parity drive in RAID 3, or the extra data space required for RAID 5 distributed parity, slows the process down.

Considering RAID 3 or RAID 5 for tape arrays does not make quite as much sense (compared to disk drives) because the factor of diminishing the transfer rates comes into play. Every drive added or every extra bit that must be recorded reduces system throughput, and the balance between all these factors becomes important.

DATA AND MEDIA INTEGRITY

Tape systems, whether in arrays or in single units, should be capable of identifying the integrity of both the media and the data on each cassette. Cassettes with high cycle rates should be backed up and replaced once they reach some preestablished number. This is accomplished as a background task that temporarily caches the data to a server, instructs the operator to replace the worn tape, and clones the data back to the tape. An incremental version number is added to the new cassette, and the old cassette is purged from the database.

The same kind of analysis should become routine when testing data integrity. If the bit error rate begins to climb, or reach some threshold, the data should be copied to a new cassette as described previously. All these fault tolerant preventative measures should, and can, be accomplished as background tasks. The scheduling should be part of routine system maintenance, established during system setup and configuration.

ARCHIVE, BACKUP AND PROTECTION

A final note before concluding our look at preventing the preventable. We've explained many different scenarios to consider related to disaster prevention and protection. Through example, we've placed definition on the three terms – but not specificity. There appears to be crossover on each of the terms to some degree, but they really are three separate items.

Archive will always include the long-term preservation of the content. Archive may be near-line (as in a robotics tape system) or it may be far-line (a term probably not dealt with, at least not yet, in the television industry). Far-line is equivalent to having the entire original MGM movie library kept off-site and protected in an underground, thermally controlled environment—while at the same time having a process that transfers the library to a future proofed digital media for use and preservation. Eventually, the film media will no longer be usable. We've already witnessed this in the 2" tape format–where thousands of hours of quad tape has been transferred.

Backup is a routine and regularly scheduled process that is generally site- and operations-specific. It could be as simple as making routine copies of the digital media to another form, including digital. It may include never discarding the native format material that was delivered to the facility prior to digital transfer. It might include something as exotic as a parallel control center, operating in complete parallel—just in case.

Backup schemes probably cover the widest range of choices of all activities for the digital domain.

Protection is just that, guarding the environment, the air product, and the revenue stream from disaster. Protection is redundancy in complete systems and in subsystems, and involves weighing the economics with the feasibility balanced by reality.

CHAPTER 36

VOD: NEW OUTLETS FOR CONTENT PRODUCERS

An integral part of interactive television (ITV) includes the ability to selectively "order" information at will. Video-on-demand (VOD) was to be *the* technology for ITV. Born first from laserdisc and the VCR, VOD transformed to the magnetic disk drive as server technologies emerged from the cradle.

Video-on-demand, like ITV, made marketing promises about technology that was still in diapers. What emerged from this technological concept, when coupled with compression, networking, and computer adapted software implementation, has found exceptional acceptance in hotels, cruise ships, hospitals and educational segments of industry.

Eventually, perhaps via the Internet, perhaps as wider pipes to the home and business are deployed, many of the promises of VOD will extend to the content producer.

This chapter, which was number four in the series column, essentially remains as it was originally published in October 1994 for *TV Technology*. Ironically, many of the concepts developed for the creation of material aimed at VOD have already been extended to the World Wide Web. The Web and the Internet are forming supplemental and competitive models for auxiliary services for DTV. Certainly the software giants see DTV as an avenue to add features and capabilities to broadcast television. As such, the wish lists of VOD should challenge us all in the coming new millennium.

No doubt about it, the third generation of television is becoming a reality. Since its origin, television has been driven first by technology, then later by the profitability of its business worth. Now [October 1994] it is about to be driven by the availability and diversity of its content as it merges with the capabilities of the computer platform.

As the delivery medium has transgressed from strictly "over the air" to a hybrid of cable and over the air; it is now faced with how to address its single biggest change and challenge in communications history.

Video-on-demand (VOD) is making that transition happen. It's a complex combination of content, user-definable timing of delivery and the independent capabilities and selection of both. Technology has been a big factor in making VOD happen. However, as this column will explain, it is not just the most recent changes in disk media storage that have set the course for VOD; it is a conglomerate of many factors, desires and capabilities.

PPV OR VOD?

Video-on-demand has already been around the block in a number of different and already established forums. Ordering a pay-per-view (PPV) movie in a hotel utilizes a form of VOD, as does the more mechanical and direct selection of movies from the local rental outlet.

Broadly speaking, video kiosks, game selections, even the channel changer and instructional laserdiscs could be considered forms of VOD. Combine a massive number of choices, integrate continuous media server technology, and place the whole system into a delivery scheme composed of networked paths to individual viewers, and you have the rudiments of the VOD systems of tomorrow.

Focusing on new technologies is not enough. With these new storage and delivery methods comes an entirely new outlet for content creators, producers and distributors.

Today, there are still only a limited number of methods available for getting specific messages out to a defined mass audience. With the future delivery method improvements, programming directly from the "source" may become the norm.

Audiences continue to fragment. By their own nature, they seek out alternative programming and therefore become more involved. Producers that once competed for a single time slot, one that might not even reach the intended audience, will soon have much greater opportunities for delivering their own concepts and ideas.

How will the VOD systems of tomorrow function? What are the basics—and what will the future bring? These are some of the questions to be answered.

NEAR VOD (NVOD)	Client selects from predetermined schedule and start times. Typical to PPV in hotels and cable systems. Navigation is by video menu screens or printed schedules.
INSTANTANEOUS VOD	A subset of NVOD. Offers predetermined media with user-selected start times. Navigation by video menu screens and remote TV controllers.
LIVE INTERACTIVE VOD	User feedback, channel to mass audiences. Requires special video servers. Stunt features like those found on VCRs (fast forward, rewind, pause, freeze). Integrates live programming: home shopping, ordering, remote ticket processing.
TRUE INTERACTIVE VOD	Most advanced of delivery systems. Clients have control over all media delivery. No time slotting–media is delivered randomly. Content employs program branching (like CD-ROM),

Table 36-1: Characteristics of video-on-demand.

Video-on-demand has been broadly divided into four conceptual forms, described in Table 36-1. We will begin by focusing on and giving an overview of these different types of video-on-demand services. VOD basic technology, terminology, and methodology will be discussed so that a framework of the methods available for getting a message to a desired audience, in the intended time frame, might be better understood.

As mentioned, certain types of VOD systems have been in place and functional for years; they've just never been called that. Tomorrow's VOD will involve more than just the addition of server technology for instantaneous storage and retrieval of content. VOD will require new navigational controls that will allow the client to interact with the system in the same way they might visit the movie rental store, go shopping at the mall or order pizza for delivery at home.

Feature sets and options will be available, giving video the flexibility and diversity of the VCR or laserdisc. Add the integration with computer platforms and you have more choices and functions, all potentially merged into one vast visual communications system.

NEAR VOD

The baseline, elementary VOD system is called "near-VOD" service. Most noted are the hotel room PPV systems that appear to take an "order" and produce a program. In the early days, they really summoned a human or machine to locate a tape and mechanically find an appropriate VCR, although they made it appear as though you just pressed a button and had instant movies.

The navigational aids included video menu screens, keypad set-top devices, printed schedules, or any combination thereof. Selection of the content was still on a client by client [i.e., room by room] basis. An option generally included is the regularly preprogrammed movies and features, usually satellite-delivered, that function the same as the cable "premium" channels except on a pay by event or view basis.

What makes "near VOD" (NVOD) distinguishable is that these systems provide a wide variety of media choices that begin at predetermined start times and with a relatively simple structure.

Prominent players in the hotel PPV business, such as Spectradyne, have been improving their NVOD systems for some time. Until recently, multiple programs were all fed directly, in real time, by a number of satellite transponders.

In the not too distant future, "data" (compressed digital video) will be downloaded at off hours via satellite from a master library—direct to a video server's disk array located on the hotel's premises.

Several streams (or channels) of video will be available for switching within the hotel when called up by the room's occupant.

The server's video engine would decode the data to analog video, modulate it and feed upwards of 32 channels to various rooms using the existing cable distribution network.

By using digital storage and transmission technology, this first-step transition in NVOD will give better quality purely by eliminating noisy satellite feeds. Systems will offer a larger selection of programming, be more reliable and have lower costs all around. In turn, the same satellite transponders would then be open to offer live programming because the media library will already be on site, digitally stored on the server for near-instant retrieval.

Indeed, this offers an economical solution to manually loaded or direct satellite fed programming. VCRs or laserdiscs may no longer be required and, of course, the human operational element (and its many costs) are all but eliminated.

INSTANTANEOUS AND INTERACTIVE

"Instantaneous VOD" is another link in the chain, providing a subset of NVOD and offering no-predetermined start times with a limited number of other options. Such options are VCR-like feature sets such as pause, replay, or fast forward.

Almost every VOD system is planned to be scalable. Systems can start out modest and then grow as capacity demands or features dictate. In addition, the level of quality (usually gauged by the degree of compression) will be scalable, providing various resolutions depending upon program content.

"Live interactive VOD" media delivery is expected to provide some level of user feedback and can be channeled to the masses. A service of this type requires a special type of server and is intended for such programming as live interactive shopping and other special venues such as remote ticket orders or vacation tour planning.

Not unlike the cable shopping channels, live programming can be in real (linear) time, with the viewer being offered the ability to break away from live programming to select more detailed information on specific products being offered for sale. Since the live portion continues to be recorded, nothing is lost and the viewer can return to the live show at will without missing a single product. The programmer gets to show more products and the customer (viewer) gets their money's worth by seeing detailed, specific product information or services that interest just them—a kind of "VOD kiosk."

"True interactive VOD" is the most advanced of all these previously described systems. In this service, each client/viewer has control over any or all of the programming material. Nothing will be time slotted, as interactive VOD will provide for instantaneous and random delivery of media [forward, backward, restarted or paused in real time].

This is the most demanding of the system designs and will most likely be offered only on a larger scale or for metropolitan services.

The interactive VOD service will provide for branching of programs, including the selection of in-depth video sidebars that can be viewed any time during the programming without missing a single portion of the mainstream content. You'll be able to interactively comment or take a side trip, then return without missing the main program content.

The software, in this case the content creation, has the potential to extend a 10-minute short subject into 30 minutes or more—at the full control and complete discretion of the viewer. If the service is metered, it just keeps on running, providing programmability and endless profitability. Ideally, a system of this sort could be billed out just like CompuServe or any other fee-based on-line service.

Both the live and the interactive VOD services are being pilot studied in several markets over the next year or so [circa 1994–95]. In each case, technical delivery, navigational, and program content capabilities are all being tested. Trials are necessary from every perspective, as no one really knows what it will take to attract and hold the interest of the widely varying audiences that these systems intend to entertain.

In terms of structure, there may be many hybrid systems composed of services bridging any of these described systems and services. From these various types of systems it is easy to see how rates (or "tiers" in cable speak) will adjust based upon the service or system functionality.

It is also fairly easy to visualize how advertising, which we only think of as commercials or paid programming, will self adjust and in turn may continue to provide new avenues for revenue generation.

Two Way Exchanges

With VOD delivery comes two way response rendered systems that will incorporate sophisticated database and demographic tracking capabilities. These systems will eventually have advertising funneled directly to you, the specific viewer, in the same fashion that you'll have program selectability[62]. You still won't escape the commercials, but they will be more targeted to your interests— you'll probably watch them.

Controls can be monitored so advertising data will be tracked, following and monitoring your response and individual profiles in a profound and almost uneasy way. You'll only see what the advertiser thinks you want to see. Advertisers will not only hope they guessed right —it's quite possible they'll *know* they guessed right.

Content Creation Alternatives

With the basics of the various services and systems theoretically conceptualized, the structure of content creation will begin to change dramatically. Eventually, almost anyone with an interest in production will have the avenue to distribute their creative material directly to local servers, or they may indeed provide services directly from their own studios or homes.

Software and delivery (i.e., distribution networking) packages will enable the creators of content to directly serve the audience they wish to target. It is quite possible that the desktop applications now being used for interactive content creation will be further expanded or customized to particular VOD systems so that producers can develop applications specific for client demographics.

Incorporating direct distribution network software and two-way interaction, producers would connect to the video media server network on a local system and reverse broadcast to the masses. Producers may

---

[62] Ironically, the purveyors of ITV have found another avenue for their concepts, that of the Internet, using "push" technology.

wish to target focus audiences, allowing them to effectively collect revenue directly from the particular client that is viewing their program.

Art, music, literature, or movie viewing may generate revenue just like regional long distance calling, at a few pennies per minute. For example, a 1 percent penetration in a 25,000 viewer system, at a dollar a showing, equates to $250 for programming that may very well have been produced on your home PC (which, of course, is now equipped with video editing, audio production, library access and compression).

Seem far-fetched? Not really! These systems can and will happen. Billions of dollars aren't being invested on a just "wouldn't it be neat if . . ." concept. All this is possible when the systems are planned for scalable architectures, affordability, and mass distribution.

While most will not want to produce full content out of their homes, one could start small and grow as the system capabilities do. It is inevitable that the dividing line between one on one connections (such as Internet video-electronic mail) and the VOD server network of the future will begin to blur.

What this all means is that the economics of distribution will be something all producers (large or small) will no longer view as an obstacle to getting their individual messages out. Even though the paths will be many and the selection of which network to use will be challenging, the result will be more demand for these services, and in turn more need for content. It really can be a two-way street.

(Free) Home Delivery

Getting the programming to the various clients is different subject altogether. Software content, whether it resides in a local server or on a main system library, will arrive at the home by various methods.

Video fiber would be the ideal delivery method to each client household, but for the present, the cost is prohibitive. Conventional coaxial cable, as currently installed by the cable TV industry, will be the most likely direct connection for now.

In the eyes of some of the Telcos, twisted copper wire is still viable—but may not provide the depth of service capabilities that a combination of coax and fiber can offer. Moreover, of course, satellite delivery (DBS) is already here.

As we've demonstrated throughout this book, the alternatives for distribution, and the technical properties required for the disk systems of continuous media servers and networks, are complex and varied. We can see how by using conventional Internetworking principles, applied to the modern continuous media server technology, content might eventually be delivered from any source to any destination. The Internet, with personal Web pages and push/pull technology, is a shining example.

To achieve this, software, hardware, and networking technologies will continue to evolve such that they will no longer discriminate between data and media in form of storage, assembly, archival or distribution.

> Digital television is the next promise for the content producer. The broadcaster, although they use the word "broadcast", is really a content producer. They've all been given a second chance. The avenues available for the delivery of their services have just been granted a full scale Los Angeles freeway on ramp. We expect to see congestion, but we also expect to see eight lanes of available bandwidth. Let's hope it gets used wisely!

CHAPTER 37

APPROACHES TO VIDEO ON DEMAND

When employing disk drive technologies for video-on-demand (VOD), or for any media-oriented streaming video and audio, several factors need to be considered, many of which were unavailable from a technology standpoint in 1994. We will now focus on the state of the art in drive technologies, as we saw them at that time.

In some cases, the examples that follow were the perspectives that leading developers of new hardware and software systems felt, at the time, were the necessary directions to take in order to meet the demands for VOD.

Many of these directions have changed significantly, driven by the change in drive technologies, the reduced price of storage (RAM and HDD), and the change in market demands over a relatively short period of time. The comparison of what was available then and what is available now gives a true example of how fast this technology has changed over a relatively short period of time.

Keep in mind that this chapter is printed as it appeared in November 1994, the fourth article in the *Media Server Technology* column. You will find some astounding statements that now, in early-1998, make no sense at all—but that was the way industry viewed the future of VOD, computers, CD-ROM, and even MPEG. It is almost frightening to look back and see that most of the ideas have either been overwhelmingly surpassed or totally abandoned.

CONTINUOUS HOME DELIVERY

The drive for VOD is centering on home delivery and the potential for consumers to want and pay for these services. Already, they are paying for cable TV, satellite services, and movie rentals. Add in on-line services and the expansion of the commercial Internet, plus the growth of CD-ROM software and games, and you'll find there's a lot of consumer spending on entertainment leisure.

Yet some believe video-on-demand's success in the home will be limited. Predictions [made in mid-1994] stated that VOD would be less successful than existing cable, and others expected confusion, poor program selections, and overly high-priced services. Some skeptics expressed concern over complicated navigational systems that would deter all but the overtly curious from getting the full benefit of the still undefined and untested capabilities.

If these negative predictions were true, then try to explain why there's been an explosion in development and technology directly aimed at VOD.

This technology appears to be on a natural course, one that's headed in the direction of the continuous media server as the delivery platform of the future.

So what is a "continuous media server" and what does it take to make one? Several species of these servers are now being developed by some well-known players in the broadcast, computer and software industries. However, one common thread to all these systems lies in the storage medium—which for now is principally the hard disk drive (HDD) or combinations of drives assembled into massive arrays that can store hundreds of gigabytes of data.

Drive manufacturers are reacting to the increased demands for greater storage, faster data rates, and continuous, uninterrupted transfer of media-oriented data. Companies are creating new breeds of drives, adding the "AV" (audio/video) suffix to their model names. This new class of drive becomes critical to the continuous media server.

To better understand what makes these new drives important for continuous media delivery, one should take a brief look at the primary inroads in drive technology improvements over the past couple of years.

In just one year's time, disk drive seek time and rotational latency has improved 50 to 60 percent. Random and sequential data throughput has increased 300 to 500 percent. Performance, overall, has excelled due to new technologies in high-speed data channels and sophisticated command strategies.

Some manufacturers are employing smart thermal recalibration systems. Through the integral design of both the drive and peripheral video on board preprocessors, recalibration schemes become more closely matched to the data transfer rates, resulting in continuous, smooth delivery of audio/video-type media.

A Data Advantage

Data for applications such as word processors, spreadsheets or databases have the distinct advantage of being able to spread out, or fragment, data all over the platters in a hard drive assembly without significantly hindering the baseline performance of the application.

When compilers or database servers require continuous duty from a single hard drive assembly, the drives incorporate an automatic process called "thermal recalibration" (TCAL) to maintain performance. TCAL is a balanced sequence that optimizes the drive's proper head shift as it moves from track to track. The TCAL process insures that the head will be properly positioned directly over the track so that the maximum signal is extracted and the number of errors from recovered data is minimized. This can cause delays in delivery of continuous data from the output processor circuitry. However, in the world of data-only delivery, this is insignificant when compared to the compute processes of the CPU or the application's latency factors.

Drives primarily used for data applications recalibrate frequently. However, when used for streaming video applications, these recalibrations need not happen as often. This is because the media data, in the form of digitized video, is generally written to the drive sequentially and therefore is read back in the same or nearly the same linear order.

In a "true VOD" system, it is impossible to predict the demands on the read process. Getting a disk array ready for linear video is fairly straightforward, but in the continuous demands of the randomized nonstop video server, this is far more complicated.

For digital video, the process requires reading and immediately processing the data. The critical performance metric now becomes the sustained data throughput rate. In 1987 that rate (from buffer to host) was only about 800 kb/s. In 1994, this rate has climbed to 5 Mb/s or more.

DRIVE TECHNIQUES EXPAND

Advancements in PRML (partial response-maximum likelihood), which is a sophisticated data detection technique, have further contributed to drive improvements. First used in digital communications during the 1970s, PRML senses magnetic transitions recorded onto the drive surface during the write cycle and correspondingly allows data to be written much closer and thus more accurately.

Another area of improved drive performance is called OLTP (on-line transaction processing). This is the ability of a drive to handle random read/write requests efficiently. Generally, the single-threaded versus multithreaded operating systems (OS) place different demands on systems, which necessitates optimization of the "command execution sequence."

In a single-threaded OS, execution only occurs in the order that the commands are received, an almost linear type of approach. In the multithreaded OS (i.e., the Novell, SCO UNIX or NT operating systems), commands are routinely reordered and executed in any sequence (non-linear). One might now begin to see interesting parallels between video-on-demand and today's network operating systems. This is one reason why some manufacturers believe that with modern drive technology it will be possible to implement VOD in software rather than strictly with complicated hardware-based structures.

Intelligence in hard drives has also made continuous media servers possible. Now there are algorithms such as the "elevator seeking algorithm", used in high-end drives of greater than 500 MB, that are used to minimize seek times. In this process, the drive head always moves to the next track in sequence. Therefore, the latency in deciding which way to move the head arm is reduced. Once this is accomplished, it becomes possible to increase rotational speeds (from, say, 3600 to 5400 rpm or above), and the result becomes increased random throughput.

Couple these techniques with new firmware, which has the ability to overlap execution of two or more read or write commands in a queue, and video playback or continuous recording of the higher data rates for video makes higher quality imaging a reality.

LARGE SCALE – SMALL SCALE

Will the VOD system of tomorrow require sophisticated supercomputers, smart set top box (STB) devices, mainframe functionality with mass-storage capabilities, or a scalable bank of PC-type servers integrated through software? The answer is "yes"—and depending upon who is promoting which type of system, there will be varying opinions and striking differences in architectures.

Typically, if you can identify the specific companies that have formed alliances with other companies to develop their systems, you can pretty much tell what that company's philosophical architecture will be. In the case of a system that is primarily developed as a software approach, you'll find there are generally statements about "off the shelf" component solutions for hardware. For example, Microsoft is betting on their Tiger, based around their NT operating system. Coupled with Intel's Pentium technology and aligned with Compaq for hardware, it is generally stated that supercomputer-based technology will not be necessary to implement VOD. This is the opposite of the Oracle approach, which is betting on nCube's massively parallel processing technology to supply similar functionality.

For comparison, Microsoft further believes that Tiger will not require RAID arrays or other specialized fault tolerant hardware. They state you will not needs lots of RAM on the client or the server side; that by using low-cost hard drives, the same capabilities are possible.

ESSENTIAL SERVANTS

Video servers are seen as an essential component in VOD. It is postulated that no one single "standard" video server will be anticipated for the implementation of video-on-demand. Video servers in cable systems must, by necessity, deal with both memory (in bytes) and transmission (in bits). The issues of off-line and secondary storage may become a Herculean task, and the economics in storage could be monumental.

Today, a CD-ROM costs about five dollars per gigabyte. Reaching back to our Greek electronic prefixes, a terabyte (one thousand gigabytes) is 10^{12}, a petabyte (a hundred thousand gigabytes) is 10^{15} and an exabyte (a million gigabytes) is 10^{18}.

The movie *Jurassic Park* requires a dinosaur-sized 1.44 terabytes of uncompressed storage. If a server library maintains 700 movies of similar length (a very small video rental store indeed), uncompressed disk storage alone would consume a whopping petabyte of disk space.

You can linger over the costs on that one, but remember that storage on CD-ROM is one thing, but true VOD, in an uncompressed form, must be equated to hard disk drives at a 1994-cost of around $500 per gigabyte. Without employing some level of compression, a service provider would need to cough up $500 million to store 700 movies—just for one video server system! By 1994 standards, you could certainly buy a lot of VHS tape for half a billion dollars.

The tasks we're discussing require a constant integration of economics and technology. That's what good engineering is all about. Moreover, compression will make many improvements over the next few years that will begin to reduce the incredible task of storing all this data.

As of mid-1994, Compaq and Intel both expect to ship with Tiger software by early 1995. Besides the tests already in process, Microsoft and TCI will begin full market tests in Seattle and Denver by mid-1995, with major rollouts in early 1995.

In order for this to happen, Microsoft is relying on two new technologies that still aren't quite ready for prime time. MPEG-2 and full ATM (on wide area fiber) will be required in order for Tiger systems to be successful. In the meantime, several tests and systems are being planned and developed right now—the results should be known over the next 18 months or so.

A Compact Approach

Another delivery technology is also a possibility. The progress in CD-ROM storage is continuing to take off. Still, though, there is a feeling that "movies on ROM" won't be practical until the CD-ROM can deliver at data rates above the 3 to 4.5 Mb/s.

Video jukeboxes using MPEG-2 compression will also be constrained until the bit rate is fully accepted at CD speeds in excess of "triple speed" (which now has a practical data rate limit of 4.5-Mb/s). IBM is in the process of developing the 6.5-GB CD-ROM by effectively stacking 10 ROMs together and reading them as one. The "blue laser" will be surfacing soon, with its promise of much higher recording densities. The 4X-ROM is around the corner, and that will help as well. In addition, if we have a 4X-ROM, why not a 10X, and so on?

Fortunately, MPEG-2 development is also coming along. Even though some believe MPEG-2 will not run on the ISA PC bus (which is limited by its slower internal bus-speed), MPEG-2 encoders will be ready by the end of 1994, and many that are already implementing MPEG-1.5 chip sets will be upgrading to implement MPEG-2 on the PC platform. (Note: MPEG-1.5 is "unconstrained" MPEG-1 running at higher data rates than 1.5 Mb/s.)

If you blend all these technologies, you can see that higher hurdles are still being leaped. The race is continuing, and video-on-demand will be technically ready just about the time the broadcaster is faced with the challenge of "what exactly will I have to do with ATV?" (For the answer to that question, we'll all have to wait and see).

CHAPTER 38

INTERACTIVE TELEVISION: ONCE AN EMERGING INDUSTRY

The broadcast facility faces new challenges in the wake of the computer industry's involvement. As these new technologies were thrust upon the market, the Telcos[63] began looking for their entrance, and the expectations of the yet to be defined digital broadcasting era. One new venue was experiencing both its definition and its epitaph. As hundreds of millions of dollars were being poured into a science fair project without a market, other technologies evolved. The results of those experiments are the fruits of the future for video content delivery.

This, our concluding chapter, will look at the philosophical expectations of what was once called interactive television—a concept not yet ready for prime time. It is ironic that we close with this early and nearly vanishing paradigm for the television medium. There is so much to do and there are so many ways to do it.

The excerpt that follows this chapter is provided as a reflection—some things just haven't changed much yet.

In May of 1995, the emerging video and media server industries had a lot to think about in what some felt was a moderately short time frame. The

---

[63] Telcos refers generally to the "telephone companies," RBOCs, and other telephony carriers.

media server was to be a requirement for the delivery of interactive TV. It was also expected that media servers would become part of either the TV or the PC or both.

At just about this time, something happened that nearly derailed the entire prospect of putting media servers into action on a consumer level. The technology needed to implement interactive television was slower to develop than expected. Prospective and active market trials began to drop like flies. The costs for deployment were predicted as staggering. There was no uniformity in plan and no direction for implementation.

The sophisticated development cycle of bringing any product to market is a complicated one. When you think about how fast compute oriented markets were changing, and how long people will expect their investment to be of value, the viability of changing the entire television market just because a new technology comes along is pretty thin.

In the mid-1990s, with the rapid growth in media technology, business was changing. Many existing companies recognized the potential for new sources of revenue, driven by technology. However, in business, by the time a plan is proposed and financing is put in place, the project is at risk of being placed in jeopardy. Something new comes along that demands a complete rethinking of the original plan. Often it is competition. Routinely it is a lack of anticipated demand. Sometimes, when the reliance upon another frame of technology just doesn't come about, the new idea doesn't even get a chance.

There was, in May 1995, a frenzy going on in the area of video media delivery technology. This new child was only in its infancy. The big hype in 1995 was still the information highway or "I-way," as it was affectionately called. This was to be the answer to all the questions.

The questions were plentiful. "Which way will the I-way go?" "Will there actually be a meaningful offramp into the living room?" "If so, how will it get there?" "Once it arrives in the front room, which piece or pieces of hardware will it terminate in?"

The broadcast industry sat and watched in the wings. They had watched the seemingly endless technical and political battle over Advanced Television (ATV) continue. Broadcasters watched interactive TV trials go up in smoke. They had also watched the beginning of a new dawn in the delivery of information—over the Internet.

Servers for the broadcast industry had already become their new technological salvation. Like many of the manufacturers, the broadcast industry thought that moving toward a tapeless operating system might actually happen. Even though major manufacturers stood poised with new products, development of new professional and even consumer videotape transports took a deep breath and paused.

A new stepchild was about to be born—the emphasis was about to be placed upon the digital video disk or DVD[64]. Although the impact on the broadcaster will not be felt for several more years, if at all, the constant change in storage and delivery infrastructure was making a serious impact on just what direction to take.

During tape technology's pause and regroup phase, the broadcasters continued to ponder just how far they must continue to explore, evaluate, and define where their business might be headed. They had decisions to make. Transports were (and still are) in need of replacement; the D-2 format was dying an evolutionary death. D-1 was far too expensive and the acceptance of Digital Betacam in other than post production applications or satellite digital delivery was just not that widespread in the daily on-air broadcast community.

As continues to be the case in consumer electronics through the mid-'90s, consumer media delivery products still dominated the living room. Even with all the new developments in the PC, a computer replacement for the conventional television set just hadn't made it there yet, and it was doubtful that it would.

ENTERTAINMENT EVOLUTION

The evolution of entertainment delivery devices shows marked and different perspective when comparing the PC to the TV. Historically, we recognize the many changes that have occurred in the relatively short time frame of just over 20 years from the mid-1970s.

CD players had already replaced the phonograph. The eight-track vanished in favor of the cassette deck. The BetaMax, which never quite caught on at all, can hardly be found.

Laserdiscs, even with their stunt feature sets, still have only a small market penetration. Big screen projection TVs are still not the

---

[64] DVD was later generalized to digital *versatile* disk, covering all the aspects of DVD-ROM, DVD-RAM, DVD-video, etc.

predominant form for delivering moving images, but that may change with DTV.

Still, there was one question that loomed on the horizon even to this day. "How long will it be before they're going to turn the living room over to the Apples and the Microsofts of the world?"

With the life cycle on most new computer products being only a few years, how much longer would people put up with having to continually buy larger hard drives, faster processors, and more advanced display drivers before they would holler "enough is enough?" There is no clear answer to why that momentum hasn't changed much clear into early '90s, and history will show that the growth continues even stronger.

THE PARODY IN MEDIA

Looking back upon the television industry, a similar parallel might be applied to media delivery methods. When looking at the avenues for media delivery, the broadcaster, cablecaster and satellite delivery industries all seem to be wrestling with similar questions: "What will be the predominant method of delivery and how much should we invest and when?" The similarities in delivery methods have boiled down to a few primary and a couple of secondary approaches. Delivery methods remain basically the same; some now simply employ different formats for the actual transmission and delivery or reception.

From a form-factor standpoint, when you look at the TV set of 1980, it is still just as functional as the TV set of 1995. You can't say the same when you consider the evolution of the PC—just compare the differences from the PC-XT or the Apple IIe to today's personal computers.

Still, the biggest difference so far is in how you view the medium. You can view a TV set from most anywhere in the average living room, but you can't really play *Myst* from eight-feet away on a 15" screen?

Television audio hasn't moved much since 1980 either. Sure BTSC stereo broadcasting was added in the mid-1980s, but look further at the multichannel audio capabilities that were also a part of that format. The SAP (secondary audio program) channel has never really caught on. In the name of marketing, some manufacturers took a short cut and actually mixed SAP and the primary stereo channel together, rendering the broadcaster's attempts to provide additional information virtually worthless.

Delivery of SAP would be further complicated by the cable companies in ability to minimize cross-talk and bandwidth protection. Add in the complications of scrambling, converters, addressable converters, and the infamous A/B switch fiasco, and you have no better guarantee for consumers that the same set of programming would be received locally or regionally without undue complications and duress.

THE PROBLEM'S SOLVED!

It would be shown that interactive television would probably create the same types of complications for consumers as the many attempts to add feature sets to televisions already had.

Interactive television was just about at its peak of interest in mid-1995. It was not only a technological flop, but the marketing impact that most purveyors expected to emerge was just plain unbelievable.

Look at all the promises, all the possibilities, all the expectations—some actually thought that adding interactive television into the household would solve so many problems that consumers would pay *anything* to get it.

Thinking through the process and past the hype, you had to consider the impact of what interactive TV was *supposed* to do and ask: "Just how many people will use Quicken to balance the checkbook on a home TV screen sitting on the other side of the room using a four button remote control as the sole navigation device?" Was this to be a part of the interactive TV experience or was this truly to become the dividing line between the PC and the TV?

Even as we head toward the year 2000, video remains one of the only scenarios that has seems to have actually bridged the PC/TV dividing line. Streaming video on a dedicated PC is still only marginal in performance compared with over the air broadcast, cable, or satellite. Yet even with the phenomenal growth of the Internet, held back primarily by the bandwidth of the delivery pipe, video still has not taken its place in a satisfactory way on the wired PC.

If true interactive television is to be successful, one of the necessary and essential elements will be the navigational feature set. The baseline TV set has two basic controls, a volume control and a channel changer. The PC is a different animal altogether, offering a keyboard, a mouse, and a joystick as a control interface set.

Forget for a moment the VCR; the actual interactivity of a TV set is pretty straightforward, and most any couch potato can handle it. Not so for the PC. Even though every effort is being made to simplify the interface, or expose more people to operations, the PC is still a complicated device to those who don't (or won't) spend time with it every single day.

Television add-on devices, such as Nintendo or Sega, use game controllers, the predecessor to the handheld interactive controller of tomorrow. Still, these devices offer only knee-jerk reactionary controls that steer with a very limited functionality at any given moment.

Some feel that the game machine will remain a separate interactive strategy altogether. Game machines will not be positioned somewhere between the PC and the TV, as they will strive to be a totally unique experience in a completely different package. This is probably why games will continue to survive on both the TV and the PC screens.

To understand this logic, just observe how gamers really interact during "game-play." They sit on the floor, crouched directly in front of the screen, volume turned up high, twitching and flailing about, repeating the same moves over and over again until they get it right. They learn and adapt the interface to the experience by continual repetition.

The gamer interacting on the PC, with the possible exception of Doom-like action games, is definitely more focused, more direct and generally does not tolerate distractions well at all. Not that Doom wouldn't be outstanding on a 27" screen, but Doom is a one on one type of interaction. Taking over the family gathering place to watch somebody blast the daylights out of a fictitious set of characters will probably result in everyone, except the player, looking for another screen to watch.

A similar scenario might be placed on true interactive TV. Unless a market is conceived where more than one individual can interact effectively in a group environment (as in the family living room), interactive TV is doomed to fail, and you might as well sit in front of the computer screen one on one.

As for the TV, we've been accustomed to the same degree of pure "noise" for so long that it doesn't dignify any real level of interaction (or attention) much beyond what happens at the top or bottom of the hour when it's time to change channels.

TV continues to offer a variety of levels of concentration and attention. One or many can enjoy it whether focused on the program or not. Other questions remain, and what happens should the PC become an integral part of the television?

Even if progressive scanning wins and film level resolution to the home becomes real, how successful will interactive television be if there can only be one person at a time interacting? What happens when that functionality gets boring and the user wants something more? Or the advertiser or local station deems it is time to add a new piece of pizzazz!

OPPORTUNISTIC OPTIMISM

The optimistic software developers think they've figured this one out. A great deal of research has been put into the psychological and technological perspectives of interactive television. If that is the answer, who will be expected to pay for the improvements—for a medium that to date has been virtually without fees for general over the air broadcast services?

If you're in the broadcast business and you're delivering for interactive TV and conventional TV, try to imagine the reaction if every time the ratings sweeps were about to begin, that in order to get the true directed value of the programs, you had to download and install a new software set. Consider the retraining for an entire new suite of commands where before you could just "point and click". While this sounds absurd, the interactive PC/TV of the future may just force that level of interaction in order to get the additional value or even to get the program at all.

The PC still remains in a whole different domain. Granted, more users are becoming tuned in to the PC, but will the PC be built into the TV any sooner than the TV is built into the automobile? Probably not—and today it appears that the TV will be built into the PC long before the opposite is true.

Culture has accepted that each appliance has its own place and its own function. Even adding the sophistication of "Bob," the now ancient Microsoft interface, to your screen still suggests a well-established "PC" interface. The mass market of this decade doesn't seem to want that level of activity in the TV, so what makes us think that adding sophisticated media server technologies either remotely or locally will provide any more degree of pleasure than there is right now?

Looking back on this brief history, the 1995 Consumer Electronics Show (CES) displayed at least six new set top boxes (STB) for the emerging interactive TV industry. Each STB was destined to redefine the functionality of television over the next year or so. According to the industry trades, by 1996 that number was expected to be much higher. Ironically, at the same time, the interpretation of the TV versus the PC was expected to remain just as undefined.

Even today, with the rise of the Internet and the apparent demise of interactive TV for tomorrow, we've already seen some routes take a dead end. The Telco vision was of adding telephones, fax, voicemail, and banking to the channel changer's remote control. The big screen remains the entertainment device it is today (with an expected change with the introduction of DVD). Wide screen displays, supplemented by more advanced features (possibly 3-D), better sound, and different imaging for movies, education, soap operas, and news will have real promise.

A Final Perspective

In 1995, we asked a multitude of questions. We were bombarded by new concepts embedded in infant technology. The industry remained cynical and suspicious, all the while asking if it would all be worth it.

The following excerpt has been left as it was from its original *TV Technology* printing in May 1995. The comments shed a rather slanted, but reflective, light on our development to date.

> For imaging the question continues—will NTSC still be the most acceptable method of video reproduction for years to come, supplemented only by the inevitable inclusion of the digital tapeless transport? The set-top box people might understand best because the delivered output from the back of the box is still, for the most part, plain old NTSC.
>
> Digital or not, the image is still the same—albeit there might be an RGB or S-Video spigot—but for the most part, plan on looking at NTSC for some time to come.

Unfortunately, and only until recently, this was still what the server market saw as "good," as it still references its quality to that of BetacamSP. Why? Because most people accept it, just like most people accept VHS.

Will wide screen or ATV be a part of the fit [referring to the number of analog 4:3 TVs that will be discarded]? Only in that part of the market that has a landfill tradeout ready and a CD (that's "certificate of deposit") ready to be redeemed. For other types of "compute level" power, stay close to your SVGA monitor because that's still, for quite some time, the best place to look and to interact!

Before the truly interactive video server system of tomorrow can be fully defined, refined, produced, and accepted by the consumer—at any level that hopes to make a profit for the hundreds of players expected in the game— the real soul searching must begin. As video engineers or cable head end operators, all this talk about [interactive] video server technology is wonderful news. There have already been some tests; Bell Atlantic is [was] into this stuff big time! Monster-size software companies are betting on the future before any of us really know how well it will be accepted. Research continues, silicon is in the oven, and still we don't really know what all this will look like or really bring to market.

For the set-top box it looks like MPEG-2 will still be the choice. So where does this leave MPEG-1 which is just now being adopted for the PC at the motherboard level? When will MPEG-1 replace the Video Graphics Controller (VGC) for multimedia, and if so, how will this trade-off fit into repurposing for set-top or new-TV delivery? Does this mean the server of tomorrow must be MPEG ready as well?

As evidenced by this unaltered writing, printed about the time we heard about, but seldom saw, this marvelous wonder called ITV, much has changed. It is only three years later and we've seen a new direction with a lot of new technology. To wit, we've also seen a cultural shift in what interaction means; just visit the Internet.

Video and media servers, with all their technology and applications, promises and capabilities, will be as everyday as videotape transports in less than five years. What will change most is the *real* business—which will refocus us all!

ACRONYMS AND ABBREVIATIONS

AAL	ATM Adaptation Layer
ABR	Available Bit Rate
ACPI	Advanced Configuration and Power Interface
ADC	Analog to Digital Converter
A/D	Analog to Digital Converter
ADO	Ampex Digital Optics (model number)
ADH	Asynchronous Digital Hierarchy
ADSL	Asynchronous Digital Subscriber Line/Loop
AES	Audio Engineers Society
AGC	Automatic Gain Control
AMS	Array Management Software
ANSI	American National Standards Institute
API	Application Programming Interface
APM	Advanced Power Management
ASCII	American Standard Code for Information Interchange
ASIC	Application Specific Integrated Circuit
ATSC	Advanced Television Systems Committee
ATA	AT Attachment
ATM	Asynchronous Transfer Mode
ATRC	Advanced Television Research Consortium
ATTC	Advanced Television Test Center
ATV	Advanced Television
AVI	Audio-Video Interleave (compression file format)
AVR	Ampex Video Recorder (model number)
AUI	Attachment Unit Interface
BER	Bit Error Rate
BISDN	Broadband - Integrated Services Network
BNC	British Naval Connector
bps or b/s	bits per second
BVS	Broadcast Video Server
BW	Bandwidth
CAD	Computer Aided/Assisted Design
CAM	Constant Access Method

CATV	Community Antenna Television
CBR	Constant Bit Rate
CCD	Charge Coupled Device
CCIR	Comite Consultatif International de Radio-communications
CCITT	Consultative Committee on International Telephone & Telegraph
CCS	Common Command Set
CD	Compact Disc
CDDI	Copper Data Distributed Interface
CDMA	Code Division Multiplex Access
CDTV	Conventional Definition Television
CDV	Compressed Digital Video
CES	Consumer Electronics Show (exhibition and conference)
CISC	Complex Instruction Set Computing
CMOS	Complementary Metal Oxide Semiconductor
CMX	CBS-MemoreX
CODFM	Coded Orthogonal Frequency Division Multiplexing
COMDEX	Computer Dealers Expo (exhibition and conference)
COS	Class of Service
CPE	Customer Premises Equipment
CPU	Central Processing Unit
CRC	Cyclic Redundancy Code
CS	Convergence Sublayer (AAL)
CSDI	Compressed Serial Digital Interface (see SDTI)
CSU	Channel Service Unit
CVBS	Color Video Blanking Sync
D-1	Component Digital Video 19mm tape format
D-2	Composite Digital Video 19mm tape format
D-3	Composite Digital Video
D-5	Component Digital Video
DAC	Digital to Analog Converter
D/A	Digital to Analog Converter
DAT	Digital Audio Tape
DCC	Digital Compact Cassette
DCE	Data Communications Equipment
DCT	Discrete Cosine Transform (compression)
DCT	Digital Component Technology (video equipment)
DDR	Digital Disk Recorder
DDS	Digital Data Services
DLCI	Data Link Connection Identifier
DLT	Digital Linear Tape

DMF	Distribution Media Format
DNG	Digital News Gathering
DOS	Disk Operating System
DRAM	Dynamic Random Access Memory
DS-0,1,2,3	Digital Signal Layer (Level 0-3)
DSP	Digital Signal Processor
DSS	Digital Satellite Service
DST	Digital Streaming Tape (by Ampex)
DSU	Data Services Unit
DTTB	Digital Terrestrial Television Broadcasting
DTDS	Disaster Tolerant Disk Systems
DTE	Data Terminal Equipment
DTH	Direct to Home
DTS	Decode Timestamp (MPEG-2)
DTV	Digital Television (previously Advanced Television – ATV)
DV	Digital Video (recording standard)
DVC	Digital Video Cassette
DVD	Digital Video/Versatile Disc
DVE	Digital Video Effects
DVR	Digital Video Recorder
DVTR	Digital Video Tape Recorder
EBU	European Broadcasting Union
ECC	Error Correction Code/Coding
ECP	Extended Capabilities Port
EDL	Edit Decision List
EDO	Extended Data Out (memory)
EDTV	Extended Definition Television
E:E	Electronics to Electronics
EIA	Electronics Industries Association
EISA	Extended Industry Standard Architecture
EMI	Electro-magnetic Interference
ENG	Electronic News Gathering
EOM	End of Message
EPG	Electronic Program Guide
FAT	File Allocation Table
FC	Fibre Channel
FC-AL	Fibre Channel-Arbitrated Loop
FC-EL	Fibre Channel-Enhanced Loop
FC-AV	Fibre Channel-Audio Video
FCC	Federal Communications Commission

FDD	Floppy Disk Drive
FDDI	Fiber Distributed Data Interface
FDM	Frequency Division Multiplexing
FDPT	Fixed Disk Parameter Table
FEC	Forward Error Correction
FIFO	First In First Out
fps	frames per second
FRDS	Failure Resistant Disk Systems
FRS	Frame Relay Services
FTDS	Failure Tolerant Disk Systems
FTP	File Transfer Protocol
GA	Grand Alliance
Gb	Gigabit (one billion bits)
GB	Gigabyte (one billion bytes)
GFC	General Flow Control
GLM	Global Link Module
GOP	Group of Pictures
GPI	General Purpose Input (interface)
GPO	General Purpose Output (interface)
GUI	Graphical User Interface
HDD	Hard Disk Drive
HD-DDR	High Definition-Digital Disk Recorder
HDLC	High Level Data Link Control
HDSL	High-Speed Digital Subscriber Line
HDTV	High Definition Television
HEC	Header Error Control
HiPPI	High Performance Parallel Interface
HMSF	Hours:Minutes:Seconds:Frames
HP	Hewlett-Packard
HSM	Hierarchical Storage Management
HTML	Hypertext Markup Language
HVD	High Voltage Differential (drivers)
IDE	Integrated Drive Electronics
IDU	Interface Data Unit
IEC	International Electro-Technical Commission
IEEE	Institute of Electrical and Electronics Engineers
IEEE 1394	FireWire Interface
IETF	Internet Engineering Task Force
IOP	Interoperability
IP	Internet Protocol

Acronyms and Abbreviations 439

IPI	Intelligent Peripheral Interface
IPX	Novell Internetwork Packet Exchange
IS	Information Systems
ISI	Inter-Symbol Interference
ISO	International Organization for Standardization
ISP	Internet Service Provider
IT	Information Technology
ITS	International Teleproductions Society
ITU	International Telecommunications Union
ITV	Interactive Television
JBOD	Just A Bunch of Disks
JPEG	Joint Photographic Experts Group
kb	Kilobits (one thousand bits)
kB	Kilobytes (one thousand bytes)
LAN	Local Area Network
LBN	Logical Block Number
LLC	Logical Link Control
LMS	Library Management System
LRC	Loop Resiliency Circuits (Fibre Channel)
LSB	Least Significant Bit
LTC	Longitudinal Timecode
LUN	Logical Unit
LVD	Low Voltage Differential (drivers)
MAC	Media Access Control
MAC	Multiplexed Analog Components
Mac	Macintosh Computer (from Apple)
MAN	Metropolitan Area Network
Mb	Megabits (one million bits)
MB	Megabytes (one million bytes)
MIPS	Million Instructions Per Second
MDCT	Modified Discrete Cosine Transform
MIS	Manager of Information Services
MO	Magneto-Optical
MP@ML	Main Profile at Main Level (see MPEG)
MP@HL	Main Profile at High Level
MPEG	Motion Picture Experts Group
MR	Magneto-Resistive
MSB	Most Significant Bit
MTBF	Mean Time Between Failure
MTF	Modulation Transfer Function

NAB	National Association of Broadcasters
NDBC	National Data Broadcasting Committee
NFS	Network File System
NIC	Network Interface Card
NICAM	Near Instantaneous Companded Audio Multiplex
NIST	National Institute of Standards and Technology
N-ISDN	Narrowband Integrated Services Digital Network
NPRM	Notice of Proposed Rule Making
NRZ	Non-return-to-zero (coding)
NRZI	Non-return-to zero, inverting (coding)
NTFS	New Technology File System (WindowsNT)
NTSC	National Television Systems Committee
NVOD	Near Video on Demand
OFDM	Orthogonal Frequency Division Multiplexing
OSI	Open Systems Interconnection
PAL	Phase Alternating Line
PAT	Program Association Table (MPEG-2)
PC	Personal Computer
PCD	Phase Change Disc
PCI	Peripheral Component Interconnect
PCM	Pulse Code Modulation
PCMCIA	Personal Computer Memory Card Industry Association
PCR	Program Clock Reference
PCVS	Point-to-point Switched Virtual Connections
PD	Packetization Delay
PD	Phase Disc
PDU	Packet Data Unit
PDU	Protocol Data Unit
PES	Packetized Elementary Streams (MPEG-2)
PHY	Physical Layer of the OSI Model
PID	Program Identification
PMD	Physical Medium Dependent (ATM sublayer)
PMT	Program Map Table (MPEG-2)
POP	Point of Presence
POTS	Plain Old Telephone Service
PPP	Point-to-point Protocol
PPV	Pay-per-view
PRML	Partial Response Maximum Likelihood
PSI	Program Specific Information (MPEG-2)
PTS	Presentation Timestamp (MPEG-2)

PVC	Permanent Virtual Circuit	
QAM	Quadrature Amplitude Modulation	
QIC	Quarter Inch Cartridge	
QoS	Quality of Service	
RAB	RAID Advisory Board	
RAID	Redundant Arrays of Independent Disks	
RAM	Random Access Memory	
RBOC	Regional Bell Operating Company	
RCA	Radio Corporation of America	
RFI	Radio Frequency Interference	
RGB	Red-Green-Blue	
RISC	Reduced Instruction Set Computing	
R-S	Reed-Solomon (coding)	
SAP	Secondary Audio Program	
SAR	Segmentation And Reassembly	
SAW	Surface Acoustic Wave (filter)	
SCA	Single Connector/Contact Assembly	
SCAM	SCSI Configured Automatically	
SCPC	Single Channel per Carrier	
SCSI	Small Computer Systems Interface	
SDDI	Serial Digital Data Interface (see SDTI)	
SDH	Synchronous Digital Hierarchy	
SDI	Serial Digital Interface	
SDTI	Serial Data Transport Interface	
SDTV	Standard Definition Television	
SEC	Single Edge Contact	
SECAM	Sequential Couleur avec Memoire	
SGI	Silicon Graphics Incorporated	
SIF	Source Input Format (sometimes Standard Input Format)	
SMPTE	Society of Motion Picture and Television Engineers	
SNA	Systems Network Architecture	
SNMP	Simple Network Management Protocol	
SONET	Synchronous Optical Network	
SOT	Start of Tape	
SPI	SCSI Parallel Interface	
SRAM	Static Random Access Memory	
SSA	Serial Storage Architecture	
SSS	Still Store Servers	
STA	SCSI Trade Association	
STB	Set Top Box	

STP	Shielded Twisted Pair
SUT	System Under Test
SVC	Switched Virtual Circuit
SVGA	SuperVGA (Video Graphics Array)
T1S1	ANSI T1 Subcommittee
TBC	Time Base Corrector
TC	Transmission Convergence (ATM sublayer)
TCP	Transmission Control Protocol
TCP/IP	Transmission Control Protocol/Internet Protocol
TDM	Time Division Multiplexing
TRS	Timing Reference Signal
TS	Time Stamp, Transport Stream
TS	Transport Streams (MPEG-2)
UDP	User Datagram Protocol
UHF	Ultra High Frequency
ULP	Upper Layer Protocol
UNI	User Network Interface
UPS	Uninterruptible Power System
USB	Universal Serial Bus
UTP	Unshielded Twisted Pair
VAR	Value Added Reseller
VBI	Vertical Blanking Interval
VBR	Variable Bit Rate
VDR	Video Disk Recorder
VCI	Virtual Circuit Identifiers
VGA	Video Graphics Array
VHDCI	Very High Density Cabling Interconnect
VHF	Very High Frequency
VITC	Vertical Interval Timecode
VLAN	Virtual Local Area Network
VOD	Video on Demand
VPI	Virtual Path Identifiers
VSB	Vestigial Sideband
VTR	Video Tape Recorder
WAN	Wide Area Network
WORM	Write Once Read Many
YUV	Luma and two color difference signals

INDEX

0
0,4 encoding 383

1
1,7 encoding 383
1.062 Gb/s 351
100BaseT 258
10Base2 256
10Base5 256
10BaseT 39, 220, 222, 256, 259, 260
131 Mb/s 351
14.3 MHz 269
143 Mb/s 22, 64, 249
16:9 .... 133, 134
16-Mb/s token ring 280
177.3 Mb/s 64
19.39 Mb/s 135, 139, 156, 165, 283
1920 × 1080 138
19mm 400, 436

2
24 Mb/s ....
  103, 142, 157, 210, 211, 397, 400, 401
25.6 Mb/s ATM 280
27 MHz 92, 269

270 Mb/s 26, 55, 64, 89, 189, 198

3
3/2 pull-down 19, 139, 140
360 Mb/s 55, 64, 198

4
4:1:1 136
4:2:0 .... 136, 138, 142
4:2:2 28
4:3 .... 133, 134
45-Mb/s 3
48 kHz 69, 94
48 kHz sampling 94
48 Mb/s 60, 142, 154
4B/5B 280
4ƒSC 22, 27, 64, 90, 92, 269

5
50 Mb/s ....
  .... 56, 57, 137, 138, 142, 156, 222
540 Mb/s 189

7
742.5 Mb/s 189

8
8B/10B encoding 351
8mm 17, 23, 110, 398, 399, 401, 404
8mm tape 110

A
A cable 324

443

AAL ....
    xiii, 275, 277, 278, 279, 283, 435,
    436
Abekas ....
    1, 21, 24, 27, 28, 29, 31, 32, 42, 104,
    105, 106
Abekas A60 .... 28, 29, 104, 105, 106
Abekas A62 .... 24, 27, 28
Abekas A64 .... 24, 27, 28
Abekas A65 29
Abekas A66 29
Abekas TouchUp 29
access time ....
    65, 102, 143, 179, 376, 377, 387,
    401
Accom ....
    ...... 17, 31, 32, 37, 42, 105, 106, 107
Accom RTD-4224 .... 31, 32, 42, 107
Accom Workstation Disk (WSD) ....
    .... 32, 107
active termination 326
adaptation layer, AL-1 278
adaptation layer, AL-5 278
adaptive filtering 91
ADO 35, 232, 435
advanced digital television 198
advanced SCSI architecture (ASA II)
    388
Advanced Television ....
    .... 4, 7, 59, 426, 435, 437
Advanced Television (ATV) ....
    .. 4, 25, 131, 423, 426, 433, 435, 437
AES ....
    68, 69, 70, 80, 85, 93, 94, 108, 112,
    198, 256, 319, 435
AES3 .... 69, 96, 283
AES3 audio bit streams 96
Al Kovalick .... xv, 111, 113
algorithms ....
    82, 91, 121, 145, 146, 297, 299, 301,
    307, 308, 335, 377, 420
aliasing 91
Amdahl's Law 236
Ampex HS-100 .... 12, 15, 16
Ampex, ACR-25 .... 130, 159
analog peak-detect 381, 382, 385
analog-to-digital (ADC) 92, 435
angular speed 386

animation ....
    xviii, 12, 13, 19, 29, 43, 107, 130,
    296, 313, 314
ANSI ....
    xv, 64, 70, 80, 96, 106, 146, 197,
    315, 317, 318, 319, 328, 332, 343,
    344, 355, 362, 369, 401, 435, 442
ANSI X3.131-1986 317
application program interface (API) ......
    .... 62, 148, 195, 435
arbitrated loop 348
architecture (RAID) 290
archive ....
    x, xi, xviii, 28, 62, 110, 151, 154,
    155, 163, 168, 170, 174, 178, 179,
    199, 208, 209, 212, 213, 223, 225,
    226, 399, 402, 403
areal density 384, 385
artifacts 27, 55, 91, 139, 146, 156
ASA II 388
ASC Virtual Recorder 127
ASCII 98, 252, 435
asynchronous ....
    xvii, 109, 114, 232, 254, 259, 272,
    273, 274, 321, 324, 328, 402
AT Attachment 332
AT Attachment (ATA) 435
ATM ....
    vii, x, xiii, 63, 64, 112, 114, 148,
    189, 200, 237, 260, 261, 263, 266,
    272, 273, 274, 275, 276, 277, 278,
    279, 280, 283, 284, 316, 343, 350,
    352, 367, 422, 435, 440, 442
ATM base layers 275
ATM cell 274, 277
ATM cell header 279
ATM Forum 272
ATM framing 114
ATM payloads 283
ATM traffic classifications 274
ATM transport packet 274
ATM, AAL-5 279
ATM/AAL5 63
attachment unit interface (AUI) 257
ATV .... See Advanced Television
audio compression 24
audio/video-on-demand (AVOD) ... 129
AUI 435
AVI 24, 40, 225, 435

Index   445

Avid AirPlay 128
Avid Technologies 1
Avid Technology35, 36, 122, 128, 175, 215
Axial RAVE 37

B

B cable 323
backup tape 178, 296
bandwidth ....
  xvii, xviii, 3, 7, 19, 24, 29, 33, 38, 39, 41, 42, 44, 45, 48, 60, 65, 68, 72, 80, 81, 86, 92, 101, 103, 104, 105, 107, 121, 125, 128, 132, 134, 136, 141, 143, 144, 154, 165, 189, 218, 219, 221, 222, 223, 224, 225, 227, 234, 235, 236, 256, 257, 260, 266, 268, 269, 271, 273, 280, 316, 328, 329, 332, 344, 347, 348, 352, 353, 354, 358, 359, 360, 362, 363, 364, 366, 367, 368, 378, 400, 415, 429
baseband ....
  43, 44, 45, 54, 55, 56, 97, 122, 135, 139, 140, 143, 144, 146, 149, 156, 196, 224, 225, 402
Bell Atlantic 433
Berkeley papers 291
Betacam ....
  22, 23, 27, 30, 60, 65, 136, 144, 176, 177, 269, 303, 427
BetacamSP ....
  22, 54, 79, 154, 157, 177, 211, 303, 361, 398, 433
BetacamSX 135, 136, 143
bit error rate 405
bit stream server 57
bit streams ....
  48, 54, 55, 56, 57, 61, 69, 109, 115, 135, 137
broadband delivery 132
broadcast ....
  x, xiii, xv, xviii, 1, 3, 5, 6, 12, 15, 17, 23, 26, 32, 33, 40, 43, 49, 51, 54, 57, 59, 60, 67, 68, 69, 72, 82, 83, 86, 89, 90, 91, 94, 95, 96, 100, 108, 109, 124, 125, 126, 128, 129, 130, 132, 147, 151, 155, 157, 159, 160, 161, 162, 163, 164, 165, 173, 175, 179,

180, 181, 183, 184, 185, 186, 187, 188, 190, 192, 194, 195, 196, 197, 198, 199, 200, 201, 203, 215, 222, 227, 229, 234, 240, 263, 265, 271, 283, 285, 286, 291, 320, 341, 344, 359, 386, 389, 394, 395, 407, 413, 415, 418, 425, 426, 427, 429, 431
broadcast video server 5, 90
broadcaster ....
  xvii, 26, 64, 160, 162, 163, 185, 187, 190, 192, 193, 194, 415, 423, 427, 428, 429
broadcasting ....
  xvii, xviii, 15, 98, 99, 185, 190, 194, 201, 229, 255, 352, 355, 425, 428
browse server ....
  ix, 165, 166, 167, 169, 222, 223, 224
browser 165, 166, 168, 169, 170, 225
buffer ....
  15, 20, 81, 115, 134, 143, 208, 213, 271, 299, 308, 369, 403, 419
buffer, anticipatory 308
bursty 220, 269, 270
bus architecture 82
BVS 5, 6, 7, 435
BVW-75 protocol 69

C

cable ....
  xiii, 6, 86, 90, 120, 127, 128, 129, 132, 162, 163, 184, 198, 204, 234, 237, 241, 256, 257, 323, 324, 325, 326, 330, 331, 332, 335, 336, 337, 342, 364, 365, 408, 409, 410, 411, 412, 414, 418, 421, 429, 433
cache ....
  25, 67, 126, 127, 181, 183, 201, 208, 307, 308, 309, 369, 374, 375, 376, 377, 389, 404
cache hit ratio 376
cache hit ratios 375
cache hits, imbalance 376
camera 56, 106, 224
cassette, linear tape 400
cassettes, data 162, 397, 399, 402
CAT-5 .... 258, 259
Category 5 (CAT-5) wiring 258
CAV 18

CCIR-601 .... 3, 21
CCIR-656 .... 19, 21
CCITT .... x, 253, 254, 436
CCITT X.25 .... x, 253
CCS 317, 318, 328, 334, 436
CDDI 260, 436
cell 274
cell delineation 276
cell switching 273
central file server 230
channel service unit/data service unit
.... 235
character generators 13, 35
Charles Poyton 99
Chatter Disk Management 20
chroma .. 17, 28, 60, 136, 144, 269, 295
chroma channel 60
Chyron 35
Cinepak 40
Ciprico 32
circuit switching 268, 269, 270, 273
circuit switching transmission 273
class of service (COS) 353, 354
closed-architecture 29
CLV 18
CMX 16, 17, 29, 30, 70, 436
coax 79, 147, 249, 414
codec ....
ix, 39, 40, 45, 55, 56, 59, 62, 72, 75,
82, 83, 84, 85, 86, 87, 90, 92, 121,
124, 125, 131, 138, 139, 141, 142,
143, 144, 149, 174, 214
coding 3
color space 105
color subcarrier 91
color-under 17
COMDEX ....
.... 331, 358, 368, 382, 389, 436
COMDEX, Fall-93 382
COMDEX, Fall-96 .... 331, 368
COMDEX, Fall-97 358
Common Command Set (CCS) ....
.... 317, 328, 334, 436
commutator 73
Compaq Computer Corporation ....
.... 289, 421, 422
component digital ....
xviii, 12, 20, 21, 22, 23, 28, 30, 32,
33, 38, 41, 43, 44, 55, 68, 80, 90, 91,
106, 109, 155, 190, 263, 269
component digital interface 90
component serial digital ....
.... 54, 79, 156, 198
component video 90, 93, 154
composite analog decoder 91
composite digital ....
.... 22, 23, 27, 30, 43, 64, 90, 269
composite video 17, 19
compositing ....
.... 18, 20, 21, 23, 27, 28, 32, 44
compressed digital streams 149
compressed digital video ....
xviii, 3, 23, 62, 109, 114, 120, 133,
144, 198, 254, 280, 410
compression ....
2, 3, 5, 9, 12, 14, 24, 32, 33, 40, 43,
45, 48, 49, 51, 52, 53, 54, 55, 57, 59,
60, 62, 68, 71, 72, 80, 81, 82, 85, 86,
87, 90, 91, 102, 108, 117, 121, 122,
124, 125, 131, 133, 134, 135, 136,
137, 139, 141, 142, 143, 144, 145,
146, 148, 149, 152, 153, 156, 159,
162, 166, 189, 195, 198, 199, 211,
220, 221, 253, 255, 257, 264, 266,
267, 269, 281, 284, 361, 378, 387,
392, 396, 397, 398, 399, 401, 404,
407, 411, 414, 422, 435, 436
compressor 59, 81, 82, 83, 144, 290
concentrator 73
conditional access 163
connectionless 278, 279, 280, 353
connection-oriented 275, 354
constraint 56, 111
Consumer Electronics Show (CES) ......
.... 193, 432, 436
convergence sublayer (CS) 278, 436
copper data distributed interface
(CDDI) 260
copper pairs 71
copy protection 100
CRC 250, 276, 277, 366, 436
CSU 235, 236, 436
cycle time ....
.... 210, 211, 322, 323, 324, 400

D

D-13, 20, 21, 22, 23, 26, 28, 30, 38, 40, 60, 144, 153, 154, 249, 427, 436
D-2 22, 23, 25, 26, 27, 30, 40, 90, 249, 427, 436
D-3 .... 26, 90, 436
D-5 .... 26, 38, 436
DAT 398, 399, 436
data frame .... x, 250, 251
data integrity ....
.... 119, 211, 307, 311, 373, 388, 405
data link protocol 273
data loss 309, 311, 368, 394
data network 196, 199, 254, 361
data packets 283, 363
data protection ....
xviii, 43, 48, 54, 151, 171, 286, 289, 294, 300, 393
data protocols 314
data rate ....
65, 87, 102, 132, 137, 156, 165, 198, 210, 211, 221, 222, 314, 316, 322, 329, 330, 337, 386, 401, 422
data reconstruction .... x, 307, 308
data segment 211
data stream ....
3, 64, 97, 99, 100, 136, 198, 224, 252, 298, 299, 300, 301
data transport stream 283
database ....
8, 37, 62, 110, 160, 161, 162, 164, 166, 167, 168, 169, 170, 171, 177, 180, 210, 211, 213, 216, 218, 219, 220, 223, 224, 226, 239, 367, 374, 401, 403, 405, 413, 419
datagram service, Fibre Channel 353
DC300A 399
DCT 40, 55, 59, 86, 436
DDR ....
7, 14, 27, 29, 30, 31, 38, 39, 45, 105, 106, 107, 108, 109, 314
DDR, component 27
DDR, composite 27
decode ....
82, 84, 135, 137, 143, 145, 233, 267, 282, 411
decode timestamp (DTS) 282, 437

decoder ....
27, 40, 56, 59, 81, 82, 83, 95, 139, 145, 230, 281
differential fast-SCSI 337
differential SCSI 337
digital audio tape 398
digital broadcasting 283
digital effects 31, 32, 232, 233
digital linear tape ....
.... 142, 167, 182, 211, 399, 400, 401
digital news gathering (DNG) 215
digital optics 21
Digital Post and Graphics xviii
digital signal processor (DSP) ....
.... 374, 380, 437
digital streaming tape ....
.... x, 156, 209, 212, 402
digital transmission 274
digital versatile disk 24
digital video disk ....
.... 7, 11, 21, 24, 28, 29, 427
digital videotape ....
.... 7, 23, 30, 38, 153, 400
Digital-S 60, 136
direct recording 17
discrete cosine transform 40
disk cache 375
disk drive, magnetic ....
.... 286, 377, 379, 407
disk mirroring 291
disk recorder ....
7, 12, 13, 14, 15, 17, 18, 19, 21, 23, 24, 27, 30, 31, 32, 37, 38, 39, 40, 41, 42, 43, 45, 81, 82, 101, 104, 105, 106, 108, 115, 295, 403
disk striping 291
distributed file server 238
DLT ....
142, 156, 167, 178, 209, 211, 399, 400, 403, 436
DNG 437
Dolby AC-3 95
drive array ....
66, 73, 101, 103, 143, 213, 288, 293, 295, 302, 307, 337, 360, 392
drive failure .... x, 289, 307
drop frame 98
DS0 235
DS1 235

DS2 235
DS3 235
DS4 235
DTV ....
   ix, 47, 55, 57, 62, 89, 95, 112, 113,
   114, 118, 120, 125, 131, 132, 133,
   135, 139, 147, 153, 155, 162, 163,
   164, 176, 183, 186, 187, 188, 189,
   190, 191, 193, 194, 195, 229, 234,
   261, 263, 303, 360, 407, 428, 437
duplex/mirrored drive 299
DV 60, 65, 283, 370, 437
DVC 60, 283, 437
DVCPro-25 136
DVCPro-50 .... 56, 136, 137
DVD 24, 427, 432, 437
DVE 32, 208, 437
DVR-1000 21
Dynamic Rounding 20, 33

E

EBU .... xv, 96, 437
edit decision list ....
   .... 17, 36, 224, 225, 252, 437
electro-magnetic interference (EMI) .....
   258
electronic program guide (EPG) ....
   .... 164, 437
electronic still store (ESS) 6, 7, 268
EMI 258, 259, 331, 437
encode 52, 54, 82, 84, 137, 142, 267
encoder 40, 44, 81, 82, 83, 165
encoder-decoder 84
end of active video (EAV) 401
entry-level server 53
error concealment 308
error correction ....
   3, 5, 105, 112, 211, 231, 233, 234,
   248, 250, 271, 274, 294, 297, 305,
   306, 307, 308, 340, 364, 399
error correction coding 105, 233
error correction coding (ECC) ....
   .... 105, 307, 340, 437
error detection. 103, 105, 277, 351, 362
error recovery 307
Ethernet ....
   28, 31, 39, 42, 63, 71, 104, 106, 107,
   108, 149, 189, 197, 200, 223, 234,

   247, 254, 255, 256, 257, 258, 259,
   261, 279, 280, 334, 346
Exabyte ....
   28, 105, 156, 167, 209, 210, 211,
   213, 401
Exabyte 8500 series 105
extensibility ....
   31, 61, 66, 85, 131, 191, 197, 200,
   330, 335, 337, 346, 361

F

fabric switch 348
failure tolerant disk systems (FTDS)
   .... 394, 438
fast packet switching 273, 274
Fast SCSI ....
   321, 322, 323, 328, 329, 330, 332,
   333
FASTBREAK, Sundance 128
fault tolerance ....
   xviii, 3, 6, 43, 285, 286, 293, 297,
   302, 391, 395, 404, 405
fault tolerant .. 4, 74, 295, 310, 405, 421
FCC .. xiii, 132, 147, 184, 229, 258, 437
FCS 367
fiber distributed data interface (FDDI)
   .... 197, 237, 257, 346, 367, 438
fiber optic cabling 361
Fibre Channel ....
   viii, x, xiii, 39, 44, 53, 54, 64, 66,
   71, 72, 74, 75, 103, 110, 112, 114,
   148, 189, 190, 192, 198, 200, 206,
   208, 213, 214, 234, 237, 254, 261,
   293, 303, 313, 323, 331, 335, 339,
   340, 341, 342, 343, 344, 345, 346,
   347, 348, 349, 350, 351, 352, 353,
   354, 355, 356, 358, 360, 361, 368,
   369, 370, 371, 387, 388, 389, 404,
   437, 439
Fibre Channel Loop Community
   (FCLC) 331, 356
Fibre Channel, Class 1 .... 353, 354
Fibre Channel, Class 2 .... 353, 354
Fibre Channel, Class 3 .... 353, 354
Fibre Channel, Class 4 .... 344, 354
Fibre Channel, enhanced physical ... 341
Fibre Channel, Fabric ....
   .... 343, 349, 350, 352, 353

Index 449

Fibre Channel, FC-0 level 351
Fibre Channel, FC-1 level 351
Fibre Channel, FC-2 level 345, 352
Fibre Channel, FC-3 level 352
Fibre Channel, Upper Level Protocol
  (ULP).. 354
Fibre Channel-Arbitrated Loop
  (FC-AL)..
  75, 103, 234, 343, 347, 348, 387,
  389, 437
Fibre Channel–Arbitrated Loop (FC-
  AL) 66, 354
Fibre Channel-Audio Video (FC-AV)
  .... 344, 354, 437
Fibre Channel-Enhanced Loop (FC-
  EL).. 355, 437
Fibre Channel-Physical (FC-PH) ....
  .... 351, 352
FibreDrive........................ 103, 293, 355
file server ....
  118, 119, 122, 128, 219, 220, 238,
  239, 240, 241
file transfer protocol (FTP) 39, 108
FireWire....... 39, 64, 234, 335, 339, 438
first-birthday 310
flags.. 366
flying head height 386
force-formatting 295
fractals.. 60, 86
frame relay 269, 274
FTDS.. 394, 438
FTP 24, 108, 109, 110, 438
Fujitsu 19, 382

G

gain 38, 91, 146, 169, 295
gamma correction........................... 105
general flow control (GFC).................
  .... 276, 277, 278, 438
general purpose input or interface
  (GPI)..................................... 70, 438
Gigabit Ethernet............................... 261
global link module (GLM) ....
  .... 342, 343, 438
Grand Alliance (GA).............................
  .... 135, 142, 156, 283, 438
graphics workstation 232
Grass Valley Group........................... 32

group of pictures (GOP)........................
  .... 60, 87, 139, 143, 144, 281, 438
GUI ....
  19, 36, 54, 71, 72, 99, 202, 204, 438

H

Hamming code 297, 307
hand-shaking 322, 324
hard disk drive (HDD) ....
  .... 41, 288, 320, 417, 418, 438
Harry 18, 19, 20, 21, 24, 28, 37
HDCam 135, 136
HD-DDR.. 31
HDTV ....
  xiii, 25, 31, 60, 89, 104, 113, 131,
  134, 135, 139, 146, 147, 155, 156,
  188, 193, 222, 235, 438
headend 204
header...
  x, 86, 110, 164, 171, 198, 210, 211,
  250, 251, 267, 271, 275, 276, 277,
  278, 279, 282, 283, 340, 345, 352
HEC 276, 277, 278, 438
Henry 20, 21, 31, 32, 37
heterodyne.. 17
heterodyned....................................... 17
Hewlett-Packard.....................................
  32, 111, 113, 126, 128, 210, 403,
  438
Hewlett-Packard MediaStream ....
  .... 126, 129, 403
hierarchical storage management 163
high definition television 57, 89
high performance parallel interface
  (HiPPI)....................................... 352
high-voltage differential drivers (HVD)
  331
HiPPI.............. 341, 344, 352, 360, 438
host adapter 319, 328, 364
hot file.. 376
hot pluggable device 323
hot plugging 330
hot swap 310
hot swappable 287
hot-plugging.................................... 388
hot-standby....................... 74, 310, 347

hub ....
    71, 73, 214, 241, 242, 260, 346, 347,
    348, 370, 391
hubs 73, 75, 241, 259, 347, 367, 393
HVD 331, 333, 337, 438
hybrid ....
    37, 42, 45, 109, 126, 190, 240, 291,
    408, 412

I

I, B and P frames 86
IBM ....
    13, 192, 236, 317, 360, 362, 365,
    368, 369, 379, 382, 383, 385, 422
IBP coding 60
IEEE 1394 (FireWire) ....
    .39, 64, 112, 114, 234, 335, 339, 438
IEEE 802 279
IEEE-485 331
impedance, characteristic 336
Indeo 24, 40, 225
inductive thin film head 385
information superhighway 2, 4, 257
initiator 319, 328, 367
intelligent peripheral interface (IPI) ......
    .... 341, 352, 439
Intelligent Resources 86
Intelligent Resources VideoBahn 86
interactive PC/TV 431
interactive television ....
    6, 132, 186, 236, 407, 425, 426, 427,
    429, 430, 431, 432
interactive television (ITV) ....
    .... 407, 413, 433, 439
interfaces ....
    xviii, 27, 29, 39, 43, 61, 62, 63, 69,
    77, 90, 99, 103, 107, 108, 112, 114,
    148, 168, 191, 197, 199, 211, 234,
    250, 255, 260, 275, 276, 280, 314,
    315, 316, 330, 335, 336, 337, 342,
    343, 353, 354, 358, 359, 360, 362,
    367, 369, 387, 389
interlace 19, 133, 136
interleave 24, 247, 339
interleaved 94, 269, 299, 301
interleaving 55
Intermix 353

International Telecommunications
    Union (ITU) ....
    3, 26, 28, 64, 68, 91, 92, 106, 110,
    136, 144, 146, 154, 156, 307, 439
Internet ....
    113, 114, 118, 132, 180, 184, 185,
    186, 217, 229, 236, 237, 248, 257,
    265, 271, 352, 407, 413, 414, 415,
    418, 427, 429, 432, 434, 438, 439,
    442
Internetworking 245, 247, 415
interoperability ....
    26, 62, 191, 194, 200, 255, 347, 355
interpolation 105
intra-frame 40, 59, 99
intranet 113, 114, 217, 237, 248
IP streaming 112
IPI 317, 341, 344, 352, 439
IP-TV 112
ISO .... xiii, 98, 245, 275, 439
isochronous ....
    xvii, 114, 115, 265, 269, 283, 354,
    355
ITU R BT.601-5 64
ITU-R BT.601 ....
    3, 26, 28, 64, 68, 91, 106, 110, 136,
    144, 146, 154, 156, 307

J

jitter 114
Joint Photographic Expert Group 24
JPEG ....
    24, 29, 40, 44, 51, 54, 55, 56, 57, 59,
    62, 85, 86, 90, 99, 121, 125, 131,
    134, 135, 141, 145, 147, 154, 155,
    156, 157, 211, 222, 264, 347, 401,
    439
jukebox 121, 122, 399

K

Kadenza 32
Kaleidoscope 32

Index   451

L

LAN ....
  71, 118, 146, 167, 221, 233, 237,
  239, 247, 248, 252, 254, 274, 279,
  346, 439
laserdisc ....
  .. 16, 18, 20, 122, 407, 408, 410, 411
latency ....
  18, 102, 110, 115, 130, 142, 149,
  179, 233, 280, 295, 301, 308, 364,
  374, 375, 376, 378, 387, 402, 418,
  419, 420
Latin America Channel (FLAC) 127
legacy hardware 149
Leitch 126, 127, 293, 355, 403
levels ....
  xiii, 27, 45, 48, 53, 60, 99, 131, 138,
  141, 148, 154, 175, 180, 188, 198,
  215, 222, 226, 249, 271, 272, 274,
  278, 279, 285, 289, 291, 293, 294,
  298, 302, 308, 331, 332, 351, 352,
  353, 364, 368, 382, 395, 431
library management system (LMS) ....
  .... 47, 181, 182, 208, 297, 396, 439
Lightworks 37
Lightworks VIP 37
line rate 17
linear tape machines 14
logical connection 275
logical scheme 320
logical unit (LUN) 319, 439
long-term storage ....
  ...... 48, 151, 156, 161, 162, 187, 199
loop resiliency circuit (LRC) 75, 439
low-voltage differential drivers (LVD)
  .... 331, 333, 335
luma 17, 19, 28, 91
luminance 2, 61, 64, 136, 144, 295
LVD 331, 333, 359, 365, 439

M

machine control ....
  .... 62, 63, 69, 77, 94, 97, 98, 99, 198
Macintosh ....
  13, 36, 105, 118, 124, 128, 260, 303,
  313, 314, 317, 320, 325, 369, 439
macroblock boundary crossing 154

magnetic disk array 288
magneto-optical (MO) 122, 439
magneto-resistive head (MR) ....
  .... 383, 384, 385, 388, 439
Main Profile at Main Level 24, 439
MAN 118, 247, 439
mapping MPEG into ATM 280, 282
Matsushita Avionics Systems
  Corporation 130
mean time before failure (MTBF) ... 311
mechanical liabilities 376
media file server xix
Media Server Technology ....
  .... xvi, xviii, 379
media servers, architecture 77
Megatransfers per second 329
Micro Henry 32
microcode 41, 232, 305, 322, 380
microcontroller 380
Micropolis Corporation 130
MII format ... 60, 79, 176, 177, 303, 361
mirroring ....
  x, 207, 294, 296, 297, 302, 305, 394,
  395, 396, 405
MO-drive 338
Moore's Law 288, 373, 379
motion-JPEG ....
  29, 40, 51, 54, 145, 147, 154, 155,
  156, 157, 211, 222, 347, 401
Moving Pictures Expert Group 24
MPEG ....
  x, xix, 24, 40, 44, 55, 56, 60, 87, 99,
  128, 135, 137, 139, 141, 142, 143,
  153, 156, 236, 255, 264, 272, 280,
  281, 282, 417, 439
MPEG-1 .... 5, 24, 40, 49, 318, 423, 433
MPEG-2 ....
  ix, xiii, 5, 24, 40, 57, 85, 86, 87, 95,
  99, 110, 130, 131, 134, 137, 138,
  139, 140, 143, 144, 147, 149, 156,
  164, 198, 222, 275, 281, 282, 283,
  318, 370, 403, 422, 423, 433, 437,
  440, 442
MPEG-2 4:2:2P@ML ....
  .... 126, 138, 142, 156, 402
MPEG-2 MP@ML 126, 138
MPEG-2 transport packet 283
MPEG-4 .... 24, 318
MTBF 311, 388, 439

452 Video and Media Servers: Technology and Applications

multicasting............................. 188, 355
multichannel..
   ix, xvii, xviii, 103, 163, 180, 187,
   201, 205, 428
multiple stream............................... 129
multiple streams ....
   .... 6, 49, 82, 101, 103, 148
multiplexer 19, 139, 142
multiplexing ....
   96, 109, 114, 251, 255, 268, 269,
   278, 281

N

National Association of Broadcasters
   (NAB)................ 23, 25, 57, 389, 440
NCR 317
near-video-on-demand (NVOD) 411
net transfer rate 404
network file server 387
network interface card (NIC) ....
   .... 222, 249, 258, 280
nodes ....
   x, 50, 75, 229, 233, 268, 277, 279,
   342, 348, 349, 350, 358, 363, 364,
   370
non-linear editing ....
   13, 35, 36, 47, 59, 122, 123, 132,
   144, 190, 217, 221, 232, 264, 314,
   327, 359
non-linear editing (NLE) ....
   ix, 36, 37, 38, 122, 123, 124, 128,
   240, 386, 388
non-linear video editing 386
non-RAID 305
non-RAID mirroring........................ 297
non-return to zero, inverted............. 280
nonscalable 297
notch filter... 91
Novell 260, 420, 439
NRZI 280, 440
NTSC coding 52

O

open architecture 28, 343
open systems interconnection 275
optical storage 374

original equipment manufacturer
   (OEM) 18, 40, 51, 380
OSI ....
   x, xiii, 245, 246, 247, 248, 249, 251,
   253, 254, 255, 275, 334, 351, 353,
   440
OSI Applications Layer................... 334
overhead...
   211, 232, 233, 235, 258, 283, 310,
   316, 323, 324, 332, 359, 362, 363,
   366, 404
over-the-top................................. 82, 86

P

P cable.. 323
packet ....
   x, 112, 198, 250, 251, 254, 267, 270,
   271, 273, 274, 275, 281, 282, 324,
   335, 343, 345, 346
packet identifier (PID) 281, 282, 440
packetized elementary streams (PES)
   .... 164, 281, 282, 440
Paintbox 18, 29
PAL ....
   16, 17, 19, 43, 64, 92, 249, 295, 440
parallelism.. 304
parity ....
   x, xiii, 103, 207, 291, 294, 295, 299,
   300, 302, 305, 306, 307, 308, 310,
   355, 366, 405
parity bit.. 306
parity block 301
parity channel..................................... 74
parity data 308
parity disk 294, 299, 300, 301
parity drive ....
   x, 73, 103, 295, 298, 299, 300, 301,
   302, 306, 307, 309, 392, 405
parity drive failure........................... 306
parity generation 103
parity information 295, 299, 301
parity protection........................ 65, 305
parity RAID storage 309
parity striping........................... 103, 397
parity word....................................... 306
parity, interleaving 310
partial-response maximum liklihood
   (PRML) ....

Index 453

381, 382, 383, 384, 385, 388, 420, 440
partitioning 401, 402
passband 91
pay-per-view ...... xviii, 6, 127, 129, 408
pay-per-view (PPV) ....
.... 6, 408, 409, 410, 440
Pentium 233, 381, 421
Philips-BTS 22, 33, 128, 403
Philips-BTS, MediaPool 128, 403
physical scheme 320
Pluto File System 111
Pluto SPACE 109
Pluto Technologies International ....
.... 109, 110
point-to-multipoint .... xvii, 120, 196
point-to-point ....
112, 186, 219, 230, 232, 234, 235, 254, 259, 268, 343, 348, 350, 364
Point-to-point .... ix, x, 120, 350, 440
prefetch cache 308
preloading 56
preread 30
presentation timestamp (PTS) ....
.... 282, 440
program association table (PAT) ....
.... 282, 440
program clock reference (PCR) ....
.... 281, 440
program numbers 282
protection 180, 214, 299, 406
protocol ....
63, 66, 69, 97, 108, 109, 110, 192, 198, 211, 252, 259, 266, 273, 275, 277, 281, 314, 315, 316, 317, 318, 320, 321, 322, 327, 328, 334, 335, 341, 342, 344, 345, 351, 358, 365, 366, 367, 389
proximity recording 386
proximity recording head 384
pulse code modulation (PCM) ....
.... 94, 269, 440

Q

Q cable 323
QoS 112, 114, 441
quadruplex 14, 16, 18
quality of service 114, 441

Quantel ....
1, 18, 20, 21, 28, 29, 31, 32, 33, 37, 39
Quantel Dylan drives 20, 39
Quantum 376, 382
Quicktime 40, 121

R

Raceway 86
RAID "number" 290
RAID 0 ....
102, 103, 207, 293, 294, 295, 301, 302, 368
RAID 1 296
RAID 10 .... 293, 301
RAID 2 .... 294, 297, 298, 307
RAID 3 ....
x, 74, 103, 207, 212, 290, 294, 298, 299, 301, 302, 306, 307, 368, 392, 397, 405
RAID 4 .... 294, 299
RAID 5 ....
x, 103, 294, 300, 301, 306, 308, 310, 368, 389, 405
RAID 53 .... 291, 293, 302
RAID 6 .... 291, 293, 294, 301, 308
RAID Advisory Board (RAB) ....
.... 291, 293, 295, 394, 395, 441
RAID controller ....
.... 74, 305, 307, 310, 393, 394
RAID Level 290
RAID Level 0 .... 291, 294
RAID Level 1 .... 291, 296
RAID Level 10 301
RAID Level 2 297
RAID Level 3 .... 290, 291, 298
RAID Level 4 299
RAID Level 5 300
RAID Level 53 .... 291, 302
RAID Level 6 .... 291, 301
RAID, control architectures 373
RAID, expansion chassis ix, 73, 74, 213
RAID, in software 355
RAID, software implementation 66
RAIDbook, The 291, 293
RAIDsoft 355
RAM disk 321, 378

random access ....
   1, 13, 14, 18, 20, 29, 35, 40, 101,
   123, 167, 185, 220
RBOC 441
RCA, TCR-100 ....130, 159
RCP..109, 110
read cache 375
read channels... 381, 382, 383, 384, 385
redundancy.......................................
   43, 45, 48, 54, 75, 103, 117, 201,
   250, 277, 289, 294, 296, 297, 298,
   305, 311, 366, 391, 396, 404, 406
redundant arrays of independent disks
   (RAID)................................285, 303
redundant arrays of inexpensive disks
   (original RAID) 289
redundant bit 305
redundant copy.................................296
remote access service.......................248
RGB 19, 104, 105, 106, 433, 441
RGB, valid color..............................106
RISC 13, 441
robotics ....
   126, 131, 148, 170, 208, 209, 212,
   213, 397, 406
rotational latency............. 102, 376, 418
rotoscoping 19, 29
RS-422 .... ix, 63, 69, 71, 94, 97, 98

S

SAM........................xiii, 334, 335, 339
SASI...317
SCA-2 388
scalability....... 6, 53, 131, 191, 192, 357
SCAM 331, 358, 368, 388, 441
SCAM Level-1................................388
SCAM Level-2................................388
SCAM-1...331
SCAM-2...331
Scitex 39, 368, 370
SCSI ....
   31, 44, 104, 314, 315, 316, 317, 319,
   320, 323, 332, 333, 337, 338, 343,
   344, 350, 352, 360, 367, 368, 373,
   388
SCSI applications............................318
SCSI Architectural Model (SAM)...334
SCSI architecture, internal 314
SCSI bus..
   211, 314, 315, 318, 319, 320, 326,
   329, 336, 337, 358, 359, 404
SCSI bus bottleneck 404
SCSI commands..............................376
SCSI Configured Automatically
   (SCAM)...............................331, 441
SCSI data protocol 314
SCSI device....................................
   .... xiii, 314, 319, 330, 338, 393
SCSI ID................................. 319, 320
SCSI interface ....
   71, 313, 314, 319, 324, 326, 331,
   336, 404
SCSI mapping 334
SCSI numbers 318
SCSI parallel interface (SPI)..........331
SCSI peripherals 315
SCSI protocol.................................318
SCSI Trade Association (STA).............
   ....xiii, 331, 332, 441
SCSI, balanced...............................320
SCSI, communications............321, 322
SCSI, control..................................39
SCSI, data transfers........................321
SCSI, differential 337
SCSI, Fast + Wide..........................
   .... 322, 323, 329, 335, 337, 364
SCSI, Fast and Wide......................323
SCSI, Fast-20 ....
   .... 328, 329, 332, 336, 337
SCSI, high-level communications
   protocol...................................320
SCSI, parallel 341
SCSI, parallel Fast-40 329
SCSI, parallel interface 321
SCSI, protocols 357
SCSI, RAID controller...................66
SCSI, Ultra2............ 330, 331, 333, 365
SCSI, Ultra2 transceivers................331
SCSI, Wide Ultra 330
SCSI, Wide Ultra2 330
SCSI-1..
   313, 317, 318, 321, 322, 323, 324,
   325, 326, 327, 328, 330, 331, 332,
   333, 336, 362
SCSI-1, synchronous......................321
SCSI-2..
   31, 257, 313, 318, 321, 322, 323,

326, 327, 328, 332, 335, 336, 339, 358, 362, 363, 367, 368
SCSI-3 ....
   xiii, 313, 318, 320, 322, 323, 324, 327, 328, 329, 332, 333, 334, 335, 336, 337, 339, 344, 355, 358, 360, 363, 388, 389
SDTI ....
   ix, 55, 56, 65, 69, 90, 96, 112, 113, 114, 134, 137, 139, 197, 198, 255, 344, 361, 436, 441
Seagate 102, 382, 387, 388, 389
secondary controller 310
security ....
   32, 119, 122, 131, 199, 226, 237, 240, 267
segmentation and reassembly layer (SAR) 278, 441
serial digital interface ....
   .... 26, 64, 70, 79, 198
serial digital interface (SDI) ....
   26, 64, 65, 69, 70, 90, 112, 113, 114, 135, 136, 137, 144, 146, 189, 196, 198, 224, 256, 283, 441
serial protocols 314
Serial Storage Architecture ....
   x, xi, xiv, 44, 190, 233, 303, 313, 331, 335, 340, 343, 350, 354, 355, 358, 360, 361, 362, 363, 364, 365, 366, 367, 368, 369, 370, 371, 441
set top box (STB) 421, 432, 441
SGI 105, 190, 313, 360, 441
short-term storage 49, 199
Sierra Design 32
Silicon Graphics Incorporated (SGI) .....
   ...... 13, 104, 107, 313, 314, 360, 441
single connector assembly (SCA) ....
   .... 330, 358, 368, 388, 441
slider 385, 386
sliders, 30 percent 386
sliders, 70 percent 386
slo-mo 15, 18
SMPTE 125M 90
SMPTE 12M-1995 .... 70, 96, 401
SMPTE 259M ....
   54, 55, 56, 64, 69, 80, 90, 113, 114, 136, 146, 149, 155, 197
SMPTE 292M 114

SMPTE 305M, proposed (SDTI) ....
   .... ix, 55, 69, 90, 197
SMPTE Engineering Committees 93
SMPTE Journal 115
SMPTE, machine control protocol 29
SMPTE/EBU Task Force 111, 115
Society of Motion Picture and Television Engineers (SMPTE) ....
   .... xviii, 441
solid state disk ....
   .... xviii, 373, 374, 376, 377
SONET 65, 280, 350, 441
Sony 21, 22, 26, 29, 30, 70, 97, 208
source input format (SIF) 166, 441
spatial 24, 86, 91, 362, 363
Spectradyne 410
SPI 331, 332, 333, 335, 441
STA .... xiii, 331, 332, 441
standard definition television (SDTV)
   .... 113, 135, 188, 441
start of active video (SAV) 401
still store server (SSS) 6, 7, 441
stopband 91
StorageTek 209, 210, 403
store and forward 127, 264, 392
streaming data tape 167, 401
striped array 296, 299
striping ....
   x, 74, 101, 102, 103, 207, 289, 291, 294, 295, 298, 299, 300, 302, 304, 310, 352, 397, 404
stunt mode 166
subcarrier 17, 22, 28, 269
subsystem (RAID) 290
Sundance 128
S-VHS 23, 154
synchronous ....
   92, 108, 111, 112, 114, 220, 234, 235, 254, 270, 274, 318, 321, 322, 324, 328, 337
systems network architecture (SNA)
   .... 259, 441

T

T1 .... 112, 113, 224, 235, 442
T1C 235
T2 235
T3 112, 235

T4235
tagged command queuing 328
tape library ....
  121, 126, 152, 165, 167, 168, 170,
  179, 182, 210, 403
target 85, 175, 319, 413, 414
TCP/IP 63, 106, 343, 442
Tektronix Profile................................
  .... 37, 103, 125, 126, 210, 403
temporal 24, 86, 91, 114
termination 254, 326, 328, 336
Texas Instruments TMS320C 374
thermal calibration (TCAL) 388, 419
thermal recalibration (TCAL) 419
time division multiplexing (TDM) ....
  .... 269, 442
time-base corrector................ 15, 17, 24
timecode..
  28, 31, 36, 38, 70, 77, 89, 96, 97, 98,
  99, 102, 112, 220, 398, 401
timeline editing 31
timestamps 282
timing slew errors 336
trailer.. 283
transcoding....................................... 106
transmission time slot.................. x, 268
transport stream.................................
  .... 96, 110, 164, 198, 281, 282
transport stream (TS)164, 198, 281,
  442
TV Technology..................................
  xvi, xviii, xix, 2, 25, 183, 313, 407,
  432
TV versus PC................................... 432
Type-C 17, 22

U

Ultra SCSI...
  65, 74, 102, 190, 321, 328, 330, 331,
  332, 358, 359, 387, 388
UNI 276, 277, 279, 442
uniprocessor..................................... 380
UNIX .. 29, 62, 118, 190, 314, 369, 420
unshielded twisted pairs (UTP)....... 256
upconversion... 133, 136, 137, 139, 155
UTP........................ 256, 258, 259, 442

V

VBI 89, 92, 96, 99, 100, 442
VCI 276, 277, 278, 279, 442
vertical blanking interval............. 89, 99
very high density cabling interconnect
  (VHDCI).................................... 330
VHDCI........................... 330, 358, 442
video cart machines...........................
  .... 159, 391, 396, 397
video compression ....
  .... 12, 32, 59, 124, 199, 257, 392
video email..................................... 280
video file server....... 121, 132, 135, 142
Video Graphics Controller (VGC) ..433
video kiosks 408
video server.......................................
  ix, xi, xviii, 2, 12, 13, 37, 38, 42, 43,
  47, 48, 49, 50, 59, 60, 61, 62, 63, 64,
  65, 66, 67, 68, 70, 71, 73, 75, 77, 78,
  79, 80, 81, 82, 84, 85, 89, 90, 91, 92,
  94, 95, 96, 97, 99, 101, 102, 103,
  108, 114, 117, 118, 121, 122, 124,
  125, 126, 128, 129, 130, 131, 132,
  134, 137, 139, 141, 142, 143, 145,
  149, 152, 155, 164, 165, 169, 173,
  174, 177, 178, 182, 183, 184, 188,
  195, 199, 200, 201, 202, 203, 211,
  212, 215, 218, 219, 220, 229, 285,
  293, 297, 303, 341, 344, 347, 348,
  355, 357, 378, 381, 389, 391, 394,
  395, 396, 397, 410, 419, 421, 422,
  433
Video Still Projector 21
video streaming................................ 24
video teleconferencing 280
video-on-demand ....
  xiv, xviii, 6, 11, 38, 49, 82, 121, 124,
  127, 129, 130, 163, 183, 227, 290,
  320, 374, 388, 409, 417, 418, 420,
  421, 423
videotape transport............................
  8, 22, 37, 39, 79, 97, 132, 152, 180,
  202, 208, 223, 361
Vidifont.. 35
virtual channel connection 279
virtual circuit identifier 276
virtual connection............ 251, 252, 275
virtual contact head 384

virtual path 276, 279, 280
virtual path identifier 276
visual degradation 139
VITC 92, 96, 97, 100, 442
VOD ....
  viii, 6, 7, 129, 163, 227, 407, 408,
  409, 410, 411, 412, 413, 414, 417,
  418, 419, 420, 421, 422, 442
  VOD, live interactive 411
  VOD, true interactive 412
volume stacking 401
VPI 276, 277, 278, 279, 442

W

WAN 233, 346, 348, 442
Wavelets 60
Western Digital 382
Wide SCSI ....
  321, 323, 325, 329, 330, 332, 333,
  335, 337
Winchester 13, 21, 105

Windows NT ....
  43, 50, 62, 63, 103, 260, 293, 303,
  332, 359, 369
wireless cable 132
workstation ....
  3, 7, 13, 29, 32, 44, 105, 107, 108,
  109, 163, 165, 169, 170, 195, 216,
  218, 232, 240, 242, 261, 313, 314,
  348, 369
World Wide Web .... iv, 118, 407
write once read many (WORM) ....
  .... 338, 442
write-back cache 309

X

X3.131-1994 334
X3T11 343

Y

Y, B–Y, R–Y 54
$Y_{C_B C_R}$ 105
YUV 19, 105, 106, 442